Advanced Materials and Structures Conference (AMS 2024)

IOP Conference Series: Materials Science and Engineering
Volume 1319

Timisoara, Romania
30 May – 1 June 2024

ISBN: 979-8-3313-1005-9
ISSN: 1757-8981

Printed from e-media with permission by:

Curran Associates, Inc.
57 Morehouse Lane
Red Hook, NY 12571

Some format issues inherent in the e-media version may also appear in this print version.

Institute of Physics
Dirac House, Temple Back
Bristol BS1 6BE UK

Phone: 44 1 17 929 7481
Fax: 44 1 17 920 0979

techtracking@iop.org

Additional copies of this publication are available from:

Curran Associates, Inc.
57 Morehouse Lane
Red Hook, NY 12571 USA
Phone: 845-758-0400
Fax: 845-758-2633
Email: curran@proceedings.com
Web: www.proceedings.com

TABLE OF CONTENTS

IOP Conf. Series: Materials Science and Engineering **1319** (2024) 011001 doi:10.1088/1757-899X/1319/1/011001

Editorial

The *IOP Conference Series: Materials Science and Engineering* journal contains the peer-reviewed papers presented at the 10th International Conference on Advanced Materials and Structures - AMS 2024.

The conference took place between 30.05 to 01.06 2024 at Timisoara, Romania and it was the tenth edition of the scientific event organized by the Department of Materials Engineering and Manufacturing and Department of Mechanics and Strength of Materials from University Politehnica of Timisoara with the support of the Politehnica Foundation, Technical Sciences Academy of Romania and Romanian Academy - Timisoara Branch.

Additionally, this tenth edition continues the tradition of previous AMS conferences by providing a platform for researchers, academic staff, young PhD students, and industrial partners to present their scientific achievements in the field of:

- Advanced materials- amorphous, nano-structured materials, composites, cellular materials, biomaterials etc.;
- Surface engineering;
- Modern fabrication, additive manufacturing, joints and recycling technologies;
- Materials damage under time-dependent-actions (fatigue, creep, impact, corrosion);
- Computational techniques for advanced engineering materials and structures.

We hope that our readers find this special issue both interesting and useful.

Guest editors

Prof. Viorel-Aurel ŞERBAN, University Politehnica of Timisoara, Romania

Prof. Liviu MARŞAVINA, University Politehnica of Timisoara, Romania

Prof. Ion-Dragoş UŢU, University Politehnica of Timisoara, Romania

Assoc.Prof. Emanoil LINUL, University Politehnica of Timisoara, Romania

IOP Conf. Series: Materials Science and Engineering **1319** (2024) 011002 doi:10.1088/1757-899X/1319/1/011002

Peer Review Statement

All papers published in this volume have been reviewed through processes administered by the Editors. Reviews were conducted by expert referees to the professional and scientific standards expected of a proceedings journal published by IOP Publishing.

- **Type of peer review:** Single Anonymous
- **Conference submission management system:** Morressier
- **Number of submissions received:** 41
- **Number of submissions sent for review:** 41
- **Number of submissions accepted:** 40
- **Acceptance Rate (Submissions Accepted / Submissions Received × 100):** 97.6
- **Average number of reviews per paper:** 2.075
- **Total number of reviewers involved:** 28
- **Contact person for queries:**
  Name: Dragos Utu
  Email: dragos.utu@upt.ro
  Affiliation: University Politehnica Timisoara

IOP Conf. Series: Materials Science and Engineering **1319** (2024) 012001 doi:10.1088/1757-899X/1319/1/012001

Application and Impact of Automation in Crimping Processes

Ilca Dacian[1], Manescu Tiberiu[2], Gillich Gilbert-Rainer[3]

[1,2,3] Doctoral School of Engineering, Babeş-Bolyai University, Cluj-Napoca, Romania

[2] Department of Automation, Industrial Engineering, Textiles and Transportation, Aurel Vlaicu University, Arad, Romania

[3] Department of Engineering Science, Babeş-Bolyai University, Cluj-Napoca, Romania

[2] Corresponding author

*E-mail: [1]dacianilca@gmail.com; [2]manescu.tiberiu@gmail.com, [3]gilbert.gillich@ubbcluj.ro

Abstract. This paper investigates the Subassembly Process Automation (SPA) program's application and impact on industrial manufacturing processes, particularly underscoring its efficacy in refining production accuracy and efficiency. It explores SPA's systematic advancements and the incorporation of meticulous quality control strategies, focusing on how it streamlines operations, boosts efficiency, and ensures adherence to industry norms. The SPA program integrates automation for comprehensive process oversight, including job initiation, ongoing production, and critically, scrap reduction, thereby markedly diminishing waste and upholding rigorous quality standards. Offering an in-depth evaluation of SPA's capabilities, the paper illustrates its role in enhancing production flow and supporting environmental sustainability through effective scrap management. By analysing its implementation in crimping operations, from equipment preparation to production execution, the research highlights SPA's essential contribution to reducing errors, maximizing resource efficiency, and maintaining exceptional manufacturing quality.

1. Introduction

Crimping is a widely utilized process in the subassembly area of industrial manufacturing that involves the deformation of metal or other materials to join components together. This method is particularly favoured for its ability to create a strong, reliable connection without the need for heat, solder, or adhesives, making it ideal for assembling electrical wiring, connectors, and terminals. The process uses a specialized tool or machine to compress and reshape the material, ensuring a tight and secure fit. Crimping is essential for creating connections that can withstand vibration, temperature variations, and other environmental factors, making it a critical step in the production of a wide range of products, from automotive electrical systems to consumer electronics [1].

The significance of crimping in industrial manufacturing, especially in the subassembly area, lies in its efficiency and reliability. By allowing for quick and effective connections, crimping significantly speeds up the assembly process, reducing production times and costs. Moreover, the quality of crimped connections is consistently high, leading to reduced failure rates and longer product lifespans. This reliability is crucial in industries where safety and performance are paramount, such as in aerospace and medical devices. Additionally, the versatility of crimping tools and techniques enables their application across a broad spectrum of materials and component sizes, further underscoring the process's integral role in modern manufacturing.

The SPA (Subassembly Process Automation – internally developed software application) program significantly enhances the efficiency and quality of crimping operations, pivotal in industrial manufacturing's subassembly areas. This comprehensive program systematically addresses various stages of crimping processes, from operator identification, workstation setup, to job execution and completion. It incorporates advanced features for job setup, ensuring that the correct parameters and materials are used, and includes detailed monitoring and reporting capabilities to track productivity and identify areas for improvement. Moreover, the SPA program integrates quality control measures and provides mechanisms for immediate corrective actions, thereby reducing the likelihood of errors and rework. This holistic approach to managing crimping operations streamlines the workflow, optimizes resource utilization, and significantly improves the overall productivity and quality of the manufacturing process, making it an invaluable tool in the subassembly area of industrial manufacturing.

This paper examines the SPA Program's application and impact on crimping operations, encompassing a detailed overview of its features, historical development, and operational process. It delves into quality control measures and compliance with industry standards, illustrated through case studies and practical applications, including challenges and solutions during implementation. An impact analysis assesses the program's effects on productivity and operational efficiency, while a comparative analysis with other systems highlights the SPA's unique advantages and limitations, offering a comprehensive insight into its role in enhancing crimping processes.

2. SPA Program and Crimping Operations

The SPA program is an integrated software system designed to streamline and optimize manufacturing processes, specifically in the context of crimping, stripping, and other assembly operations in industrial settings. It encompasses a suite of functionalities aimed at improving operational efficiency, enhancing quality control, and facilitating a smooth workflow. Key features include user authentication, workstation unlocking, job loading with data entry through scanning, machine setup with specific parameters, quality control checks, scrap recording, and material ordering. The program also supports automatic measurements recording and visual inspection to ensure adherence to quality standards. Each document provided details procedures for different aspects of manufacturing operations, highlighting the SPA program's adaptability and critical role in modern industrial processes.

SPA program incorporates a wide array of features and functionalities designed to optimize manufacturing operations, particularly in the context of crimping, stripping, and assembly processes. Key features include:

User Authentication: Ensures that only authorized personnel operate machinery.

Workstation Management: Includes unlocking and locking stations, job loading, and setup based on specific operational needs.

Procedural Adherence: Guides operators through detailed procedural steps for various operations, including crimping, stripping, and assembly, to ensure quality and efficiency.

Quality Control: Implements checks and validations to maintain high-quality outputs.

IOP Conf. Series: Materials Science and Engineering **1319** (2024) 012001 doi:10.1088/1757-899X/1319/1/012001

Material Management: Facilitates direct ordering and management of materials required for jobs.

Data Management: Involves automatic and manual data entry, along with validation processes to maintain accuracy and traceability.

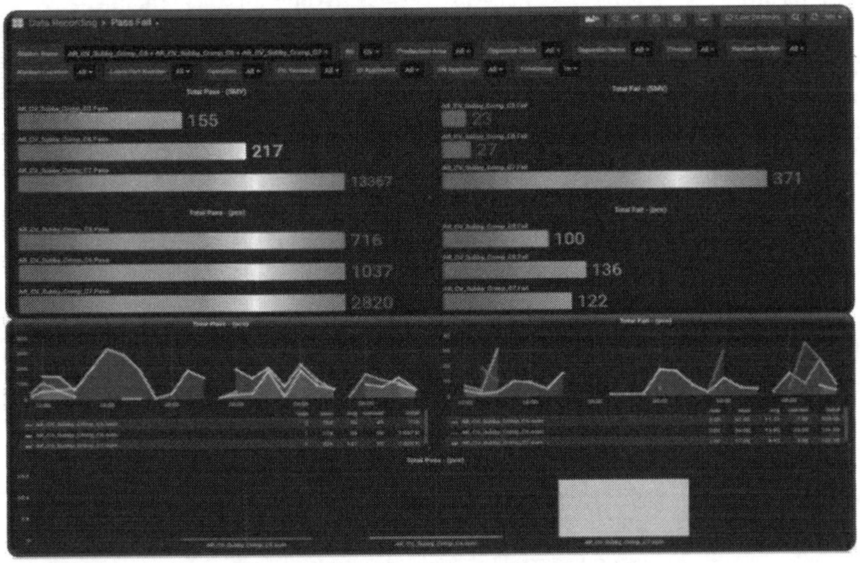

Figure 1. SPA displaying various reports.

The SPA program is utilized across various specialized workstations in the manufacturing process, including crimping, stripping, assembly, quality control, and material management stations. Each workstation leverages the program's features to streamline operations, enhance efficiency, and ensure quality compliance, tailored to the specific tasks and requirements of the station.

Crimping: Focuses on precision in connecting wires with terminals, utilizing specific parameters and quality checks to ensure secure connections.

Stripping: Automated and manual stripping processes are optimized for accuracy in removing insulation without damaging the wire.

Assembly: Guides through complex assembly tasks, ensuring parts are correctly joined according to specifications.

Quality Control: Integrates measurement and inspection tools to uphold product standards.

Material Management: Streamlines the ordering and handling of materials, reducing downtime and improving efficiency.

Soldering and Ultrasonic Welding: Applies to specialized connections, ensuring strong bonds with precise control.

3. Brief history and development of the program

This initiative had the goal of enhancing production efficiency and reliability while reducing costs. The approach included connecting equipment to leverage valuable data for live reporting, statistical process control, and utilizing automated data analysis to streamline operations and decision-making.

IOP Conf. Series: Materials Science and Engineering **1319** (2024) 012001 doi:10.1088/1757-899X/1319/1/012001

The program unfolded through strategic steps, focusing on connecting all relevant equipment to harness valuable, previously untapped data. This initiative marked the shift towards a data-driven manufacturing process. The development of live reporting mechanisms and online statistical process control systems were crucial milestones, enabling the monitoring of relevant indicators and parameters that were critical to operational excellence.

Key developments in the program involved the automation of material scanning, Kanban Cards Management, and the reduction of changeover durations, which led to significant operational efficiencies. These changes were supported by digital advancements that minimized manual errors, improved traceability, and facilitated real-time monitoring. The outcomes of these efforts were quantified in substantial cost savings, underscoring the financial and operational benefits of the automation projects.

4. Case Studies and Practical Applications

4.1 Smart Factory use cases

Our objective to achieve a Smart Factory included several areas or use cases of implementation:

a. Digitalizing the Lead Preparation Area

Objective: Improve efficiency and reliability in the lead prep. area through automation.
Practical Application: Equipment was connected to utilize previously untapped data, enhancing live reporting and online statistical process control. This led to improved production efficiency and cost savings.

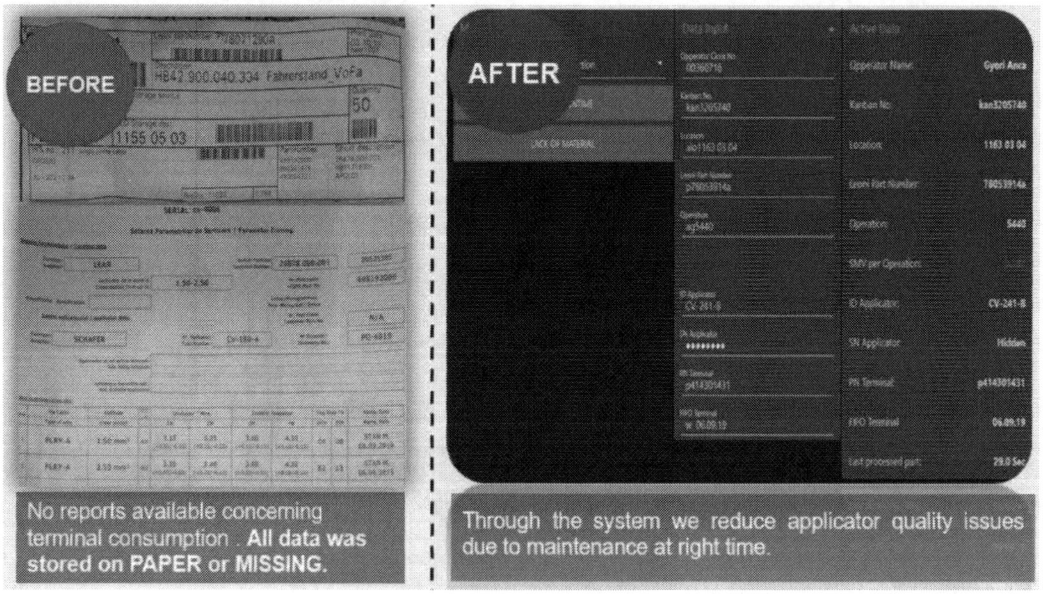

Figure 2. Material scanning before and after.

b. Material Scanning

Objective: Automate the material scanning process to reduce errors and improve efficiency.
Practical Application: The implementation of automated scanning systems led to enhanced traceability and reduced manual errors, contributing to more efficient material handling and inventory management.

c. Kanban Cards Management

Objective: Digitize the management of Kanban Cards to streamline operations.

Practical Application: By automating the input and management of Kanban Cards, our plant minimized manual data entry, improving accuracy and efficiency in production planning and scheduling.

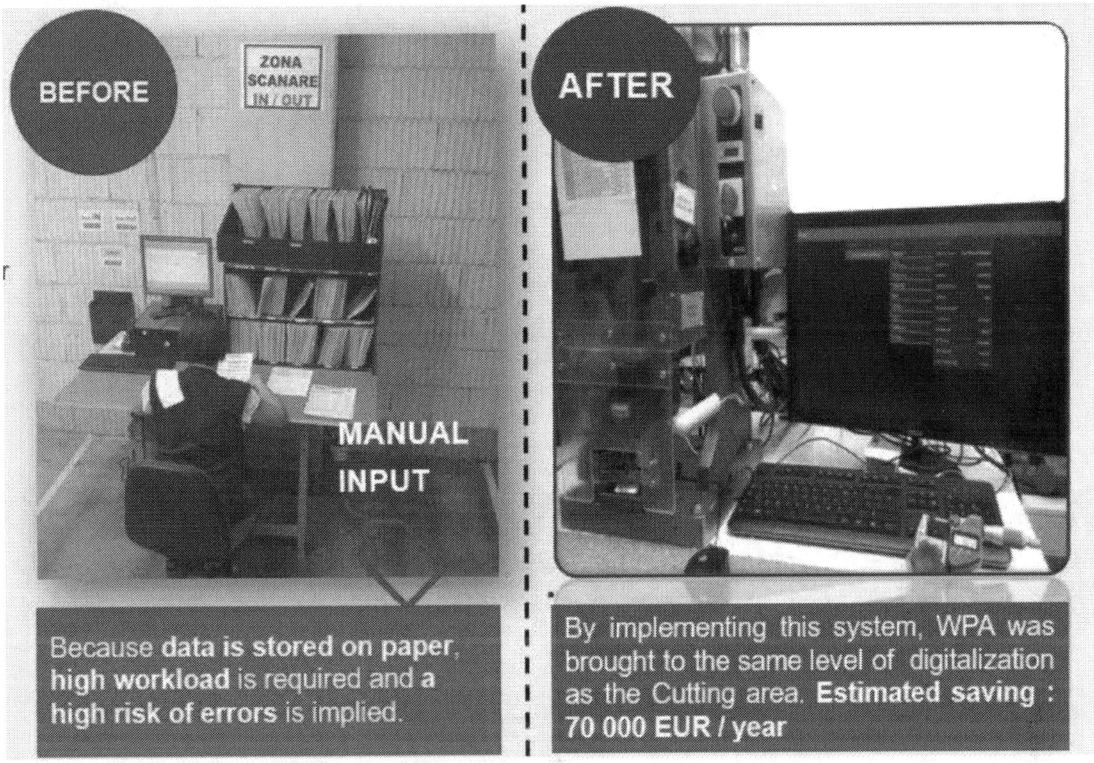

Figure 3. Kanban Cards Management before and after.

d. Reduction of Changeover Duration

Objective: Decrease the duration of changeovers to enhance production output.

Practical Application: Through digital tools and processes, the company was able to reduce the time needed for equipment changeovers, leading to increased production capacity and reduced downtime.

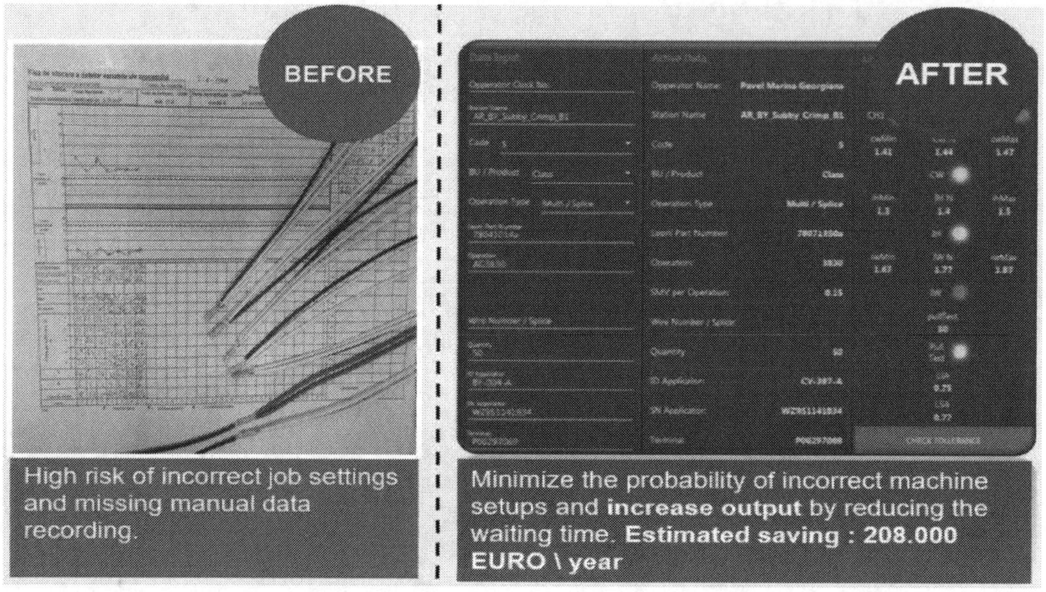

Figure 4. Reduction of changeover duration before and after.

e. Downtime Management

Objective: Minimize machine downtime and improve root cause analysis.

Practical Application: The deployment of a web interface and dashboard for management allowed for live monitoring of machine status, automatic notifications for technical service, and the generation of custom reports. This resulted in reduced reaction times to downtime events and improved analysis of underlying causes.

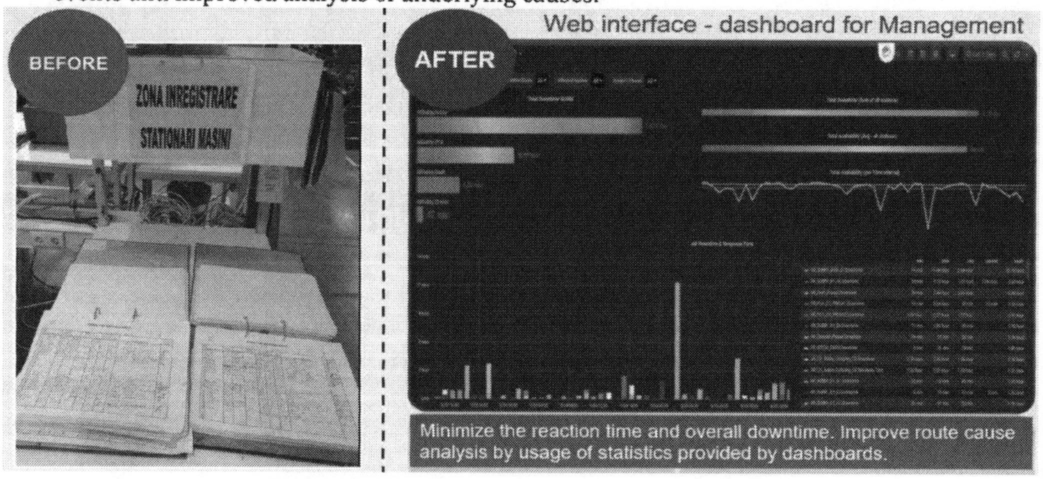

Figure 5. Downtime management before and after.

4.2 A case study of the SPA program's implementation at a designated crimping station

A step-by-step process typically involved in using a program like SPA at crimping stations:

Login and Authentication: The operator logs into the SPA system using their credentials or scanning a barcode to authenticate.

Workstation Unlocking: Once logged in, the operator unlocks their workstation, making it ready for the job.

Job Loading: The operator loads the job details into the system, which may involve scanning a job card or manually entering job specifications.

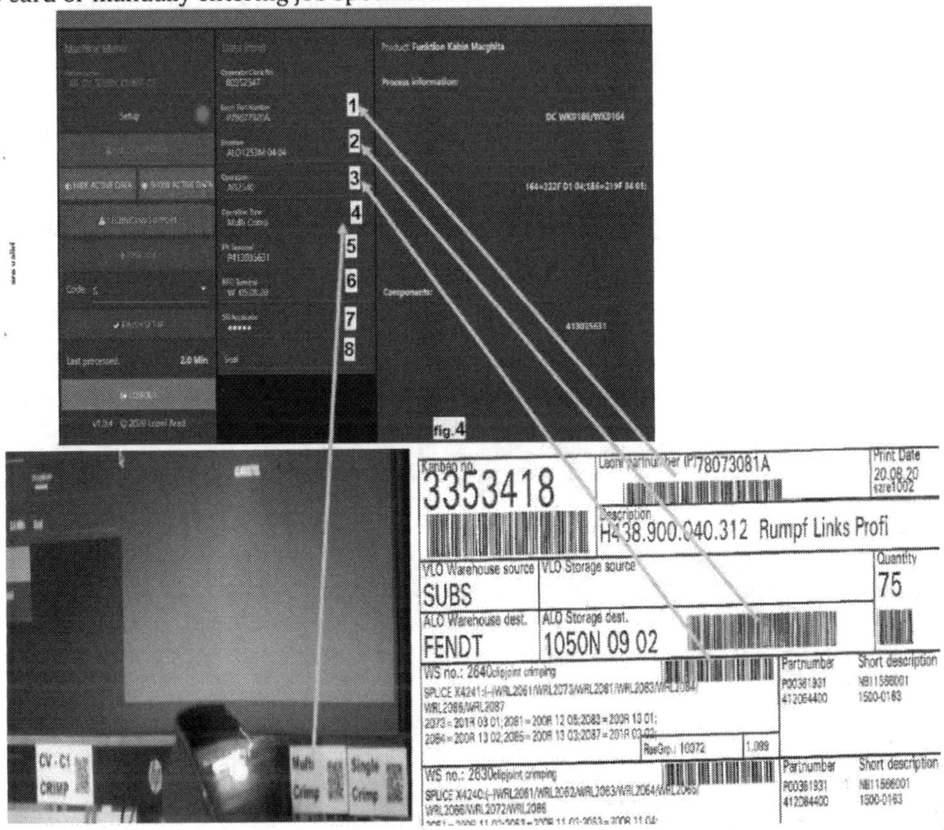

Figure 6. Job loading using scanner and Kanban card.

Machine Setup: Based on the job details, the system guides the operator through setting up the crimping machine. This includes selecting the correct crimping tools, dies, and making any necessary adjustments to the machine settings.

Sample Preparation and Testing: Before full production, the operator prepares sample crimps for quality checks to ensure the setup is correct.

Quality Control and Adjustments: The quality of the sample crimps is assessed, and any necessary adjustments are made to the machine setup to meet quality standards.

Production Start: Once the setup is validated, the operator starts the production process, crimping the wires as specified by the job details.

Ongoing Monitoring and Quality Checks: Throughout the production run, the system and operator monitor the crimping process for quality and efficiency, adjusting as needed.

Completion and Logout: After the job is completed, the operator logs out of the SPA system, and the workstation is locked until the next job.

Scrap: Scrap Recording: After completing a job, the operator clicks the "SCRAP" button, filling in relevant fields in the popup window, then sends the data by clicking "SEND". This returns them to the main window, where a new job can be started by pressing "NEW JOB".

5. Quality Control and Compliance

The SPA program implements stringent quality control measures, including detailed validation checks and protocols for handling nonconforming products. During the crimping process, operators are required to prepare samples for quality inspection, which are then measured and validated against pre-set standards. If the measurements fall within acceptable ranges, production proceeds with a "Normal Production" status. Conversely, if the samples do not meet quality standards, the system indicates a "Rejected by Quality" status, prompting necessary adjustments or rejections. This ensures that only products meeting the quality criteria proceed further in the manufacturing process.

Quality control measures implemented through the SPA program are crucial for ensuring compliance with industry standards. They guarantee that products meet stringent quality criteria before reaching the market, reducing the risk of failures that could lead to safety issues or customer dissatisfaction. Such measures ensure that manufacturing processes are consistent, reliable, and capable of producing high-quality products. By adhering to these standards, companies can maintain their reputation, avoid costly recalls, and comply with regulatory requirements, thereby ensuring customer trust and satisfaction.

6. Impact Analysis

The SPA program has significantly enhanced productivity across various operations. For instance, the material scanning and Kanban Cards Management systems have streamlined material handling and inventory management, reducing the time and errors associated with manual processes. The introduction of live monitoring and automated data input systems has led to more efficient job setups and changeovers, directly improving production throughput.

In discussing the impact of the SPA Program on crimping operations, the importance of effective scrap management, as explored by Narasimha Prasad Bhat in "Optimizing Scrap Management in Production Planning Processes", provides essential context. Bhat's research emphasizes how optimized scrap management can significantly enhance manufacturing efficiency and reduce costs, aligning with the objectives of SPA Program implementations in industrial manufacturing [2].

Moreover, the SPA program has made a substantial impact on quality control by enabling live monitoring of applicator usage, traceability of materials, and automated validation of job settings by Quality Auditors. These changes have minimized the probability of incorrect machine setups and non-conform parts production, thereby enhancing overall product quality.

Significant operational efficiency improvements, such as reduced downtime through automated downtime management systems and enhanced root cause analysis via dashboard statistics have been observed. The scrap monitoring and recording have become more accurate and less prone to errors, facilitating better scrap analysis and reducing waste.

The SPA program has led to specific positive outcomes, such as the avoidance of sending non-conforming parts to clients, exemplified by the inline test project where NOK parts were detected and rejected by the system, preventing potential claims from clients. This incident underscores the program's impact on enhancing quality control and avoiding financial and reputational damage.

The enhancement of scrap management through automation, as detailed in the SPA program, parallels advancements discussed by Schäfer, Faltings, and Glaser in their creation of the DOES - Dataset of European Scrap Classes. This dataset underpins the critical role of digitalization in scrap classification, crucial for optimizing scrap utilization in steel manufacturing. Their work demonstrates how automation and machine learning are essential in advancing green steel production by improving scrap sorting and classification, aligning with the SPA program's objectives to enhance operational efficiency and sustainability [3].

In discussing the SPA program's automation benefits, it's insightful to refer to Yadav et al.'s innovative project, "Smart Bin: A Comprehensive Solution For Scrap Management Leveraging Machine Learning for Customer Segmentation." This research introduces a web-based platform aimed at simplifying scrap management by allowing users to sell or donate scrap easily. It leverages machine learning for customer segmentation, enhancing the efficiency of scrap management similar to the objectives of the SPA program. [4].

Compared to traditional manual systems, the SPA program offers several advantages, including higher accuracy in data management, reduced manual labour, and faster reaction times to operational issues. The automation of data entry and the integration of systems for live monitoring and reporting enable a level of operational visibility and control not achievable with manual systems.

7. Challenges and Limitations

Implementing the SPA (Subassembly Process Automation) program, particularly in crimping processes, introduces several challenges and limitations across technical, organizational, and financial spheres. Technically, integrating SPA with existing systems presents complexity, especially when dealing with outdated infrastructure, necessitating meticulous planning to ensure compatibility. The transition to digitalization brings to the fore issues of data accuracy and cybersecurity, requiring robust measures to maintain data integrity and protect against breaches. Additionally, the reliance on automated systems underscores the need for high reliability and regular maintenance to avoid production interruptions.

From an organizational standpoint, the shift towards automation demands significant change management efforts to address resistance among employees who may fear job displacement or struggle with new skill requirements. Ensuring the workforce is adequately trained and upskilled for the digital landscape is imperative, yet it represents a substantial commitment of resources. Financially, the initial investment for the SPA program can be considerable, encompassing software, hardware, and training costs, with the return on investment (ROI) potentially uncertain in the short to medium term. Businesses must carefully evaluate these expenditures against the projected efficiencies and savings.

Specific to crimping operations, challenges include adapting the SPA program to meet diverse product specifications without compromising operational efficiency. Moreover, maintaining quality control in automated crimping processes demands advanced monitoring and diagnostic capabilities to ensure consistent product quality. Overcoming these challenges requires strategic planning, investment in technology, and a commitment to continuous training, positioning organizations to fully leverage the benefits of smart manufacturing solutions.

8. Conclusion

The paper discusses the significant role of crimping in industrial manufacturing and introduces the SPA (Subassembly Process Automation) program as a transformative tool for enhancing crimping operations' efficiency and quality. The SPA program, designed to optimize the crimping, stripping, and assembly processes, incorporates features such as user authentication, workstation management, quality control checks, and material management. It

IOP Conf. Series: Materials Science and Engineering **1319** (2024) 012001 doi:10.1088/1757-899X/1319/1/012001

aims to streamline workflows, improve operational efficiency, and ensure product quality, demonstrating adaptability across various specialized workstations.

The impact of the SPA program on crimping operations is profound. By automating and integrating different stages of the crimping process, the program significantly enhances productivity, ensuring high-quality connections that comply with industry standards. It minimizes the likelihood of errors and rework, thereby reducing production costs and time. The SPA program's emphasis on quality control measures and compliance with industry standards ensures that products meet stringent quality criteria, reducing the risk of failures and maintaining customer satisfaction.

However, the implementation of the SPA program in crimping processes faces technical, organizational, and financial challenges, including integration complexity, the need for change management, and high initial investments. Despite these challenges, the SPA program offers considerable advantages over traditional manual systems, such as improved data accuracy, reduced manual labour, and enhanced operational visibility.

For future research or development, exploring the integration of AI and machine learning for predictive maintenance, quality control, and process optimization could further revolutionize crimping operations. Additionally, assessing the long-term ROI of SPA implementations and developing more user-friendly interfaces for training and operation could address some of the current limitations and encourage wider adoption of automation technologies in industrial manufacturing.

References

[1] SAE International, Performance Specification for Cable-to-Terminal Electrical Crimps. SAE/USCAR-21, 4th SAE International, U.S. and Canada, 2020. https://global.ihs.com/doc_detail.cfm?document_name=SAE%20USCAR %2D21&item_s_key=00407431

[2] Bhat, Narasimha Prasad. (2023). Optimizing Scrap Management in Production Planning Processes. International Journal of Computer Trends and Technology. 71. 48-52. https://doi.org/10.14445/22312803/IJCTT-V71I5P108.

[3] Schäfer, Michael & Faltings, Ulrike & Glaser, Björn. (2023). DOES - A multimodal dataset for supervised and unsupervised analysis of steel scrap. Scientific data. https://doi.org/10. 780. 10.1038/s41597-023-02662-6.

[4] Yadav, Akash & Sharma, Aditya & Anji Reddy, Medam & Nadiadwala, Aliabbas. (2023). Smart Bin: A Comprehensive Solution For Scrap Management Leveraging Machine Learning for Customer Segmentation Revolutionizing Scrap Management: A Web-Based Platform for Efficient Scrap Management. https://doi.org/10.13140/RG.2.2.27490.61124.

IOP Conf. Series: Materials Science and Engineering **1319** (2024) 012002 doi:10.1088/1757-899X/1319/1/012002

Crimp Height influence over Resistivity in Wiring Crimping Process

Florin Dragomir[1], Manescu Tiberiu[2], Gillich Gilbert-Rainer[3]

[1 2 3] Doctoral School of Engineering, Babeş-Bolyai University, Cluj-Napoca, Romania
[2] Department of Automation, Industrial Engineering, Textiles and Transportation, Aurel Vlaicu University, Arad, Romania
[3] Department of Engineering Science, Babeş-Bolyai University, Cluj-Napoca, Romania
[2] Corresponding author
*E-mail:[1]florin.dragomir@ubbcluj.ro;[2]manescu.tiberiu@gmail.com, [3]gilbert.gillich@ubbcluj.ro

Abstract:

The main purpose of this paper is to present the influence between the crimp height and the resistivity for a crimp joint between a copper contact and a wire copper stranded cable using statistical relevant analysis. The experiment followed the analysis of 45 crimped assemblies grouped in 3 main categories to simulate a potential validation procedure according to the international validation process requirements from: DIN ("Deutsches Institut für Normung") with the corresponding Romanian standard SR EN 60352-2:2006 and SAE/USCAR-21 REVISION 3. The baseline used was the behaviour of the resistivity versus the pre-set crimp height of the crimped joints before and after thermal shock test to enable us to conclude if we have a significant relation between the crimp height of the assembly joint and the resistivity.

1. Introduction:

In today's automotive industry and especially in the wiring harness manufacturing process, crimping is considered the most important and therefore, being identified as critical, all control actions focus on achieving high stability and reliability. Current international standards regulate all suppliers by imposing process limits that are built to ensure both stability and reliability.

It is very important for the life cycle of electrical wiring to create a crimped connection capable of withstanding vibrations [1],[2],[3] temperature variations, different levels of humidity [4], and a full range of environmental factors specific to the wide range of products using electrical wiring [5],[6].

When we talk about the crimping process in industrial manufacturing, especially in wiring harnesses the most key features are efficiency and reliability. This reliability is decisive for the vehicle's life cycle.

For this reason one of the most important parameters used by either all the passenger cars manufactures [7],[8], either by truck[9],[10] or agriculture manufacturers [11] is the resistivity of the crimped connection and, consequently, the influence of the geometry and the pull force requirements needed to ensure a fully quality compliant connection.

The purpose of the current study is to analyse, using a statistical approach, how the resistivity will influence the geometry of the crimped assembly and to find the best relation between the resistivity, geometry, and the required pull force.

Table 1. Sample configuration categories

Configuration	Crimp Height Nominal CH(mm)	Termina l Type	Cable Type	Cross Section (mm^2)	Pictures	Quantity (pcs)
1	2.54	211CL3S 3120	T8930B	0.5		15
2	2.69	211CL3S 3120	T8930B	0.5		15
3	2.39	211CL3S 3120	T8930B	0.5		15

2. Methodology:

The study involves the analysis performed on 45 crimped assembly connection between a copper terminal 211 CL 3 S 3120 [12] and a stranded copper cable ACOME T8930 5.0 mm^2 B 125°C PVC [13].

As first step the 45 samples were grouped in 3 main configuration categories [Tab.1] according to crimped height configuration (CH). These configurations allowed us to monitor the

Figure 1. Schaefer EPS 2000 press

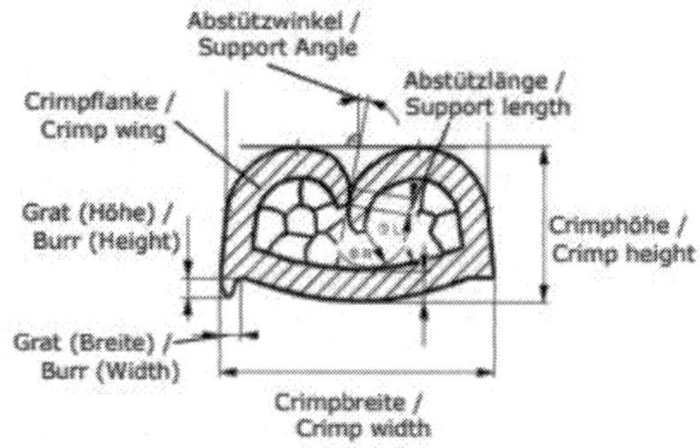

Figure 2. Crimped terminal geometry

behavior of the crimped geometry versus the pull force and resistivity.

To achieve the agreed crimp height configuration we utilized a bech press Schafer EPS 2000 [Fig.1] and the crimped dimensions were ensured using the specification requirments from the terminal data sheet FCI 211 CL 3 S 3120 [Fig. 2]

In the second step of the samples preparations for all 3 configurations groups was performed a resistivity measurements combined with termal shok cycle test with the aim of collecting the Initial resistance of the crimped assembly and the resistance after the thermal shock cycle test.

The Crimping Resistance before and after thermal shock measurment were performed according to standard RNDS-B-00029 V2.0 (paragraphs 13.1/13.2.4/13.3.1)[14].

2.1 Crimp Resistance Before/After Thermal Shock T=125°C

Test Condition:	Temperature:	23°C ± 5°C
	Humidity:	60%HR ± 15%HR
	Atm. Pressure:	960hPa ± 100hPa
Test Parameters:	Amperage:	100mA
	Voltage:	20mV
Test equipment:	Microhmmeter Sefelec/MGR 10B	

The validation requirements followed were, Initial crimped resistance for a cable with 5mm^2 cross section to be ≤ 0.3mΩ and the variation after thermal shock |ΔR1| ≤ 1mΩ and ± 0.02mΩ measurement uncertainty.

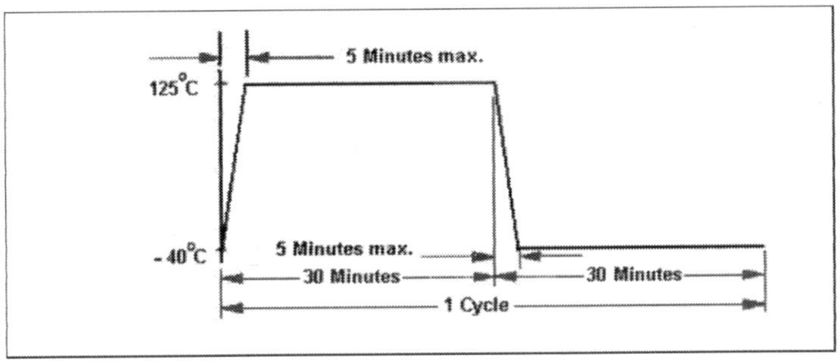

Figure 3. Thermal shock cycle diagram

2.2 Thermal Shock T=125°C

For the thermal shock cycle test all the samples were placed in a Weistechnik/T/120/V2 thermal shock chamber for 100 cycles as detailed in Figure 3, with a measurement uncertainty of ± 2°C.

3. Results and Statistical Analysis

The full range of the 45 samples measurements were recorded in a data collection plan [Tab.2], that was used as an input in the next phase of the experiment related to the statistical analyses of the recorded data using Minitab Software [15], as processing tool.

Table 2. Example from Data collection plan

Crimp Height Nominal (mm)	Pull test Nominal (N)	Resistance before T/S R(mΩ)	Resistance after T/S R(mΩ)	Resistance After-Before |ΔR1|(mΩ)
2.39	374	0.78	0.90	0.12
2.39	400	0.79	0.83	0.04
2.39	398	0.78	0.92	0.14
2.54	420	0.82	0.93	0.11
2.54	382	0.80	0.92	0.12
2.69	350	0.73	0.87	0.14
2.69	343	0.74	0.87	0.13

As specified in the beginning the main goal of this current study is in finding if during the life cycle of a crimped joint the main characteristic as the geometry of the crimped assembly which is the crimp height (CH) together with the connected characteristic as pull test values and resistance values performed before and after thermal shock cycles have a relationship between them, if these variables are influencing each other and if they do what is the degree of interdependency [16].

First parameter that we started to analyse was the resistivity versus crimped height (CH) and for this we used and interval plot diagram [Fig. 4]

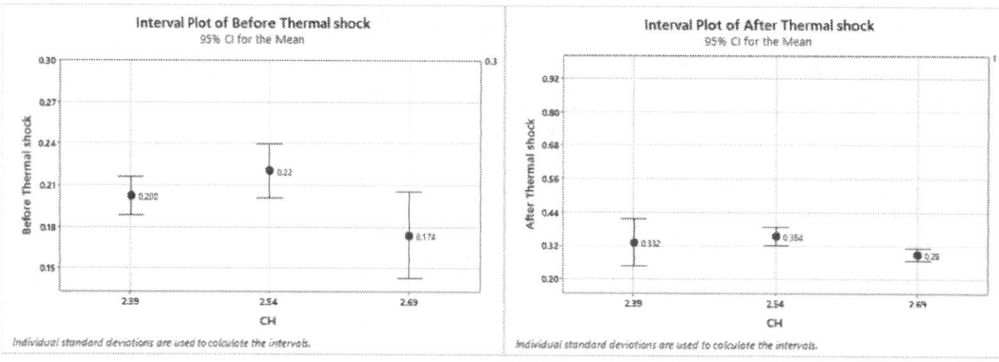

Figure 4. Interval plot diagram Before/After Thermal shock

If we look at the plotted data versus the imposed limits all the samples are inside the required specifications but if we compare the configurations between each other we can clearly observe that the assemblies with CH=2.69 mm is outperforming in both "Before" or in the "After" thermal shock test which could conclude that the best crimped connection if we look only at the resistivity parameter would be the samples from Configuration 2 with CH=2.69.

To further prove as valid our initial conclusion we put all three configurations in a Time Series Plot for a better visualisation of the comparison of resistivity values versus each configuration [Fig.5]

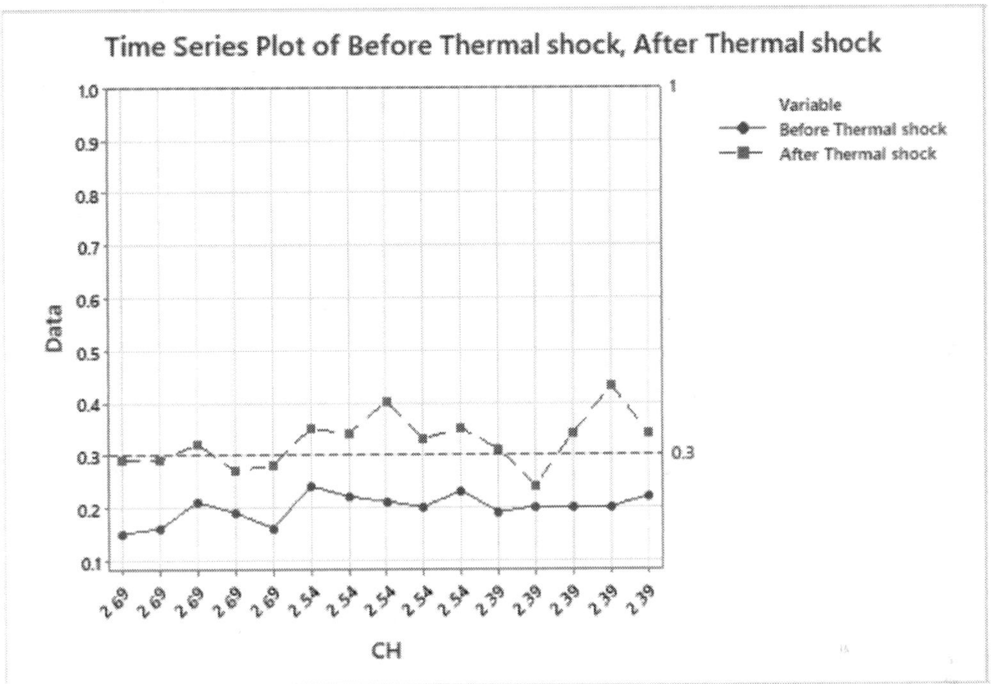

Figure 5. Time Series Plot of Before/After Thermal shock

To finalize the analyse the influence of the resistivity over the crimped geometry expressed by the crimped height (CH) we checked if the data is normal [Fig.6], and if there is a significant difference between the crimped height configurations [Fig.7]

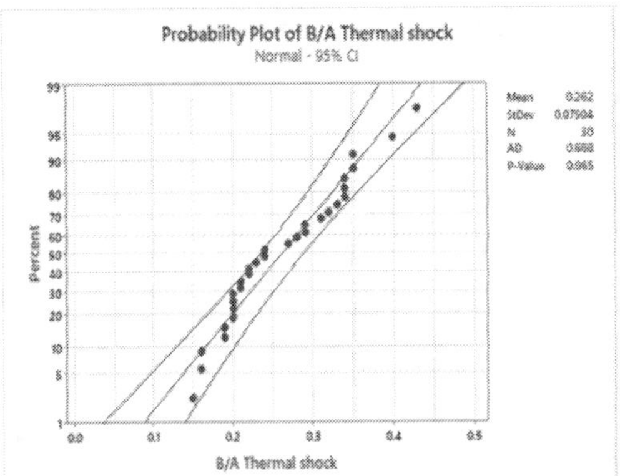

Figure 6. Probability Plot

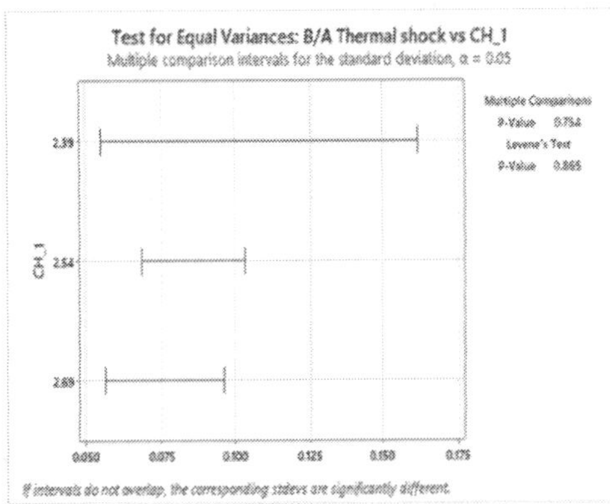

Figure 7. Test for Equal Variances Plot

IOP Conf. Series: Materials Science and Engineering **1319** (2024) 012002 doi:10.1088/1757-899X/1319/1/012002

Our data is normal with 95% confidence and because the P-Value is greater then 0.05 in both plots there is also no significat differences if we compare the standard deviations of each crimped height configuration.

To conclude the first part of the statisitical analysis performed taking in account the resistivity for each Crimped Height (CH), the CH=2.69 will be the preffered geometry to be utilized in the serial production.

Continuing the study, we began to analyse the same geometric crimp configurations expressed by the crimp height value in relation to resistivity and also the pull test values obtained. In this perspective, we started using the multiple regression tool [15] and plot all the data points to understand if we first have a clean regression model and we observe [Fig.8] to return to the data collection plan to understand the reasons behind the data points that were identified by the model as large residues or unusual data that could have a strong influence on the model. In our case, we couldn't identify any of the issues above, so we were confident that our regression model was good.

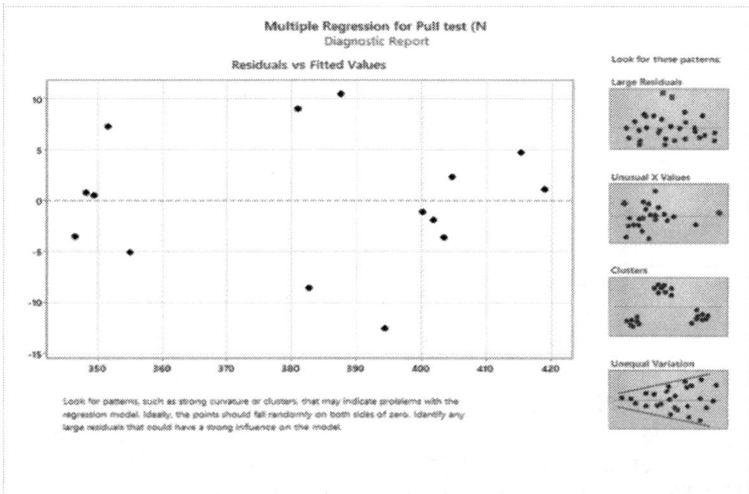

Figure 8. Multiple Regression Diagnostic Report

The second step in our multiple regression analysis was to determine if all the characteristics will have a statistically significant relation, if the percentage of the variation of the output "Y" will be explained by the regression model and if any of the "X" variables are not part of the model so consequently not relevant.

In [Figure 9], after the graphical representation of all recorded data points due to the fact that the P value is below 0.10, we can observe that the relationship between the variables graphically represented in the model is statistically significant and that 94.41% of the output variation in our case, the Pull test (N), can be explained by the regression model. The R-sq value of 94.41% shows that we have high confidence in our statistical model, and the sample size we used to build the model was sufficient for strong statistical relevance.

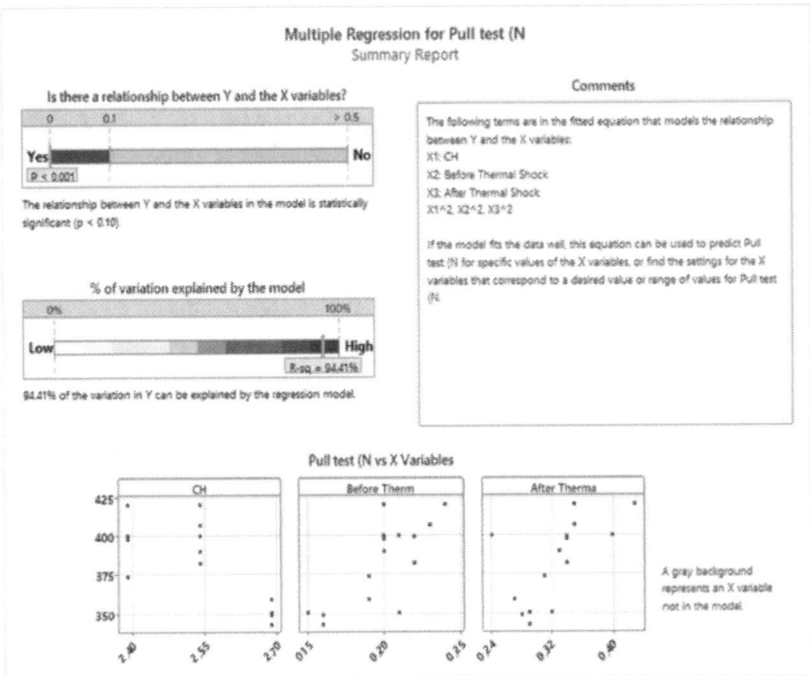

Figure 9. Multiple Regression Summary Report

Once we proved that we build a strong statistical model where there is a strong and significant relation between crimp height, resistivity and pull test values we add in our model the Pull test limit which according to standard specification for an 0.5 mm^2 cross section should be higher then 360 N.

To enable us to do so we set in our multiple regression model the goal to have the pull test values as maximized that in plain terms is to use the 360 N value as the lowest acceptable limit and to calculate an optimal solution between pull test value, crimp height, resistance before thermal shock and resistance after thermal shock

In the Prediction and Optimization Report [Fig.10], we obtain an optimal setting for CH=2.43 mm, with a Resistance Before T/S of 0.24 mΩ and Resistance After T/S of 0.43 mΩ and a Pull test value of 451 N.

Our model is also showing the next five alternative solution with pull test value range between 418N to 401N which will also give as the results around the optimal setting already sugested.

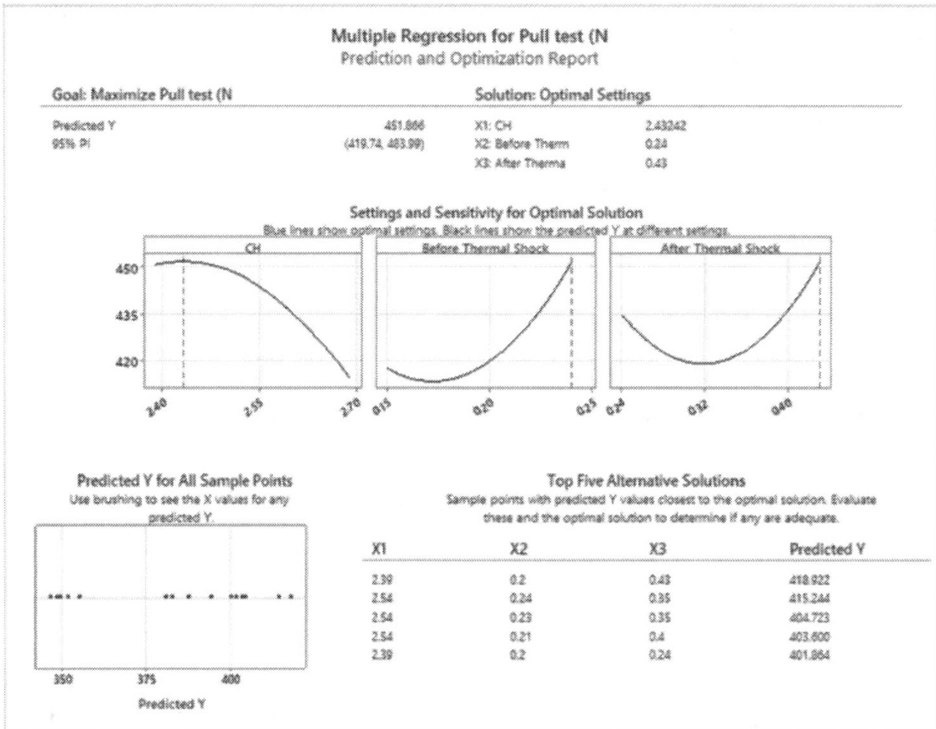

Figure 10. Prediction and Optimization Report

4. Conclusions and Final Consideration

As observed in our study, we initially achieve the optimal resistivity of the crimped joint by having a higher CH value, which in our case was 2.69 mm, but as we started to analyze the pull test value, we noticed that for this specific CH, the pull test values for a cross section of 5 mm^2 were all below the threshold imposed by the standards specifying the lower limit at 360 N.

A common practice of copper terminal suppliers to exceed this pull test limit is to reduce the crimp height, which will automatically have a positive effect on the tensile test values and place the output values above the 360 N threshold. Currenlty this process of finding the optimal Crimp Height (CH) that is fully complient to resistivity limits and one one-handed and fulfilling the requirements of tensile test values is done through trial and error with physical samples, which are time and resource dependent. Due to this time and financial resource constraints it is a common practice to directly reach the lowest possible CH and, by doing so, exceed the pull test limit and maintain the resistivity at an acceptable level. The long-time effect of this practice will have a negative impact on the tools used for the crimping application and will also shorten the life cycle of the crimped joint.

Our study is proving that by using a statistical model will be able to determine the optimal geometry express through Crimp Height (CH) with an optimal pull test values and the best outcome for Resistivity values before or after the thermal shock [17].

If initially, out of our 3-configuration setup CH= 2.39 mm, CH=2.54 mm and CH=2.69mm, the best resistance before and after thermal shock was for the highest CH=2.69, we observed the pull test values for this respective CH were under the 360N limit. According to the common current industrial approach for exceeding this pull test limit would have been to go directly to the lowest configuration CH = 2.39 should have been reached directly, which will indeed have good results in both tensile resistivity measurement and tensile testing. In our current study we showed with a 95% certainty using our statistical model that the optimal CH is CH= 2.43 followed by three alternative optimal solution on CH=2.54.

This study was performed on a crimped joint assembly for a 5mm^2 cross section which is only one type from the family of cable/terminal connections that are currently used in wiring systems manufacturing. By continuing to perform this type of statistical analysis on the entire spectrum of terminals and cables we will be able to obtain a systematic approach on what is the optimal Crimp Height and by doing this we will be able to design from the beginning the tooling to optimize from the early stage of design the overall cost linked to validation and operational performance of an wiring harness.

Data availability

The datasets generated and analysed during the current study are available from the first author on reasonable request.

Author contributions

The authors' contribution is as follows: F. Dragomir - Conceptualization, Formal Analysis, Investigation, Writing–Original Draft Preparation; T. Manescu - Conceptualization, Supervision, Writing–Review and Editing; G.-R. Gillich - Supervision;

Conflict of interest

The authors declare that they have no conflict of interest.

References

[1] G. R. Gillich, M. Tufoi, Z.I. Korka, E. Stanciu, A. Petrica,The Relations between Deflection, Stored Energy and Natural Frequencies, with Application in Damage Detection, (in Romanian), Romanian Journal of Acoustics and Vibration 13 (2), 2016, pag. 87-93.

[2] G.R. Gillich, Z.I. Praisach, V. Iancu, H. Furdui, I. Negru, Natural Frequency Changes due to Severe Corrosion in Metallic Structures, Strojniški vestnik, (in Romanian), Journal of Mechanical Engineering 61 (12), 2015, pag. 721-730.

[3] K. Krüger, H. Yuan, J. Song, The influence of the vibration test mode on the failure rate of electrical connectors, Microelectronics Reliability, Volume 135, 2022, 114567, ISSN 0026-2714, https://doi.org/10.1016/j.microrel.2022.114567.

[4] L. B. Veldscholte, S. de Beer, OpenHumidistat: Humidity - controlled experiments for everyone, Hardware X, Volume 11, 2022, e00288, ISSN 2468-0672, https://doi.org/10.1016/j.ohx.2022.e00288.

[5] SAE International, Performance Specification for Cable-to-Terminal Electrical Crimps. SAE/USCAR-21 4th,SAE International,U.S.andCanada,2020. https://global.ihs.com/doc_detail.cfm?document_name=SAE%20USCAR%2D21&item_s_key=00407431.

[6] SR EN 6032-2:2006/A1:2014, Solderless connections, Part2: Solderless crimped connections, General Requirements, test methods and practical guidance. https://magazin.asro.ro/ro/standard/224996.

[7] BMW GS 95006-1 Wiring Harnesses in Motor Vehicles, Requirements and Tests

[8] VW 60330 Issue 2013-12, Crimp Connections, Solderless electrical Connections

[9] DAF PGS 4615 Issue 1.0 Oct.2020, Vehicle Wiring Harness, Main Requirements

[10] MAN M3599 November 2020, General Requirements for cable harnesses

[11] CLAAS 2024-03-01 CN 28 0100, Requirements for cable harnesses

[12] Terminal Technical Specification FCI 211 FT 013

[13] Wire Cable Technical Specification ACOME T 8930

[14] Renault Nissan Technical Specification RNDS-B-00029 V2.0,paragraphs 13.1/13.2.4/13.3.1

[15] M. R. Delozier, S. Orlich, Discovering influential cases in linear regression with MINITAB: Peeking into multidimensions with a MINITAB macro, Statistical Methodology, Volume 2, Issue 2, 2005, Pages 71-81, ISSN 1572-3127, https://doi.org/10.1016/j.stamet.2004.11.005

[16] Mocellin, K., Petitprez, M., Experimental and Numerical Analysis of Electrical Contact Crimping to Predict Mechanical Strength. Procedia Engineering, 81, (2014), https://doi.org/10.1016/j.proeng.2014.10.274.

[17] Morgado, L., Teixeira, A., Sá, J., Pinto Ferreira, L., Almeida, F. D., Analysis and Development of a Failure Prediction Model for Electrical Terminals Used in the Automotive Industry, Procedia Manufacturing, 51, 207-214, (2020), https://doi.org/10.1016/j.promfg.2020.10.030.

Improving dynamic energy absorption performance and stability loss prediction of an all-PET sandwich structure through artificial neural networks

PRODAN Iulia Maria[1], LACHE Simona[1*] and BERARIU Andrei Ionut[1]

[1] Transilvania University of Brasov, 29 Eroilor Blvd., Brasov 500036, Romania

*E-mail: slache@unitbv.ro

Abstract. One key characteristic that is frequently studied in composite sandwich structure mechanics is low-velocity impact performance.

Single polymer composites present higher ductility when compared to traditional glass fibre and carbon fibre reinforced plastic composites (GFRP, CFRP), which makes them strong candidates for impact energy absorption. These lightweight structures are also fully recyclable due to their all-PET constituents, an important aspect when considering possible applications.

In this paper an existing second order composite SrPET and PET foam sandwich structure (self-reinforced poly-ethylene terephthalate matrix and fibre) is numerically investigated for optimal structural design configuration under out-of-plane, low velocity, dynamic loading. The study explores the parametric geometry interdependencies of the structure's configuration with respect to its energy absorption potential. The results are then compared to the quasi-static loading performance behaviour studied by M.N. Velea et all [1]. This work provides a deeper understanding of how geometry configurations can influence energy absorption capabilities highlighting structural strengths and weaknesses while exploring non-equilateral triangle configurations.

The Design Space Exploration (DSE) is carried out by using an optimization algorithm for a dynamic out of plane 3D impact simulation based on the finite element (FE) method software. The synergy between DSE and FE generates a guided simulation loop that can successfully be used to train a neural network to predict the dynamic behaviour of different geometric configurations from the exploration space.

Based on the data samples obtained, an artificial neural network (ANN) is built to predict the energy absorption capacity for a given set of structural design parameters, useful in experimental validation test cases.

Keywords: low-velocity impact optimization, artificial neural networks, Keras 3, Keras Tuner

1. Introduction

Lightweight structures impact performance studies are a significant milestone to reach in the transportation field. The role of these structures is to absorb as much kinetic energy and turn it into non-recoverable strain energy. Composite sandwich panels show higher impact characteristics in out-of-plane dynamic loading, mostly because of the way they are constructed: the upper and lower sheets provide stiffness and the core's flexibility enhances specific bending. In this regard M.N. Velea et al. [1] found that an equilateral triangular shape of a SrPET 2nd order hierarchical sandwich structure subjected to quasi-static loading possesses superior dynamic compression properties than those of a polymeric foam. The FE simulations stress-strain curves were compared to the experimental results for ten variations of the same configuration and an evaluation of the energy absorption capacity was conducted through this parametric study.

F. Tarlochan [2] showed in the review of sandwich structures designed as energy absorbers that no particular design has been identified to have superior capabilities in the vastness of data found in literature and concluded that such studies require extensive design parametric research to optimize the panel for certain applications, highlighting the need to implicate artificial intelligence/data mining coupled with topology optimization in the process. In this respect, X. Du et al. [3] used thin-walled energy-absorbing structures subjected to crash loading to develop novel data mining techniques to use as general design rules for engineering structures.

Sandwich panel optimization and nonlinear compressive response predictions using artificial neural networks (ANN) and genetic algorithm have been investigated by M. M. Sahib et al. [4] and Y. J. Wong et al. [5] and concluded that this methodology can be considered a cornerstone for reducing sandwich structures optimization problems being able to also reduce heavy experimental work efficiently for performance prediction.

2. Structural characteristics: design, materials, and numerical modelling

The baseline geometry configuration of the structure that is subjected to the dynamic load optimization process can be seen in Figure 1-a, which corresponds to "configuration III", geometric configuration in [1] (presented in detail below in Table 1) where the corrugation angle ω was kept at 60°.

Geometric parameters: l_1 – length of the sandwich section, l_2 – length of the monolithic section, h – height of the sandwich unit cell, t_c – thickness of the PET foam, t_f – thickness of the face sheets, ω – folding angle, w – width of the specimens.

The numerical model was constrained and parametrized to accurately update dynamically during the design space exploration process.

Table 1. Baseline geometric configuration carried over from [1].

Configuration	t_f [mm]	t_c [mm]	l_1 [mm]	l_2 [mm]	h [mm]	ρ [kg m^{-3}]	ω [deg]	w [mm]
III	0.45	7.1	81	80.8	70.99	23	60	40

The mechanical characteristics for the composite materials (SrPET and PET foam) used for the numerical model have been carried over from M.N. Velea et al. [1], obtained originally from Schneider et al. [6] and S. Kazemahvazi et al. [7]. A finite element (FE) Abaqus software model was created and computed in an explicit analysis environment to simulate the dynamic response of the 2nd order sandwich structure under out-of-plane impact loading. The model discretization was done using bi-dimensional four node shell elements with the same mesh densities: 5 mm for all 2D assembly constituents: SrPET sheet, impactor, and rigid support. A tri-dimensional four node element shape with the same mesh density (5 mm) was used to simulate the PET foam.

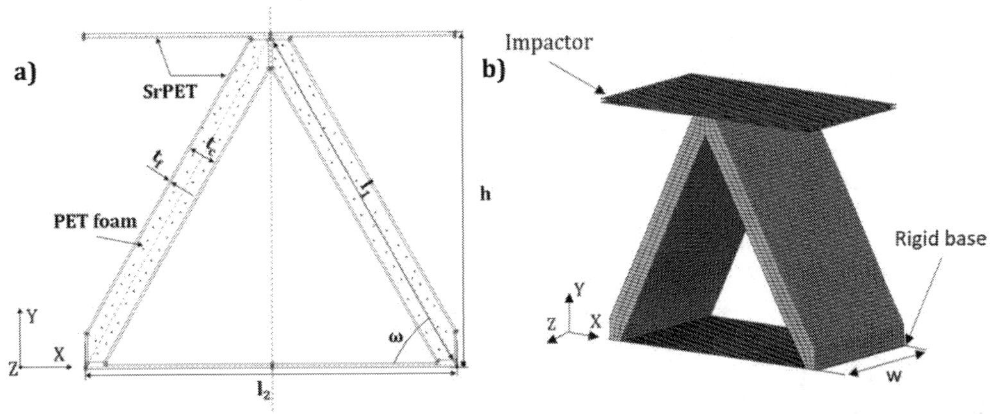

Figure 1. Baseline geometry configuration of the investigated all-PET 2nd order hierarchical sandwich structure (a); FE simulation baseline model assembly (b).

A general contact was used to avoid mesh intersections within the model and connection elements were created to simulate the bonding obtained by local heating between the PET foam and the constituent sandwich SrPET sheets. The structure is resting on a rigid base and is being impacted by a 20 Kg impactor at 2 m/s in the -Y direction with a 100% surface coverage distribution. To simulate the structure's continuity, symmetry conditions have been put in place to account for the X and Z axes continuity (Figure 1-b).

3. Multi-objective optimization: inputs, outputs, and result dissemination

The FE model was connected to HEEDS software to optimally explore the design parameter sets and get an indication as to how these can influence the structure's response to the intended load case, with the ultimate goal of identifying the optimal parameter set.

Several analyses have been carried out to meet this goal by incrementally adding variables to the baseline study. Like any optimization algorithm, the proprietary optimization algorithm of HEEDS software requires variables, objectives, and constraints to perform a successful study. This algorithm was preferred by I. M. Prodan et al. [8] because of its performance in searching large design spaces. There are no constraints applied to the studies performed, but cautionary limits have been defined for the variables. The results have been later verified for variable edge cases. Objectives were set for maximizing the internal energy, but also minimizing the structure's mass.

The baseline study parameters and boundary conditions were mentioned above. No variables or objectives were set for this case and only one evaluation was performed for which the outputs were recorded in Table 2.

Case:1 introduces one variable into the system: ω. This first scenario explores the main constant parameter that is characteristic to this specific sandwich structure, equilateral triangles shape (Figure 1-a). Keeping everything else constant, the design optimization identifies a more performant ω angle given the objective to maximize internal energy. This in turn has a positive impact on the objective value when compared to the Baseline model (Figure 2-a, Table 2).

Illustrated in Figure 2-b is the relationship between ω and the potential internal energy indicating a clear positive impact when ω is increased. The disadvantage with this result is that the crashing event occurs in a short time frame describing a steep deceleration.

Case:2 explores the design space with three variables: ω, t_c and t_f (Figure 3-a) and three objectives (Figure 3-b). t_f was introduced as a discrete value to account for 1 (0.45 mm), 2 (0.90 mm) or 3 (1.35 mm) sheet layers of SrPET. In this case the weighted sum of all objectives was considered to be the optimization challenge.

Table 2. Optimization inputs for all studies.

Study name	Variables	Objectives	No. of Evaluations
Baseline	none	None	1
Case:1	$2 < \omega < 100$	Maximize: internal energy	150
Case:2	$2 < \omega < 100$ $0 < t_c < 15$ t_f: 0.45, 0.9, 1.35	Maximize: internal energy, analysis run time Minimize: Mass	800

The FE analysis setup is conditioned to stop when the impactor speed reaches zero velocity; therefore, to detect a well-controlled velocity decrease a simulation run time maximization objective was requested (Table 2). The optimized configuration and results have been recorded in Table 3. Worth mentioning is that approximately 5% out of the 800 evaluations were discovered to have similar optimized results.

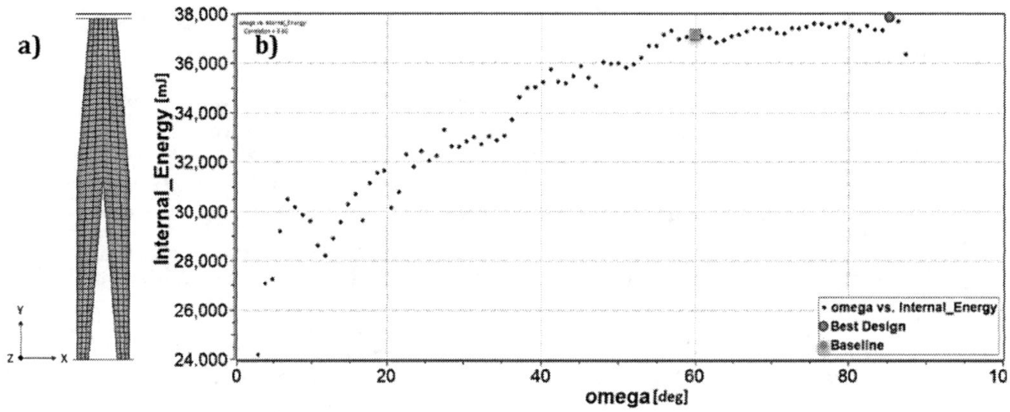

Figure 2. Case: 1 FE discretized model assembly (ω = 85.3 [deg]) (a); Variable vs. Objective (b).

When compared to the baseline model (Table 3) both cases demonstrate an increase in internal energy with a structural mass reduction of 38% (case:1) and 51% (case:2) respectively. Also, the structure's overall occupied volume logged a significant decrease by 82% (case:1) and by 87% (case:2). Another important aspect highlighted is the influence of ω in the structure's evaluated performance showing that a general isosceles tringle configuration is better suited for this kind of application, while a similar strong ω angle increase in both cases indicate a narrow base triangle configuration. Case:2 t_f parameter result indicates that additional plies do not have a significant contribution to the structure's energy absorption capacity, supporting the same conclusion reached by I. M. Prodan et all [8] while studying a novel SrPET sandwich configuration subjected to low-velocity impact.

4. Energy absorption characteristics: failure mode discussion

M. N. Velea et al. [1] showed that for "configuration III" the specific energy absorption capacity SEA [KJ/Kg] can be determined analytically in [1] by knowing the ratio between the absorbed energy E_{abs} and its mass m. Because of the quasi-static loading conditions, E_{abs} was determined with knowing the crushing force and distance which in turn meant that the SEA estimation was that of the lower boundary of absorption capacity. In the dynamic case presented in this paper the SEA can be calculated by knowing the internal energy E_{int} and mass m of the structure (Eq. 1 and Eq. 2). The determined SEA capacity for "configuration III" in [1] was 0.52.

$$SEA_{Baseline} = \frac{E_{int}}{m} = 1.86 \quad (1)$$

$$SEA_{Case:2} = \frac{E_{int}}{m} = 3.91 \quad (2)$$

By way of the optimization process the SEA indicator increased with more than double when compared to the baseline model.

Table 3. Study optimization summary.

	ω [deg]	t_f [mm]	t_c [mm]	l_1 [mm]	l_2 [mm]	h [mm]	w [mm]	KE [mJ]	E_{int} [mJ]	Mass [g]	Time [s]
Baseline	60	0.45	7.1	81	80.8	70.99	40	40000	37158	19.9	0.027
Case: 1	85.3	0.45	7.1	81	12.37	78.92	40	40000	37913	12.3	0.013
Case: 2	86.28	0.45	2.52	81	9.61	79	40	40000	38362	9.8	0.05

KE – Kinetic Energy; E_{int} – Internal Energy

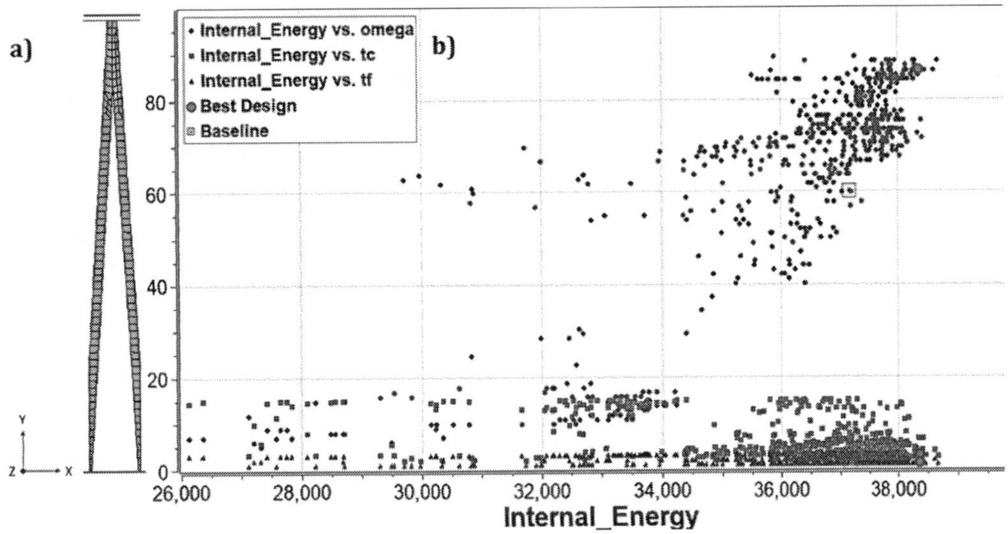

Figure 3. Case: 2 FE discretized model assembly (ω = 86.28 [deg]) (a); Variable vs. Objective (b).

IOP Conf. Series: Materials Science and Engineering **1319** (2024) 012003 doi:10.1088/1757-899X/1319/1/012003

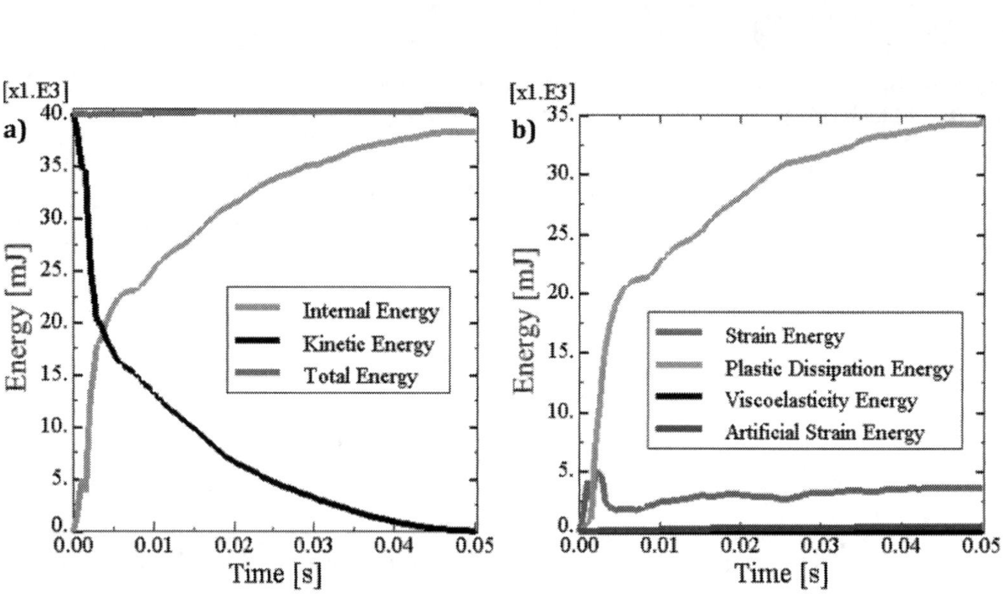

Figure 4. Case: 2 Energy balance (a); Internal Energy components (b).

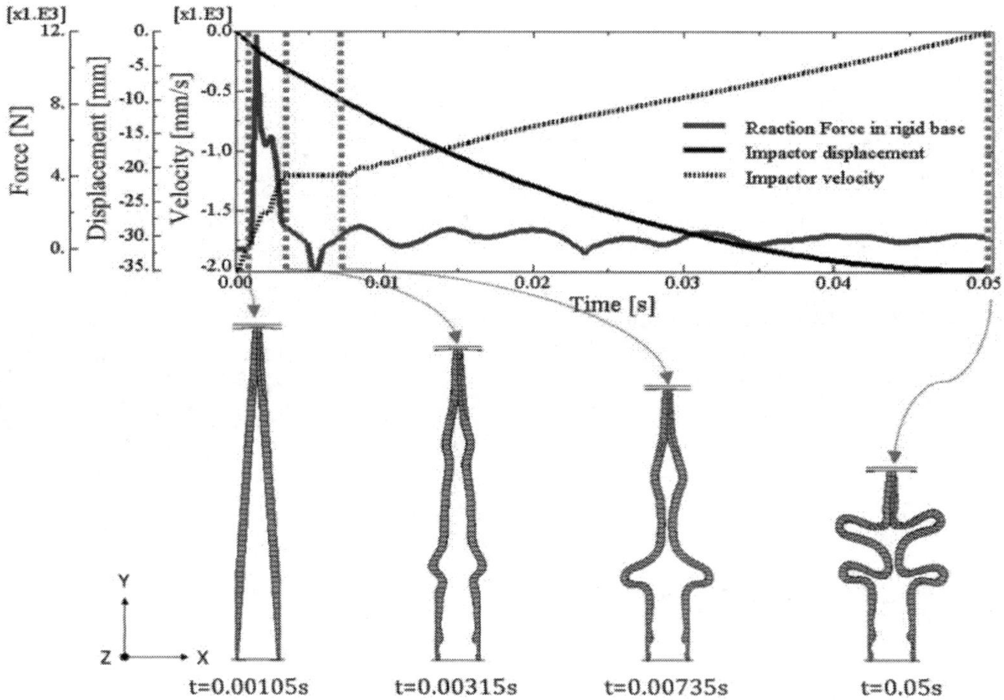

Figure 5. Evolutions of Force, Displacement and Velocity along with deformation snapshots from the impact simulation.

The energy balance seen in Figure 4-a displays how the kinetic energy is transformed into internal energy in relation to time while a slight increase of 0.87% in total energy is apparent due to the numerical method used. This percentage is acceptable being below the required 1%

threshold. Figure 4-b exposes the internal energy by its components: viscoelasticity (creep) energy is 0 [mJ], artificial strain energy (energy stored in hourglass resistances and transverse shear in shell elements) reaches a maximum of 0.37 [mJ], strain energy maximum shows 5 [mJ] while leaving plastic dissipated energy accountable for 87% of the internal energy.

Figure 5 exposes a smooth controlled descent of the impactor reaching zero velocity at 35 mm. The peak reaction force followed by constant velocity is corelated with the structure's stability loss at that time due to local buckling, which is followed by general buckling later. The shape of the force response can be corelated with the Force - Displacement curve obtained by M. N. Velea et all in [1], yielding to the same type of response for the optimized structure.

5. Artificial neural network design and training

In this study, the latest version of Keras, a multi-framework deep learning API, was utilized to construct the ANN. Keras 3 supports seamless integration with various deep learning frameworks, enabling a more streamlined development process. A significant advancement in Keras 3 is the inclusion of an integrated hyperparameter tuning tool, Keras Tuner. This tool automates the optimization of the network's hyperparameters. By employing Keras Tuner, the conventional and laborious trial-and-error approach commonly adopted in the field was circumvented, thereby enhancing the efficiency and effectiveness of the regression model development [9].

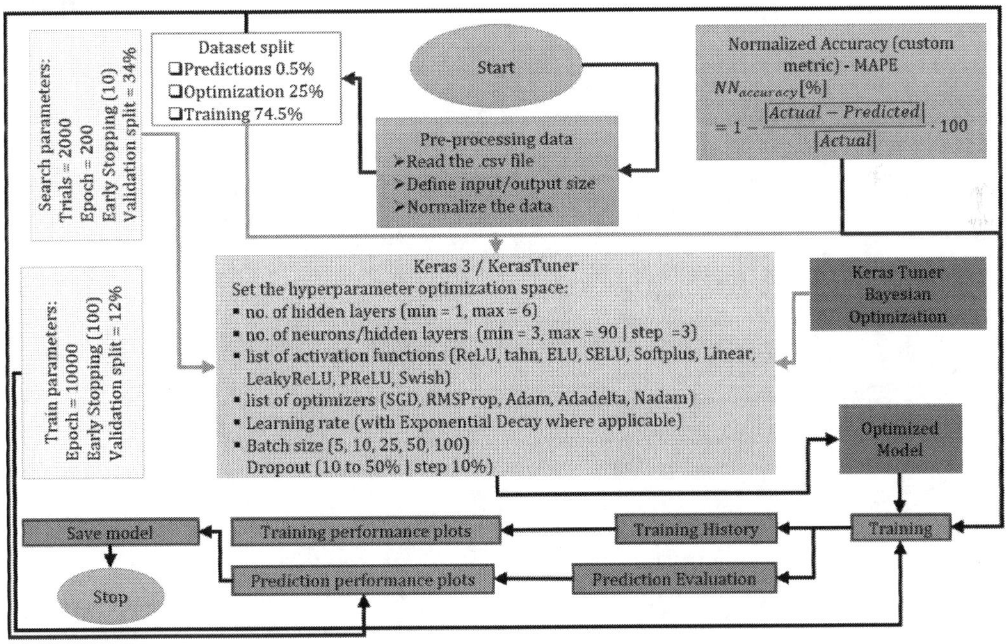

Figure 6. Flow chart for self-building ANN model with Keras 3 and Keras Tuner

Figure 6 provides a complete overview of the proposed technique for self-building of the ANN used to estimate the dynamic behaviour of the structure based on the design set obtained in section 3. The first step implied reading the data set, normalizing it per variable and divide the information as: input (X – standardized notation for the input tensor) with 3 dimensions (ω, t_c

and t_f), and output (y - standardized notation for the output tensor) also with 3 dimensions (E_{int}, Mass and Run_Time) resulting a dataset of 611 samples. The normalization is necessary mainly for the uniform scaling of the data. In the next step the original set was split into: optimization (152 samples), training (455 samples) and final prediction (4 samples) sets. A key component in this self-building ANN structure is the evaluation metric. For this research it was decided to use the simplest (but robust) way to evaluate the accuracy of the predictions thru a normalized relative error (NN_accuracy) expressed in % (in essence, a modified version of the Mean Absolute Percentage Error - MAPE). Based on this metric, the cost functions can be calculated for the different phases of the optimization/building algorithm. The dropout feature was deployed in this case to avoid overfitting the network. As for all hyperparameters, the search space is specifically defined for regression problems. At the core of the optimization step is the Bayesian Optimization, a sophisticated technique for global optimization that balances exploration and exploitation to

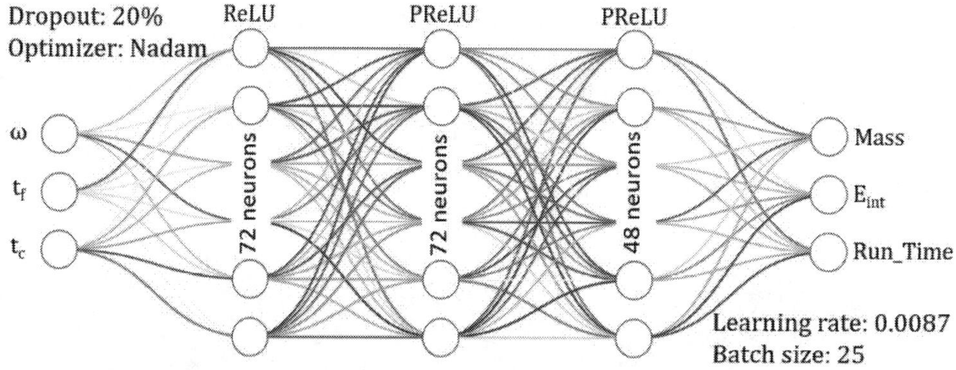

Figure 7. Optimized ANN structure for structural dynamic prediction – Keras Tuner

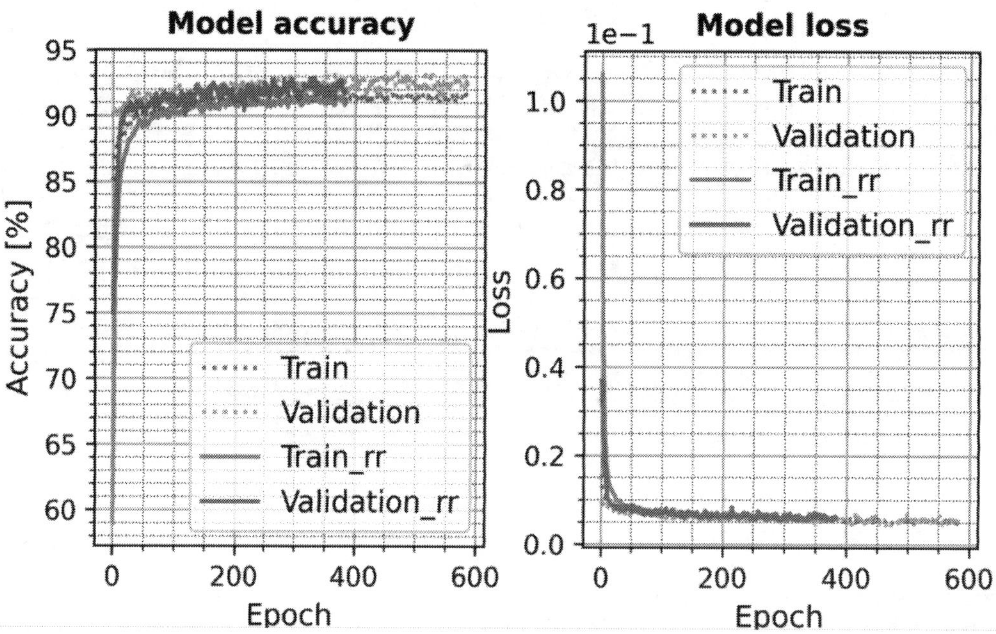

Figure 8. Training performance of the optimized ANN, model accuracy and loss

minimize the number of function queries [10]. Coupled with an early stopping condition that checks if there is any significand improvement in the last 10 epoch, generates a very efficient strategy to quickly skip trials that are not feasible. The optimization process generated after 2000 trials an ANN model exposed in Figure 7, with the following configuration: no. of hidden layers: 3; neurons in the hidden layer: 72, 72 and 48 respectively; activation per hidden layer: ReLU, PReLU for the last two; dropout: 20%; optimizer: Nadam; learning rate: 0.0087, batch size: 25. The best NN_accuracy found was 93.06% with a computation cost of 28.5 hours on a machine with the following configuration: CPU: Intel i7-8850H; RAM: 64 GB; SSD.

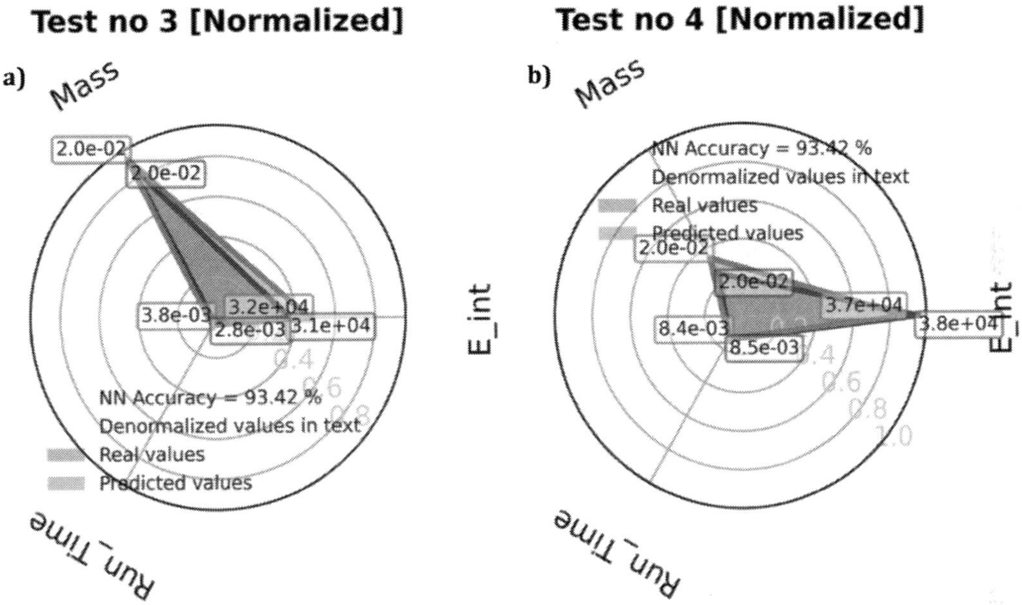

Figure 9. Prediction performance of the optimized ANN, test case 3 (a) and 4 (b)

The following step implied training the obtained ANN model/hyperparameters with the training set, applying a similar early stopping condition for the process that check if any significant improvements are present in the last 100 epoch. This condition was set mainly to avoid overfitting but also to save computation time. In order to simulate a real use case, it is also recommended to reinitialize the weights of the model, those by default are set based on the optimization set. In Figure 8, with dashed lines, are the accuracy and loss plots generated by training the network with the original weights. Both metrics are showing that the network is bias to the data and generate smoother learning behaviours thus triggering the early stop criterion later in the process.

The training process was very fast, under 2 min, triggering the early stopping criterion after ~380 epoch with a final NN_accuracy of 93.42% (Figure 8: *_rr curves). An important aspect in Figure 8 is the accuracy plot (left), where validation that both lines are increasing steadily over epochs is important. In this case a red flag behaviour would be to see the validation line plateauing while the train increasing signalling an overfitting. Similar checks apply for the loss plot (right), both lines should decrease over epoch and also the overfitting is signalled here when training loss keeps decreasing while validation increases. As demonstrated, both plots suggest steady and predictable evolutions validating the optimization step.

The best option to test the optimized and trained network is to expose it to samples that are totally new, in this case the prediction set, and compare the real values with the ones that the ANN generates. In Figure 9 there are two cases of predictions given by the network. For clarity it was decided to use normalized polar plots, where the normalized real and predicted values are used to construct the visualisation and the denormalized values are added as text boxes for each point for further evaluation.

6. Conclusions

An optimized FE model version for low-velocity impact of an all-PET 2nd order composite sandwich structure was obtained, compared to existing quasi-static studies and used for building an ANN. The improved design showed that isosceles triangle geometry configuration would be more appropriate instead of equilateral for this type of application while obtaining a significant occupied volume reduction of 87% and a mass reduction of 49% when compared to the baseline model. Similar local and general buckling failure modes were obtained by M.N. Velea et all in [1].

The use of Keras 3 and Keras Tuner was explored for creating a self-building ANN based on Bayesian Optimization. The high accuracy of 93.06% obtained in the optimization step demonstrates the effectiveness of the proposed approach in finding optimal configurations/hyperparameters. The optimized model was reinitialized before training to demonstrate the robustness of the model with impressive results, 93.42% accuracy and very good predictions for the tested cases. Overall, the work provides a strong proof of concept for using the Bayesian Optimization in self-building neural networks with applications in specific fields of engineering.

References

[1] M. N. Velea, S. Lache, Energy absorption of all-PET 2nd order hierarchical sandwich structures under quasi-static loading conditions, Thin-Walled Struc. 138 (2019) 117-123.

[2] F. Tarlochan, Sandwich Structures for Energy Absorption Applications: A Review, Materials 2021, Volume 14, Issue 16, pp.4731, https://doi.org/10.3390/ma14164731.

[3] X. Du, H. Xu, F. Zhu, A data mining method for structure design with uncertainty in design variables, Computers & Structures, Volume 244, 2021, 106457, ISSN 0045-7949.

[4] M. M. Sahib, G. Kovács, Multi-objective optimization of composite sandwich structures using Artificial Neural Networks and Genetic Algorithm, Results in Engineering, Volume 21, 2024, 101937, ISSN 2590-1230.

[5] Y. J. Wong, S. K. Arumugasamy, K.B. Mustapha, Development of a computational predictive model for the nonlinear in-plane compressive response of sandwich panels with bio-foam, Composite Structures, Volume 212, 2019, Pages 423-433, ISSN 0263-8223.

[6] C. Schneider, S. Kazemahvazi, M. Åkermo, D. Zenkert, Compression and tensile properties of self-reinforced poly (ethylene terephthalate)-composites. Polymer Testing (2013), Vol. 32(2), pp. 221–230.

[7] S. Kazemahvazi, C. Schneider, V.S. Deshpande, A constitutive model for self-reinforced ductile polymer composites. Composites Part A: Applied Science and Manufacturing (2015), Vol. 71, pp. 32–39. doi:10.1016/j.compositesa.2015.01.003 .

[8] I. M. Prodan, S. Lache, A. I. Berariu, Numerical modelling and design exploration of a novel sandwich structure designed for low velocity impact, Materials Today: Proceedings, Volume 45, Part 5, 2021, Pages 4117-4121, ISSN 2214-7853.

[9] Chollet, François and others, 2024, Keras 3, GitHub, https://github.com/keras-team/keras.

[10] O'Malley, Tom and Bursztein, Elie and Long, James and Chollet, François and Jin, Haifeng and Invernizzi, Luca and others, 2024, Keras Tuner, GitHub, https://github.com/keras-team/keras-tuner.

IOP Conf. Series: Materials Science and Engineering **1319** (2024) 012004 doi:10.1088/1757-899X/1319/1/012004

Influence of welding angle in the process of submerged arc welding of pressure vessel steel

Cioroagă Bogdan-Dorel[1*], Socalici Ana Virginia[1] Cioată Vasile George[1] Dascăl Amalia Ana[1]

[1] Politehnica University Timişoara, Faculty of Engineering in Hunedoara, Hunedoara, Romania
[2] Politehnica University of Timişoara, Faculty of Mechanical Engineering, Timişoara, Romania

*E-mail: bogdan.cioroaga@student.upt.ro

Abstract. In the industrial production of pressure vessels (PV), a very important role is played by the welding of the elements that make up the vessel. The predominant welding technology in the case of manufacturing PV is the submerged arc welding (SAW) for which different welding regimes are defined depending on the characteristics of the base material. The welding regimes are made up of various parameters easy to manipulate by the operators of the welding machines, one of them is the welding angle. In this article, aspects are discussed regarding the way in which the welding angle of SAW influences the quality of welded joints. Tensile tests were carried out on specimens taken from plates joined by SAW using different values of the electrode tilt. Aspects regarding the geometry and dimensions of the welded joints obtained as a result of the variation of the welding angle are analysed and, in the end, an optimal angle is indicated for the welding process of PV. The basic material used in the experiments is a fine-grained material intended for the manufacture of vessels that work in high pressure and high temperature regimes, named in the technical literature P355 N.

1. Introduction

In the PV manufacturing industry, it is essential to use materials that can withstand the operating conditions of these products, as well as a vessel manufacturing technology that offers a balance between the economic and the technical aspects of production. The common aspects of the materials from which the pressure vessels are made are that they are part of the category of special steels, presenting properties such as: resistance to high temperatures and pressures as well as to corrosion due to the fluids they store. In this sense, because there are PV that can function as organs of various machines and installations, they can generally be made of less demanding steels in terms of the alloys, there are PV intended for the food industry that are made of stainless steels and there are vessels used in the chemical industry that are also made of special stainless steels that have good anti-corrosion properties. A source of documentation from specialized literature that deals with aspects related to the design, operation, destination and types of PV is Ghanbari G. et al. book [1] and Moss D. R. Pressure vessel design manual [2].

A very important technological aspect in the manufacture of steel PV is the joining of the semi-finished products by welding in order to obtain the final PV. Due to the fact that PV must ensure a good seal, they are made of metal sheets, processed by plastic deformation in the rolling or embossing processes and the parts obtained are closed and joined by continuous welding seams along the entire length of the joining edges. The welding joint area should not represent a weak point of the PV product, although this area is the most prone to the appearance of material defects that can initiate cracking of the PV walls. The widely used welding method for welding PV is SAW, this is preferred because it has a high productivity and can be easily automated for straight and circular welding paths. The welding machines used during SAW processes are shown in figure 1 detail A. The equipment that has the role of ensuring the positioning and advancement of the electrode holder torch is made up of an montant that supports a mobile beam that can perform rectilinear movements in the X and Z directions. The device that supports the semi-finished product to be welded is made up of two rollers actuated by electric motors that can print a rotational movement to the cylindrical-shaped semi-finished product to carry out the welding on the inner or outer circular contour, while the electrode holder torch is kept fixed. If longitudinal welded joints are made along the semi-finished product, the electrode holder torch moves in the welding direction and the semi-finished product remains fixed.

Figure 1. The equipment used for welding pressure vessels. Submerged arc welding technology. [3]

The welding takes place by depositing a layer of granular flux material in the area of the joints, having the role of a protective and alloying environment for the metal bath, it also ensures a safer environment for the human staff who are near the welding machine due to the fact that it blocks the light emissions and the gases released after melting the metal. The machines dedicated to PV welding using the SAW method are of significant complexity and depending on the final dimensions of the PV, this equipment is adapted in some cases, massive and gauge equipment being required. PV welding involves the creation of continuous and long welded seams for which the parameters of the welding process must be kept constant throughout the welding process, therefore the use of semi-automatic SAW welding equipment is preferred. various technical aspects about process equipment and SAW technology can be found in the technical literature in books such as Ramesh Singh 's [4] or S. Nand's and Pravin Kr. Singh's [5].

Setting the parameters of the welding regime is also a decisive criterion in obtaining a good quality and productivity. This study focuses on the influence of the electrode tilt angle during the welding process, the evolution of the mechanical properties of the welded joints depending on diferent values of the welding angle is targeted.

Studies that follow the influence of the angle of inclination of the electrode during welding processes are carried out by Li, J., Li, H., Wei, H. et al. in there research [6]. The experiments carried out by them were based on welding using the MIG technology used for brazing 6061 aluminum alloy to 304 stainless steels. They discovered that the macrostructure, microstructure, mechanical properties and shape of the welding arc are affected by torch aiming position being sensitive to the changes in value of the welding angle. They obtained as a result of experiments joints of high tensile strength, which were fractured in the heat affected zone (HAZ) of Al alloys in 72% of the cases at values near to 89 MPa, using a welding angle of 20 deg. Other studies such as [7] were Deepak Pathak's et all. exposes the situation of his experiments in which they used shielded metal arc welding (SMAW) to join a base material consisting of 100x50x10mm AISI 1020 steel plates, using an electrode E6013 as filer material, observing the effects of the electrode angle during the welding process on the mechanical properties of the joint. The experiments consisted in making and testing 18 welded tensile test specimens that were welded using electrode angles of 65, 75 and 85 deg for which the welding current was also varied, which had the values of 95, 110, 125 A, different polarities of the welding current were also taken into account. Their findings were that by increasing the welding current intensity while maintaining the electrode angles in the range of 65-85 deg the tensile weld strength improved. Other authors such as J. Edwin Raja Dhas and S. Kumanan present in their disseminations [8] optimization cases of SAW processes in which aspects related to the effects of the variation of the welding parameters such as welding voltage, welding current, welding speed and arc length are presented. By going through the specialized literature, you can find other results of other research that aim at optimization aspects that take into account the welding angle and its effects.

The geometry of the joint area obtained after welding can be influenced by any of the parameters of the welding regime, indicating too low or high values of the certain parameters. The fact that the welded joints present different geometries servs as reference basis for the diagnosis of the mechanical strength from a simple visual inspection. Researches such as [9] present aspects related to the influence of process parameters on the shape and geometry of welded joints. The authors analyzed measurements made on images of different welded joints made using different welding regimes in order to determine trends in the evolution of the geometry and dimensions of the weld bed, SAW technology was used to carry out the experiments.

The presented studies are indicators that the topic of welding using different basic materials, filler materials and different welding technologies for the study of welding process parameters and especially the welding angle is a subject of interest, being an area still with potential in terms of process optimization.

2. Materials and Methods

2.1 Materials

In order to prepare welded joints using the SAW method, from which samples are taken for visual inspection and tensile testing, it is necessary to use 3 types of materials, these being the base material, the filler material and the material that constitutes the protective environment of the weld during the welding process.

The base material is a fine-grained, low-alloy carbon steel, specially designed for PV fabrication that operates under high temperature and pressure conditions. The index of this steel according to EN 10027-1:2005 in P355 N, its elaboration method being in accordance with

IOP Conf. Series: Materials Science and Engineering **1319** (2024) 012004 doi:10.1088/1757-899X/1319/1/012004

the standard SR EN 10216-3: 2003. The semi-finished form of this material is that of 6mm thick laminated tin plates. The chemical composition (maximum values) of this steel is made up of: C 0.2%, Si, 0.5%, Mn 1.7%, P 0.025%, S 0.02%, Al 0.02%, Cr 0.3%, Cu 0.3%, Mo 0.08%, Nb 0.05%, Ni 0.5%, Ti 0.04% and V 0.1%.[10][11]

The filler material is dedicated for SAW processes, being in the form of a electrode wire that has a 3.2mm. The manufacturer of this material is ESAB and according to the names given by him this material is identified as OK Autrod 12.24. This is an molybdenum-alloyed steel with copper-clothing that has a chemical composition (maximum values) of: C 0.12%, Si 0.2%, Mn 1.2%, P 0.02%, S 0.02% and Mo 0.6%.[12]

According to the carbon concentration, the base material and the addition material are close, indicating a good affinity between them for fusion by welding. Figure 2 indicates the location of the base material and the filler material on the steel equilibrium diagram. It can be observed that both materials form ferrite-pearlitic structures which must present good properties in terms of plastic deformation and weldability.

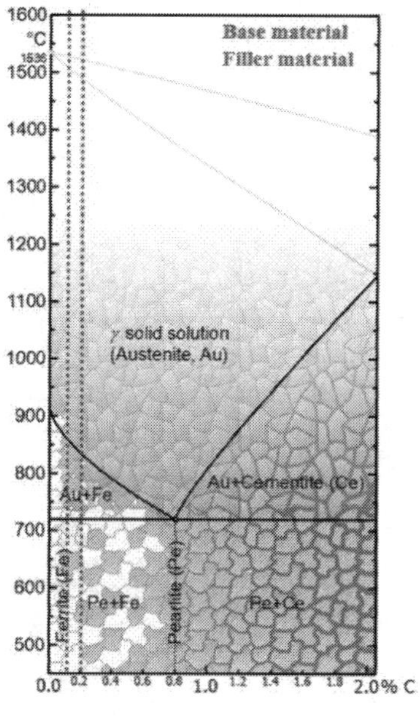

Figure 2. Basic and the filler material exposed on the steels equilibrium diagram F3C [13]

The protective environment that has the role of isolating the molten metal bath from the external environment is a granular flux-type material of granulation 0.315-2mm. This material is identified according to the manufacturer ESAB as OK Flux 10.72.

Another aspect that should be mentioned is that during the welding process a profiled rod type ceramic material was used which has the role of supporting the root of the weld so that the melted metal does not leak between the welding joints during the SAW joining process. This rod being applied along the entire length of the joint area by self-adhesive bonding, the welding being done on the opposite side, above it.

2.2 Methods

In order to carry out the experiments, steel plates were prepared to be joined by welding, having the initial dimensions of 300x125x6, becoming after the welding process 300x250x6, resulting in a weld seam of 300mm for each specimen. The preparation of the plates before welding consists in cleaning by sanding the edges that will come in the joining area, followed by positioning the edges end to end, leaving a 1 mm joint, keeping them in position by applying two welding points to each end of the plates.

Following the effects of the electrode tilt angle, the samples were made using values of 45, 60, 75 and 90 degrees of this parameter. The standard welding regime used in the industry for this type of semi-finished product is defined by the following parameter values with direct adjustment by the human operator:

- electrode tilt angle: 90 deg
- welding current intensity: 480 A
- welding current voltage: 33 V
- welding speed: 60 cm/min

During the experiments, only the electrode tilt angle (welding angle) parameter changes, the rest remain constant as in the standard regime. The welding process of the samples takes place according to figure 3 where in detail A the travel direction of the electrode and the position of the welding angle are represented. Detail B presents an image captured during the welding process of the standard welding regime sample executet using the 90 degree welding angle.

After making the welded specimens using the 4 welding regimes with different inclinations of the electrode, images were collected from which the effects of each welding regime can be visually analyzed. Measurements were also made on the welding specimens to highlight the influence of the parameter under study on the width (L) and height (H) of the seam.

After collecting the visual data and the dimensional measurements of the welded joints, the welded pate specimens are cut into strips of 37 mm width and after processed by milling to obtain 3 tensile test specimens for each welding regime under study. The dimensions and the

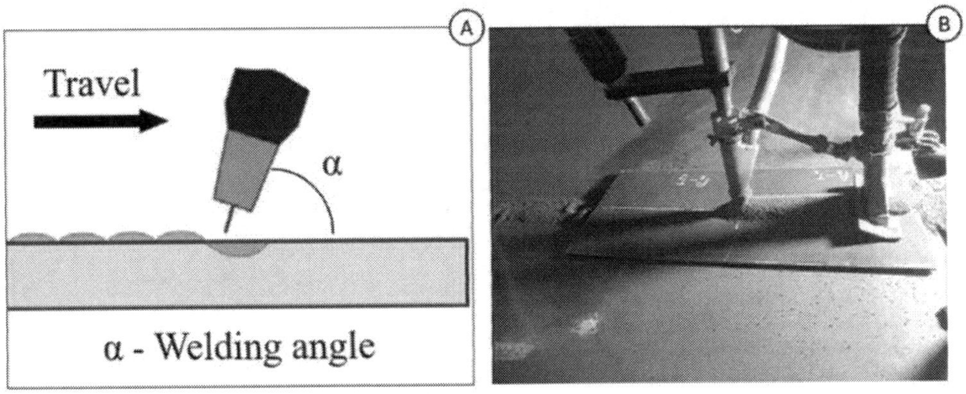

Figure 3. The SAW process of experimental samples

sampling method of the tensile test specimens from the welded specimen plate are presented in figure 4.

The sampling of the tensile test specimens was carried out according to standard BS EN 895:1995. Their processing taking place only by mechanical cutting methods, in the first phase by guillotining, then by milling with cooling in order not to thermally affect the material of the sample.[14]

The transversal tensile testing of the welded specimens was carried out with the help of a universal testing machine A009 (TC100). The tensile test specimens were tested by progressively applying a load of 50 MPa/s until the break of the specimen occurs.

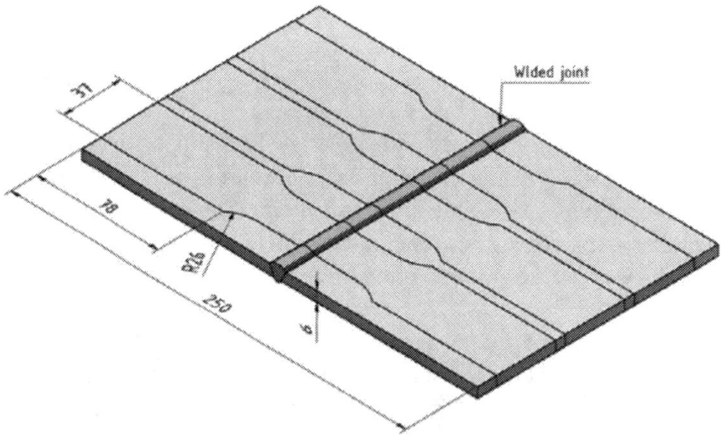

Figure 4. Specimens for tensile testing of welded joints

3. Results and discussions

3.1 Visual inspection results

The graphical data collected for visual inspection are centralized in a comparative way in figure 5. It can be observed that the images with the samples made using the welding regimes with the values of 45, 60, 75 and 90 degrees are grouped on two columns, one indicating the face on which the welding was performed (left) and the other showing the root of the welded joint (right).

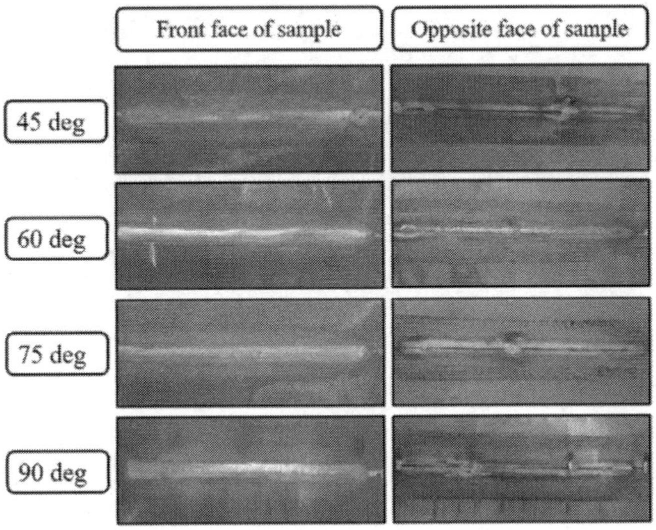

Figure 5. Welded samples using different values of the electrode tilt angle.

It is noted that from a visual point of view, the welding seams show a high degree of similarity, presenting welded joints with a continuous character and proportional width on the front face of the sample. Analysing, on the opposite face, the seam executed with an electrode inclination of 45 degrees, presents an incomplete penetration at the root of the weld. The case of the sample executed with a welding angle of 60 degrees shows a complete penetration of the filler material in the joints, forming a continuous protrusion along the entire length of the welding bead on the opposite side of the sample where the root of the weld is located. In the case of the sample welded with a welding angle of 75 degrees, it is observed the appearance on the opposite side at the root of the weld of a thermally influenced zone outside the welding joints of purple-blue color and an incomplete penetration along the entire length of the seam. The last welding regime under study, the one in which the welding angle is 90 degrees, presents on the opposite side both a thermally influenced zone and a discontinuous penetration of the filler material into the at the root of the weld.

It is found that small welding angles (45 deg) cannot provide complete penetration of the welding joint, while large welding angles (75 deg and 90 deg) cause the appearance of a pronunced HAZ due to the concentration of the heat flow and the filler material on a small area near the electrode. Small welding angles below 75 deg tend to dissipate the heat released by the electric arc along the welding joint and to push the filler material along the joint in the opposite direction to the welding advance, which favors the heat dissipation and filling of the joint on this directive, the deep penetration of of the joint is more pronounced as the welding angle tends to 90 deg.

3.2 Dimensional mesurements results

The data collected by performing direct measurements on the height (H) and width (W) dimensions of the welded seam are presented in figure 6. The values presented on the graph are average values, these being collected from 4 measurements performed on each welded sample, two measurements in the end zone of the weld and one in the middle of the sample. The measurements were performed using calipers with digital display Lumytools LT15240, 150 mm, precision 0.02mm.

Analysing the data presented in figure 6, it can be said that by increasing the welding angle, there is an upward evolution of the width of the welding bead W, presenting a directly

IOP Conf. Series: Materials Science and Engineering **1319** (2024) 012004 doi:10.1088/1757-899X/1319/1/012004

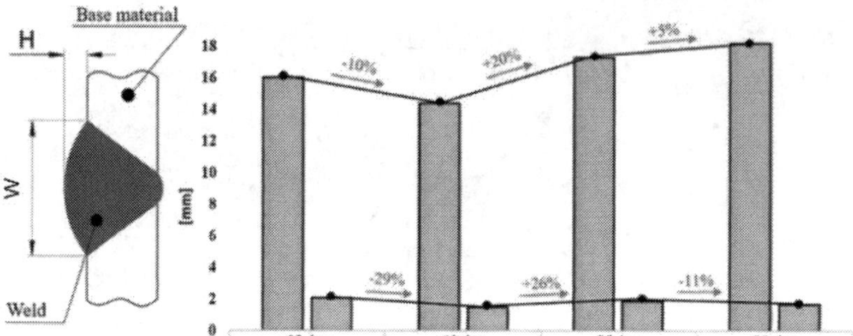

	45 deg	60 deg	75 deg	90 deg
Width welded seam [W]	16	14,4	17,3	18,2
Height of welded seam [H]	2,1	1,5	1,9	1,7

Welding angle

Figure 6. Aspects regarding the dimensions of the welded samples.

proportional evolution character. Regarding the evolution of the height of welded seams H, it has an oscillatory character with an amplitude difference between two welding regimes of 22%.

The width (W) and height (H) of the welded seams in general are influenced in welding processes by the degree of dilution and the deposition rate of the filler material, in this case the phenomenon that occurs is characterized by the increase in dilution for welding angles that tend to 90 deg, which favors the appearance of a flattened shape of the welded joint. Also due to the increase in the dilution rate, the leakage of the additive material can be observed through the joints.

3.3 Tensile test results

As a result of the tensile tests, a series of stress-strain curves resulted, 3 for each welding

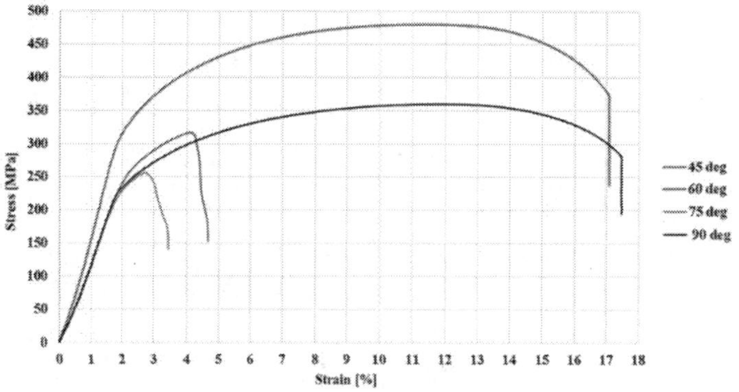

Figure 7. Stress-strain curves evolution depending on the welding angle.

regime, in total being 12 stress-strain curves, a representative curve was chosen for each presented welding regime that illustrated in the graph from figure 7. It is considered the welding regime that has the 90 degree welding angle as the criterion of comparison, since it is the one used as standard in most of the industrial production units.

It can be observed that the welding regime for which the welding angle of 60 degrees is specific, it shows a stress-strain curve with greater coverage on the stress axis, the other welding regimes present curves with lower coverages than the standard regime indicating welded joints that show poor values of the mechanical properties.

IOP Conf. Series: Materials Science and Engineering **1319** (2024) 012004 doi:10.1088/1757-899X/1319/1/012004

Following the analysis of the stress-strain curves, a series of indicators resulted, the most relevant ones were extracted and presented graphically in figures 8 and 9.

It is mentioned that the nature of the break of the tensile test specimens shows rupture through the base material for the specimens that are executed using the welding angles of 60, 75 and 90 degrees. Specimens welded using the welding angle of 45 degrees show the breakthrough the welded seam due to weak penetration of the filler material in the welding joints.

Indicators from the stress category are presented in figure 8 it is noted that for the yield tensile stress of the welded joint, the welding regimes for which welding angles of 45 and 60 degrees are representative show values higher than the standard welding regime.

The welding regime that uses the welding angle of 60 degrees shows higher values than the standard welding regime with 33% for maximum stress, 21.5% breaking stress and 34% for yield tensile stress.

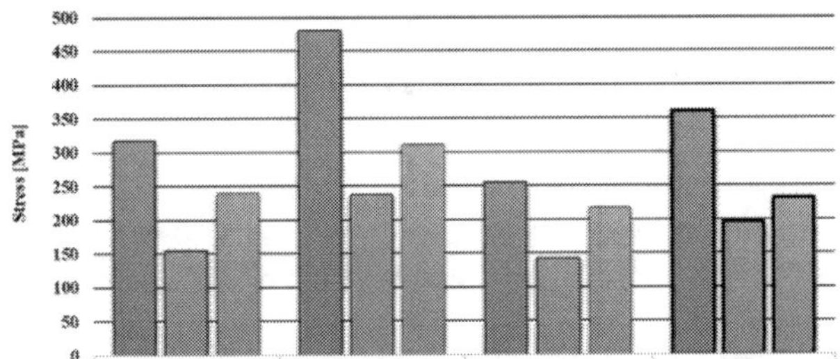

	45 deg	60 deg	75 deg	90 deg
Maximum stress	317,16	479,79	255,80	360,12
Breaking stress	154,13	237,76	141,76	195,66
Yield tensile stress	239,91	312,17	218,31	232,29

Welding angle

Figure 8. Average values of stress obtained from tensile test tress-strain curves evolution depending on the welding angle.

The situation of the indicators from the category of relative strain is shown in figure 9, where it can be seen that the maximum values are predominant for all 3 indicators within the standard welding regime. Values close to those of the standard regime are also recorded for the indicators associated with the welded specimens using the welding angle of 60 degrees.

The only strain indicator that shows significant lower values among those of the welding regime using 60 degrees compared to the standard regime is the strain corresponding to the maximum stress, being lower by 58% than the case of the standard welding regime. In the case of strain corresponding to the yield tensile stress.

The evolution of the strain corresponding to the yield tensile stress compared to the welding angle shows very small variations, being highlighted the welding regime that uses the welding angle of 45 degrees where an increase compared to the standard is recorded by 2.6% for this strain indicator.

Observing the big picture of the stress-strain curves evolution resulting from testing the tensile test specimens, it can be found that the welding regime that presents the best mechanical properties is the one in which the 60 degree welding angle is used, followed by the standard regime using 90 degree.

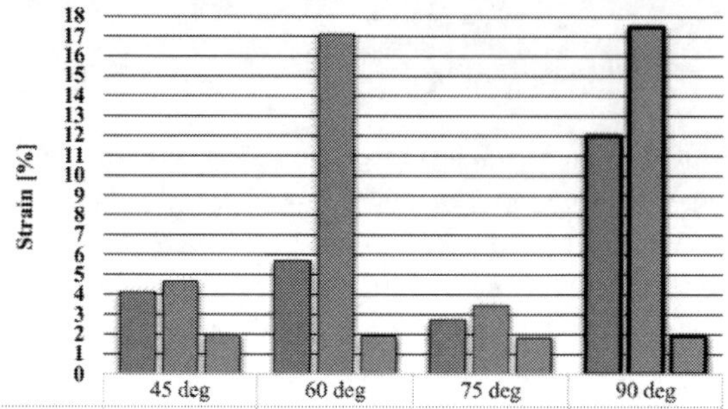

	45 deg	60 deg	75 deg	90 deg
Strain corresponding to the maximum stress	4.10	5.69	2.72	11.99
Strain corresponding to the breaking stress	4.66	17.08	3.45	17.46
Strain corresponding to the yield tensile stress	1.98	1.93	1.82	1.93

Welding angle

Figure 9. Average values of strain obtained from tensile test tress-strain curves evolution depending on the welding angle.

4. Conclusions

As a result of the experiments, it was found that the welding regime that uses the welding angle of 60 degrees presents the best mechanical properties, having the highest values of the transverse tensile strength indicators collected after drawing the stress-strain curves. The standard welding regime, which is specific to the welding angle of 90 degrees, also presents good mechanical properties, but on average 25% lower than those ensured by welding with an angle of 60 degrees.

The degree of penetration of the filler material in the welding joints is influenced by the welding angle, the best penetration being identified for the welding regime with 60 degrees and the weakest penetration specific to the regime that uses the angle of 45 degrees.

It was observed the existence of a tendency to increase the width of the welded seam with the increase of the welding angle. The rise of the welded seam is not greatly influenced by the angle at which the welded joint is executed.

References

[1] Ghanbari G et al, 2011, *PressureVessel Design – Guides&Procedure*, Navid Shiraz Publication.

[2] Moss D R, 2004, *Pressure vessel design manual 3th edition*, Gulf Professional Publishing.

[3] Web page: https://www.senlisweld.com/fit-up-rotator/, accessed on 10.04.2024

[4] Ramesh Singh, 2020, *Applied Welding Engineering, Processes, Codes, and Standards, 3rd Edition - June 2*, Elsevier Science, ISBN: 9780128213483

[5] S. Nand, Pravin Kr. Singh, 2020, *Submerged Arc Welding With Metal Powder Addition: M-SAW*, LAP LAMBERT Academic Publishing , ISBN-10 : 6202795603

[6] Li, J., Li, H., Wei, H. et al., 2016, *Effect of torch position and angle on welding quality and welding process stability in Pulse on Pulse MIG welding–brazing of aluminum alloy to stainless steel*, Int J Adv Manuf Technol 84, 705–716. https://doi.org/10.1007/s00170-015-7734-6.

[7] Deepak Pathak, Rudra Pratap Singh, Sanjeev Gaur, Vincent Balu, 2020, *Experimental investigation of effects of welding current and electrode angle on tensile strength of shielded metal arc welded low carbon steel plates*, Materials Today: Proceedings, Volume 26, Part 2, Pages 929-931, ISSN 2214-7853, https://doi.org/10.1016/j.matpr.2020.01.146.

[8] J. Edwin Raja Dhas, S. Kumanan, 2011, *Optimization of parameters of submerged arc weld using non conventional techniques*, Applied Soft Computing, Volume 11, Issue 8, Pages 5198-5204, ISSN 1568-4946, https://doi.org/10.1016/j.asoc.2011.05.041.

[9] Murugan N, Gunarajb V, 2005, *Prediction and control of weld bead geometry and shape relationships in submerged arc welding of pipes*, Journal of Materials Processing Technology, Volume 168, Issue 3, 478–487.

IOP Conf. Series: Materials Science and Engineering **1319** (2024) 012004 doi:10.1088/1757-899X/1319/1/012004

[10] *** European standard EN 10027-1:2005, *Designation systems for steels – Part 1: Steel names.*
[11] *** European standard EN 10216-3:2003, *Seamless steel tubes for pressure purposes. Technical delivery conditions Alloy fine grain steel tubes.*
[12] ESAB - Technical Handbook, *Submerged Arc Welding - Fluxes and wires for joining non and low-alloyed Steels, stainless steels and nickel-base alloys.*
[13] Web page: https://www.tec-science.com/material-science/iron-carbon-phase-diagram/summary-of-phase-transformations/, accessed on 10.04.2024.
[14] *** Standard BS EN 895:1995, *Destructive tests on welds in metallic materials – transverse tensile test.*

IOP Conf. Series: Materials Science and Engineering **1319** (2024) 012005 doi:10.1088/1757-899X/1319/1/012005

A study of the effects of welding parameters on macrostructure properties and heat distribution in FSW of aluminium metal matrix composite

Marius Pop-Calimanu[1*]

[1] Mechanical Machines, Equipment and Transportation, Politehnica University of Timisoara, Timisoara, Romania

*E-mail: marius.calimanu@upt.ro

Abstract. The objective of this study is to present the effects of welding variables, specifically traverse speed and rotational speed, on the macrostructure of an aluminium metal matrix composite. This investigation includes comprehensive thermal monitoring and measurements within the weld joint area, utilizing infrared thermography to track the primary stages of friction stir welding (FSW). By examining different areas of the weld, to study aims to elucidate how temperature is distributed across the aluminium metal matrix composite in relation to the specific parameters of the FSW process, and to determine the subsequent impact on weld quality. Using infrared thermography, the researcher not only maps the thermal profile during welding, but also correlates these thermal variations with macrostructural changes in the aluminium metal matrix composite. This method allows for precise detection of temperature fluctuations and provides a detailed understanding of how traverse speed and rotational speed influence thermal distribution. Moreover, the study demonstrates the efficacy of thermography as a valuable tool for assessing weld quality. By capturing real-time thermal images, this technique enables the identification of potential defects and inconsistencies within the weld zone, facilitating a more thorough investigation into the integrity and performance of the welded joints. Ultimately, the findings aim to contribute to optimizing welding parameters on enhance the structural integrity and reliability of aluminium metal matrix composites in various applications.

1. Introduction

Aluminium metal matrix composite (MMC's) has been attracting great interest because of their attractive properties [1], [2] and cost-efficient, and these have led to their use in various fields, including the aerospace, automotive and marine industries [3][4].

Aluminium metal matrix composites have emerged as a vital class of high-performance materials for various applications, [5] primarily due to their enhanced strength, high elastic modulus compared to conventional base alloys,[6] fatigue resistance, [7] specific stiffness, and excellent wear resistance [8].

An effective way of welding metal matrix composites is by friction stir welding (FSW), which is a solid-state joining method patented by the Welding Institute (TWI) [9][10].

In recent years, many researchers have studied the effects of the FSW welding parameters of aluminium metal matrix composite [11] on their microstructure, and only a few of them have studied these welding parameters on the temperature distribution in welded joint [12], [13].

To obtain good quality welds, with a well consolidated stirring zone over the entire thickness of the material meant to be welded, it is mandatory to reach the optimum temperature level for their plasticization [14]. Equipping the FSW welding machine with an infrared thermography camera to monitor and control the FSW process provides valuable information on the stability of the welding process and parameter and provides information on the occurrence of imperfections or defects, as well as analysis of the quality of welding by thermal imaging [15].

2. Experimental section

The composite material employed in this study consists of an AA2124 aluminium alloy, reinforced with silicon carbide, particles, at a concentration of 25%. The silicon carbide particles measure 2-3 microns in size. The microstructure of the AA2124-25% SiC particle metal matrix composite and its distribution can be observed in **Figure 1.** The AA2124-25%SiC plates have

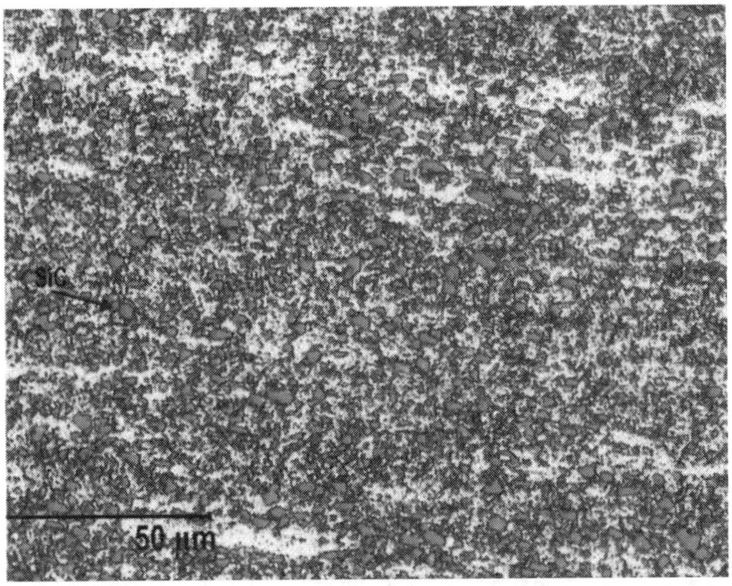

Figure 1. Microstructure of AA2124/SiC/25p-T4 metal matrix composite

been trimmed at the sizes of 100 mm × 50 mm × 5 mm and 200 mm × 50 mm × 5 mm from a forging plate of 275 mm × 220 mm × 15 mm by electro-discharge machining (EDM).

The AA2124-25%SiC aluminium metal matrix composite components were joined by butt welding using a friction stir welding (FSW) dedicated welding machine **Figure 2,** which is part of the National Institute for Welding and Materials Testing Timisoara.

In **Table 1** is presented the chemical composition of AA2124. **Table 4** also shows the mechanical properties of FSW welding such as tensile strength and elongation.

Table 1. The chemical composition wt% of AA2124

Alloy	Major elements (wt%)								
	Cu	Mg	Mn	Si	Fe	Zn	Ti	Other	Al
2124	3.8-4.9	1.2-1.8	0.3-0.9	0.5	0.5	0.25	0.15	0.05	Bal

a

The friction stir welding machine and experimental set-up consist of: 1 – driving unit no. 1, 2- driving unit no. 2, 3- driving unit no. 3, - 4 welding equipment, 5- infrared camera, 6- multimeter, 7- control panel, 8- data acquisition system of the temperature, 9- data acquisition system for energy consumption.

There are two monoblock categories of tools used in this investigation, one is made of tempered steel X38CrMoV5 with a cylindrical threaded pin **Figure 2a** and the second tool is made of sintered tungsten carbide P20S, which have a tapered pin with four flared sides **Figure 2b**. The FSW welding specimens were created using various combinations of tool rotation speed and tool traverse speed parameters, employing different welding tool setups as outlined in **Table 2**.

All welded joints were carefully evaluated to identify the macrostructure and to check for defects or other effects resulting from FSW solid state welding process. The specimens were cut

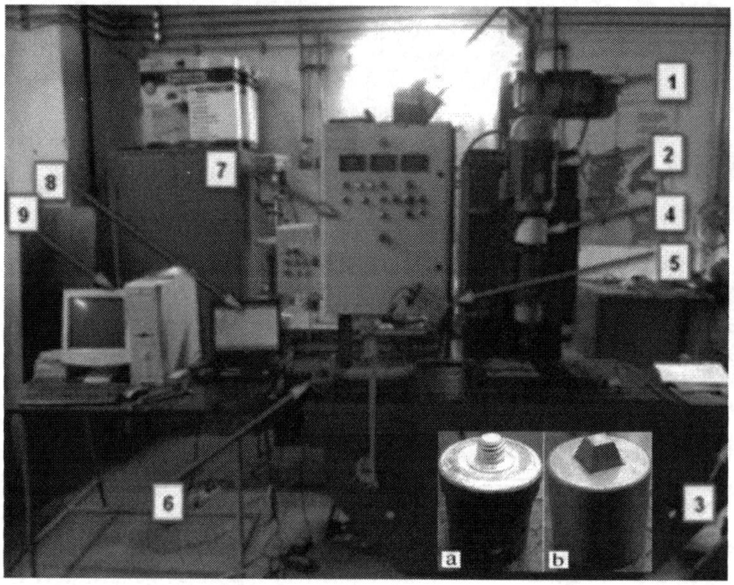

Figure 2. Friction stir welding machine and experimental set-up with a - X38CrMoV5 tool with a cylindrical threaded; b - pin sintered tungsten carbide P20S.

perpendicular to the weld joint and were arranged in accordance with ISO 17369 standardized metallographic procedures. Macroscopic analysis consists in choosing two samples from

Table 2. Welding parameters set-up.

Specimen	Material	Traverse speed [mm/min]	Rotation speed [rpm]	Welding tools			
				Type	Φpin [mm]	Lpin [mm]	Φshoulder [mm]
1	AA2124-25%SiC	50	1000	P20S	4.85	4.8	20
2	AA2124-25%SiC	50	1000	X38CrMoV5	6	4.5	20
3	AA2124-25%SiC	70/100	1100	X38CrMoV5	6	4.5	20
4	AA2124-25%SiC	150	800	X38CrMoV5	6	4.5	20

different areas of the friction stir welding joint to evaluate the influence of temperature and process parameters on the weld. The surface preparation procedure involves mechanically polishing with garnet paper ranging from 180 to 2000 grit, polishing with felt and chemical etching.

The temperature was recorded in real time using a Thermo-Vision A40M thermographic camera, at a capture rate of 25 images per second. The thermographic care was placed on the welding equipment to track the intersecting area between the tool shoulder and the welding surface **Figure 3**.

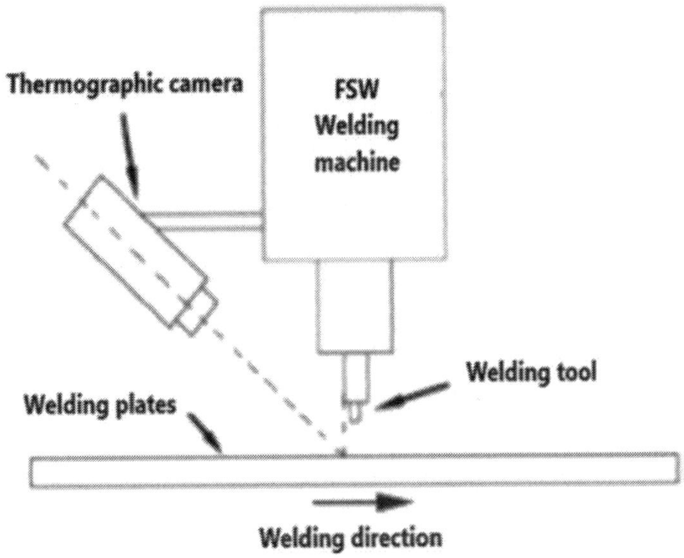

Figure 3. Thermographic camera positioning scheme on the FSW welding machine.

IOP Conf. Series: Materials Science and Engineering **1319** (2024) 012005 doi:10.1088/1757-899X/1319/1/012005

To ascertain temperature fluctuations in the material during welding, measurements were taken perpendicular to the joint line at various points from the weld origin, specifically 1 mm behind the welding tool shoulder, in the stirring zone, respectively at 6 mm from the joint line both on the advance side and on the retreating side in the thermo-mechanically affected zone, **Figure 4.**

3. Results and discussion

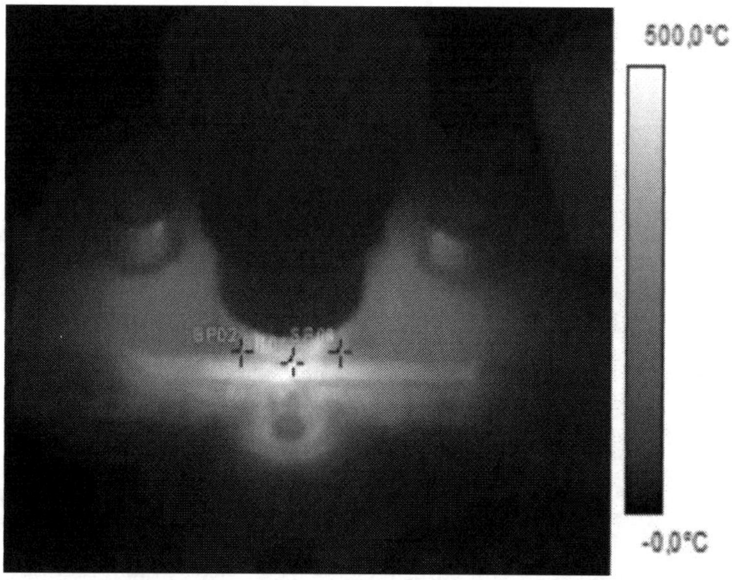

Figure 4. The measuring points seen on the infrared image SP1 – center line, SP2 – retreating side, SP3 – advancing side.

3.1 Measuring temperature during the FSW welding process in real time by infrared thermography
During the FSW welding of specimen **Figure 5**, the maximum temperatures were captured at the centreline of the joint, with the temperature values on the retreating side of the tool welding were much higher than on the advance side of the tool. The temperature discrepancy is explained by the fact that a tunnel defect has been formed on the advance side **Table 2**, which prevented the propagation of heat on this side. It is observed that at the start of the process, in the centre of the joint, the temperature rises linearly (associated with the penetration of the pin into the material), and at the contact of material with the shoulder of the tool welding, the temperature rises very rapidly under the shoulder from 150°C to 470°C in seconds, due too much axial pressure exerted by the tool on the surface of the welded material.

At this point welding begins. At the beginning of the process there is a suddenly decreases of the temperature at 300°C, followed by an irregular linear increase of temperature during the welding process, up to 460°C (from 40 to 65% of the metal matrix composite melting temperature).

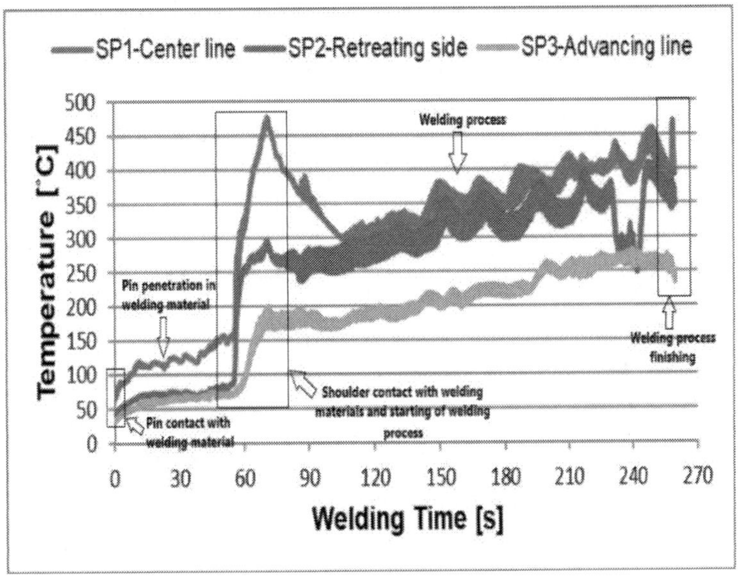

Figure 5. Temperature changes during the FSW welding process in various regions of welded specimen 1

The process parameters are not stable, and this has led to a temperature fluctuation from the beginning until the end of the weld.

Analyzing the temperature evolution of Specimen 2, **Figure 6**, it is observed that the welding process shows instability in the initial phase, specifically within the first 10-15 mm,

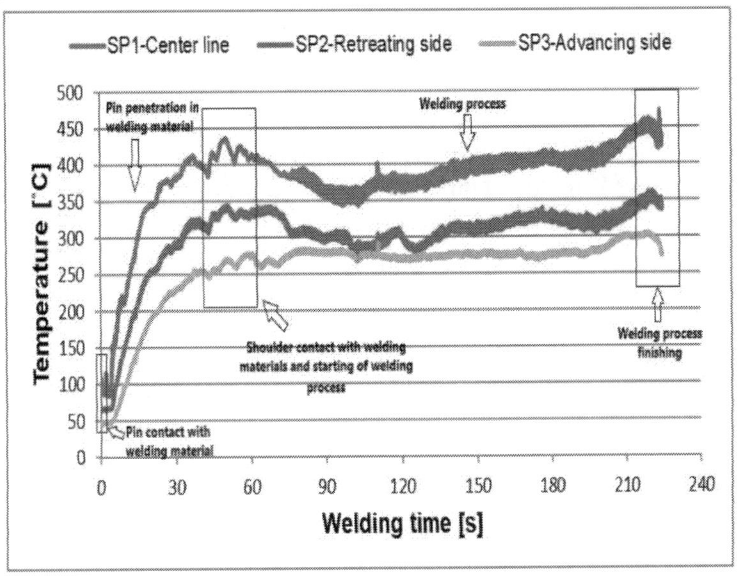

Figure 6. Temperature changes during the FSW welding process in various regions of welded specimen 2

from the beginning of the weld. After this initial segment, the welding process stabilizes, and the temperature increases steadily by approximately 100°C along the entire length of the weld. After 100 mm of welding, which is the end of the weld, the temperature reaches approximately 450°C.

Furthermore, it is noted that the temperature on the retracting side of the weld is consistently higher than on the advancing side, with a temperature difference ranging between 35-45°C. Once the welding process stabilizes, there are no significant temperature fluctuations, indicating a uniform and stable welding process for the remainder of the weld.

At the beginning of welding process for Specimen 3, a traverse speed of 70 mm/min was

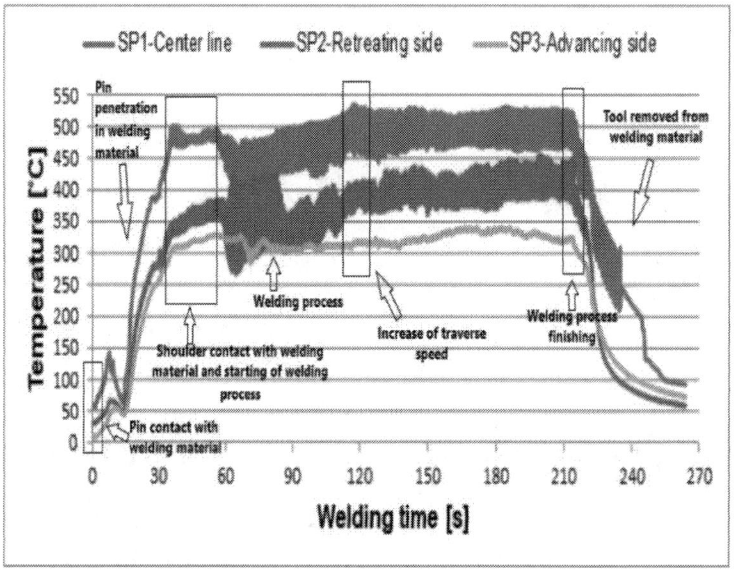

Figure 7. Temperature changes during the FSW welding process in various regions of welded specimens 3

used over the first 70 mm. After this initial distance, the traverse speed was increased to 100 mm/min for the remaining 130 mm, Figure 7. The highest recorded temperatures are found in the centre of the joint, succeeded by the retreating side.

At the onset of the process, the temperature drops by approximately 30°C within a few seconds. Subsequently, the temperature rises continuously, reaching a maximum of about 530°C after welding the initial 70 mm. Upon increasing the traverse speed to 100 mm/min, the welding process stabilizes, maintaining to the same temperature, around 530°C util the end of the weld. At the beginning of welding process for Specimen 4, the temperature increases steadily over the first 50 mm, reaching a maximum of approximately 520°C, **Figure 8**. After riching this peak, the temperature stabilizes and maintains a consistent level throughout the remainder of the jointing process. Observations indicates that the jointing process remains stable and free form disturbances after the initial rise in temperature, ensuring uniform heat distribution.

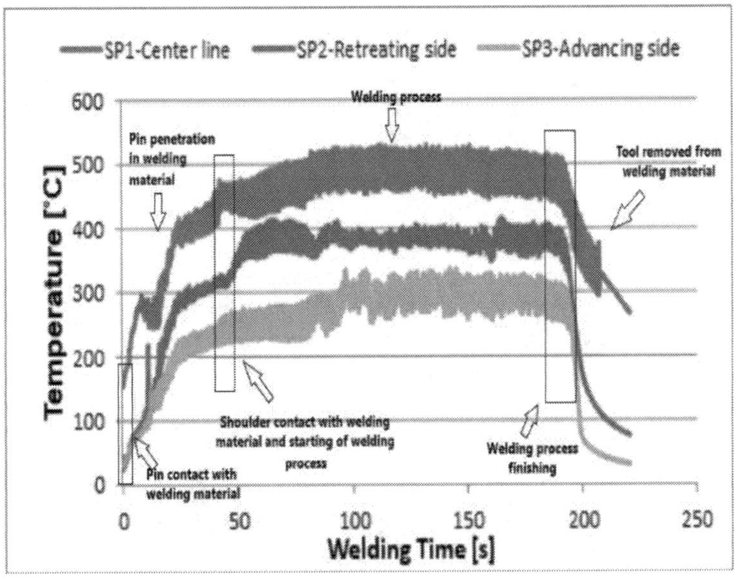

Figure 8. Temperature changes during the FSW welding process in various regions of welded specimens 4

3.2 Macrostructural analysis

Macrostructural examination was conducted to assess the quality of the weld **Table 4**. Welding of Specimen 1, after 30 mm of the weld a tunnel defect with sizes of approximately 0.5 x 0.7 mm on the advance side of the welding tool it appeard.

The situation was due to improper mixing of the material. The tapered pin with four flared sides tool was not able to ensure a sufficient fluidity of the material in a short time, which is also confirmed on the temperature evolution diagram, where the temperature registered the lowest value. After 50 mm of welding the size of defect has visibly decreased, a sign that the process begins to approach the optimum plasticizing temperature of the material. Near the end of the weld the defect almost disappeared.

At Specimen 2 it is found that changing the welding tool had a beneficial effect on welded joint. The welded region stands out distinctly owing to a distinct macrostructural contrast. No

Table 3. Mechanical properties of AA2124-25%SiC of FSW joints at different parameters

Specimen	Ultimate tensile strenght [MPa]	Elongation [%]	Joint efficiency [%]
Base material	323	2.05	-
Specimen 1	114.7	0.55	90.6
Specimen 2	292	0.48	88.5
Specimen 3	168	0.22	51
Specimen 4	286.5	0.48	88.5

Table 4. Macrostructure and defect of the weld

Specimen	Distance form beginning of the weld	Macrosructure	Name of defects	Quality of weld
1	30 mm		Tunnel defect	Very poor
	75 mm	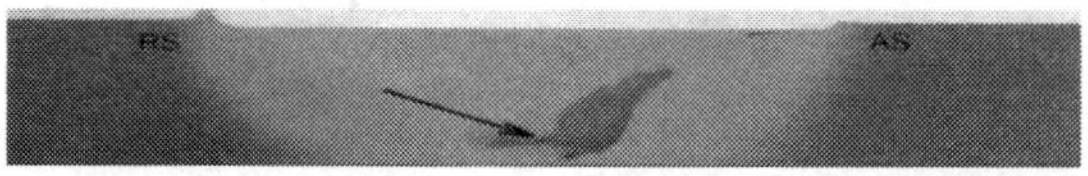		
2	30 mm		No defect	Very good
	75 mm			
3	30 mm	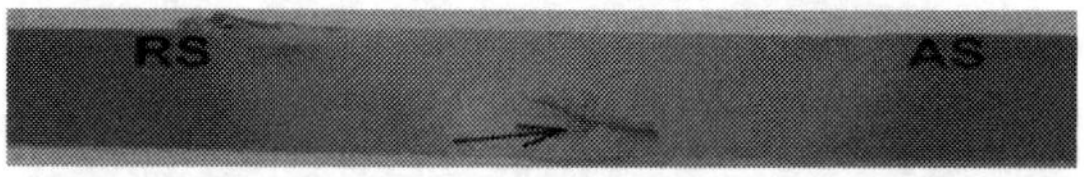	Zigzag defect and crack	Poor
	175 mm	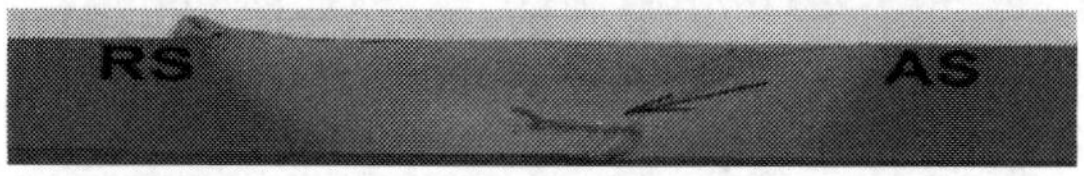		
4	30 mm		No defect	Good
	175 mm			

defects or imperfections are identified along the entire length of the weld, with a very well consolidated stirring zone in the centre of the weld, in form of concentric circles (onions rings).

Welding of specimen 3 at the beginning of the weld a zigzag defect and a crack on the advancing side occurred until the end of the weld. After increasing of the traverse speed, the size of the defect increases. The defect is a result of higher rotational speed of the tool, which produces extra heat during FSW welding. This leads to too much rise of the fluidity of the material, which results in a turbulent flow in the welding area. It is found that a higher rotational speed of the tool has the effect of producing a larger defect.

Welding of specimen 5 show us a good weld with no defects or imperfections. Towards to the end of the welding the area of the weld is reduced, a sign that the material is slightly below the plasticizing temperature.

4. Conclusion

Stable welding processes typically exhibit a steady temperature increase followed by stabilization. Instabilities and fluctuations often correlate with defects.

Defects such as tunnel defects, cracks and zigzag patterns are linked to improper mixing, excessive heat, and high rotational speeds causing turbulent flow.

The choice of welding tool and parameters significantly affects the temperature profile and quality of the weld, with optimized conditions leading to fewer defects and better material consolidation.

Temperature differences between the advancing and retreating sides highlight the importance of balanced heat distribution to prevent defects. The possibility that the defects were caused by the amount of plasticized material cannot be ruled out. Additionally, if the pin of tool had been positioned at an angle of a few degrees, there would have been a greater chance for the material to mix more efficiently. This adjustment could have prevented the appearance of defects, such as tunnel formation, by ensuring better homogenization of the material during the process.

References

[1] Kumar N., Das A., Prasad S.B., An analysis of friction stir welding (FSW) of metal matrix composites (MMCs). Materials Today: Proceedings. 2020, vol. 26, pp. 2650-2656

[2] Madhavarao, S., Raju, C.R.B., Kumar, G.S.V.S., Investigation of Friction Stir Welding of Metal Matrix Composites Using a Coated Tool. Materials Today: Proceedings. 2018, vol. 5, pp. 7735-7742.

[3] Mad Patthi, U.B, Awang, M., Friction Stir Welding on Aluminium 6092/sic/25p/t6 Metal Matrix Composite: its Microstucture Evolution and Mechanical Properties, IEEE. 2011, pp. 1-6.

[4] Peryasamy, P., Mohan, B., Balasubramanian, V., Effect of Heat Input on Mechanical and Metallurgical Properties of Friction Stir Welded AA6061-10% SICp MMCs. JMEPEG. 2012, vol. 21, pp. 2417- 2428.

[5] Pop-Calimanu, M., Cojocaru, R., Ciuca, C., Fleser, T., Ciorpac, P.L., Welding of AA2124/25p-T4 by FSW at high welding speed and monitoring the process by real-time control of infrared thermography and energy consumption. The 5th International Conference on Advanced Materials and Structures-AMS'13. Timisoara: Solid State Phenomena, 2013, pp. 288-293.

[6] Babu, S. D. D., Sevvel, P., Kumar, R. S., Simulation of heat transfer and analysis of impact of tool pin geometry and tool speed during friction stir welding of AZ80A Mg alloy plates. Journal of Mechanical Science and Technology. 2020. vol. 34 (10), pp. 4239-4250.

[7] Sadoun, A.M., Wagih, A., Fathy, A., Essa, A.R.S., Effect of tool pin side ratio on temperature distribution in friction stir welding. Results in Physics. 2019, vol. 15, pp. 1-8.

[8] Jayaseelan, P., Christy, T.V., Vijay, S.J., Nelson, R., Effect of tool material, profile and D/d ratio in friction stir welding of aluminium metal matrix composites. Material Research Express. 2019, vol. 6, pp. 1-8.

[9] Sabry, I., El-Kassas, A.M., An appraisal of characteristic mechanical properties and microstructure of friction stir welding for Aluminium 6061 alloy-Silicon Carbide (SiCp) metal matrix composite. Journal of Mechanical Engineering and Sciences. 2019. vol. 13, Issue 4, pp. 5804-5817.

[10] Bozkurt, Y., Boumerzoug, Z., Tool material effect on the friction stir butt welding of AA2124-T4 Alloy Matrix MMC. Journal of Materials Research and Technology. 2018, vol.7, pp. 29-38.

[11] Karloopia, J. MozammiL, S. Jha. P.K. An experimental study on friction stir welding of Al-Si-TiB2 metal matrix composite. Materials Today:Proceedings. 2018, vol. 5, pp. 17260-17269.

[12] Pardeep G, Narinder K., Naveen H., Sachin G., Microstructural assessment of friction stir joined AA6063/10.5%SiC Al-matrix composite. Materials Today: Proceedings. 2023, vol. 78, part 3, pp. 662-666

[13] Sangeetha K., Prabhu L., Thiagarajan C., Optimization on the process parameters and mechanical behavior of dissimilar AA6061/AA5083 metal matrix composites by FSW. AIP Conference Proceedings, 2024, vol. 3042, Issue 1

[14] Fayaz A. M., Noor Z. K., Saad P., Joining of aluminium matrix composites using friction stir welding: A review. Journal of Materials: Design and Applications, 2022, Vol. 236, Issue 5.

[15] Chen W., Beibei W., Zongyi M., Effect of tool offsetting on friction stir welding of dissimilar aluminiu matrix composite and aluminium alloy. Journal Science and Technology of Welding and Joining. 2022, vol. 27, Issue 7.

IOP Conf. Series: Materials Science and Engineering **1319** (2024) 012006 doi:10.1088/1757-899X/1319/1/012006

Strain analysis in a 3D printed car brake pedal by numerical and experimental methods

Mircea Cristian DUDESCU[1*], Laszlo RACZ[1] and Cristian VILĂU[1]

[1] Department of Mechanical Engineering, Technical University of Cluj-Napoca, 103-105 Muncii Boulevard, 400641 Cluj-Napoca, Romania

*E-mail: mircea.dudescu@rezi.utcluj.ro

Abstract. This paper presents the strain analysis in a brake pedal manufactured by additive technologies. A commercial plastic pedal made by injection moulding was optically scanned using a scanning system based on fringes projection. The result was a CAD model of the pedal that can be used in both numerical simulations and 3D printing. Two models of the pedal with 100% infill rate and a grid infill pattern were produced by 3D printing on an industrial printer, the difference between the two models is laying in the orientation of the printed part in the printer workspace. Resistive strain gauges were applied to determine the mechanical behaviour of the printed pedals regarding the injection moulded version. To perform a comparative study of the experimentally obtained results, pedals with similar mechanical characteristics to those measured experimentally were analysed numerically by the finite element method. The numerical results showed a good convergence with the experimental results, with relative deviations of 13% for the printed pedal with outer filaments oriented along its longitudinal axis, 7.3% for the printed pedal with outer filaments oriented transversely and 2.8% for the homogeneous (injected) pedal. The results also demonstrated the possibility of replacing in the future complex, parts made today by conventional technologies with 3D printed elements having mechanical characteristics allowing them to fulfil a functional role.

1. Introduction

Every industry is influenced by additive manufacturing. In recent years, technology has profoundly transformed the way new things are designed, developed, and manufactured. In the automotive industry [1] these technologies have an impact in developing new forms, fabricate more complex parts and lightweight structures, all at the lowest possible production costs. Although 3D printing technology is still mostly used only for rapid prototyping in the development of new car models or in concept cars, the technology has advanced to the point that it is also used to make certain final components. New materials, innovative finishes and faster production times will support 3D printing technology to be integrated more closely into the manufacturing process and perhaps, in the future, also into the spare parts supply chain. In the automotive industry, the most common applications of 3D printing are printed parts used in Motorsport, printed vehicles, 3D printed car prototypes, 3D printed spare parts (for production and sale), etc.

IOP Conf. Series: Materials Science and Engineering **1319** (2024) 012006 doi:10.1088/1757-899X/1319/1/012006

Williams Racing, together with EOS, one of their manufacturing solutions partners specialized on Direct Metal Laser Sintering are working on making 3D printed brake pedals from titanium [2]. The challenge was to develop a titanium part that exceeded the stiffness of standard aluminium, leading to a significant decrease in stress on the component. This was achieved through a lightweight, hollow design that saved 80g in weight.

Brake pedals are critical safety components subject to rigorous engineering principles, demanding quality standards, and stringent inspection procedures. These rigorous requirements make them ideal candidates for exploring AM's capabilities. Successfully producing a complex safety part like a brake pedal with AM would demonstrate its potential for less critical car components such as brackets, hinges, and supports. Few studies regarding the application of AM technologies in optimisation or production of a brake were identified. Some investigations aimed usage of different materials for manufacturing the brake pedal and other have employed a highly strength metal for mass reduction. Dhande et al. [3] achieved significant weight reduction in a brake pedal by applying topology optimization to its geometry. Polyamide (Nylon 6,6) with 30% glass fibre was then chosen as the material for manufacturing the pedal using a traditional injection molding process. Sapuan presents in [4] a study related the use of polyamide (nylon) with short glass fibres as a potential material for brake pedals. Sudin et al. [5] employed topology optimization to redesign a brake pedal, achieving a significant 22% weight reduction. To ensure the redesigned pedal's functionality, they validated its performance through a stress-strain test. In their research on a Formula Student racing car's accelerator pedal, Prada et al. [6] explored replacing the traditionally used 7075 aluminum with Additive Manufacturing (AM) techniques. They investigated two specific AM processes: Poly-jet, utilizing Fullcure 720 material, and FDM, utilizing ABS-plus (P430) filament. Building on prior research, Sargini et al. [7] investigated the potential of using Fused Deposition Modelling (FDM), a cost-effective additive manufacturing (AM) technology, for creating metallic automotive parts. They compared the results of finite element analysis (FEA) with experimental testing to assess the performance and economic viability of an FDM-produced metal brake pedal.

This research investigates the application of FDM technology for designing and producing a brake pedal of polymeric material. Firs part of the paper presents the model creation considering a real pedal of a commercial car. The procedure involves reverse engineering based on 3D optical scanning. The second part presents the analysis of two different brake pedals with different build orientations realized by FDM procedure. Strain analyses were performed by strain gauge technique, the obtained measured values being compared with the numerical simulations. The paper ends with the presentation of comparative results and formulate the conclusions.

2. Materials and methods

2.1 Reverse-engineering the pedal model

The major advantage of the FDM technology is that it is possible to produce parts in a short time and at low cost starting from a CAD model or using the advantages of reverse engineering. To demonstrate these concepts, the paper presents the study of a brake pedal from a Dacia Duster car produced using additive technologies (FDM). A commercial plastic pedal (made of ABS) made by injection moulding was purchased from the Dacia network. Based on this, a reverse engineering process was performed [8]. The pedal was optically scanned using a scanning system based on the Atos Core fringe projection system produced by the German company GOM (www.gom.de). To ensure the black pedal could be scanned by the optical sensor, a special white anti-reflection powder was used to eliminate glare and improve projected fringe detection.

Following the scan, a 3D points cloud was obtained. A triangulation process is necessary to obtain a polygonal grid containing information about the surface and dimension of the brake pedal. The grid is a polygonal model representing the real pedal that can be exported in a CAD format. The process then involves filling in the remaining gaps from the scanning procedure to create a 3D model, some shaded area being not reachable by the optical system. This procedure was carried out using GOM 3D scanning software, subsequently to generation of some surfaces using Catia V16 software. The result was a CAD model of the pedal (Figure 1) that can be used in both numerical simulations and 3D printing.

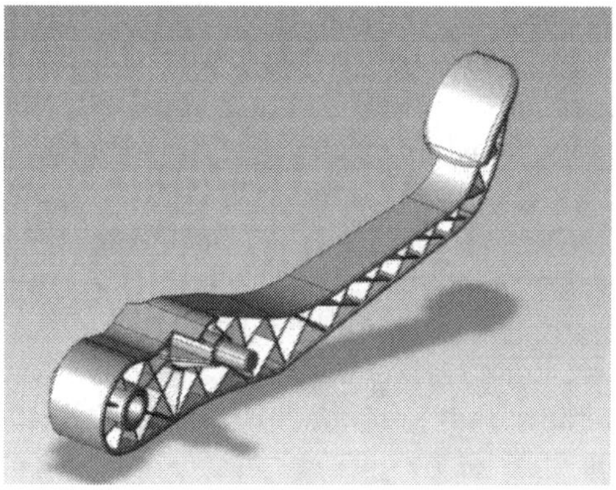

Figure 1. CAD model of the pedal obtained by reverse engineering of the commercial pedal

On the basis of the CAD models, two models of the pedal with 100% infill rate and the infill pattern (grid +45°-45°) recommended by the printer's producer to get the maximum strength, were produced by 3D printing on an industrial printer model STRATASYS Fortus 450mc. The material used was ABS-M30 produced by Stratasys. This infill pattern; also investigated by us in previous works [10,11]; was found to provide good static properties but more importantly the highest fatigue strength [12]. The difference between the two models lies in the orientation of the printed part in the printer workspace, resulting in a model with the outer filaments oriented transversely to the longitudinal axis of the pedal (Figure 2.a) and a model with the filaments oriented along the longitudinal axis of the pedal (Figure 2.b).

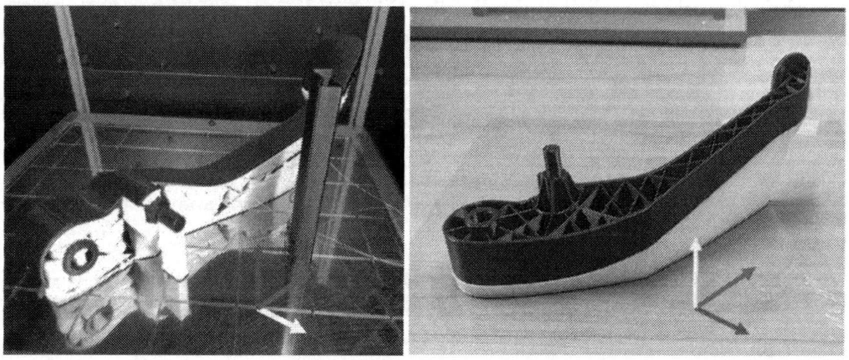

Figure 2. Brake pedals printed in different printing planes: a) outer filaments oriented transversal to the pedal's axis, b) outer filaments oriented along the longitudinal axis of the pedal

Outer filaments direction is indicated by the light axis. In the two images it can also be seen how the printer builds the model (supports respectively supporting elements). The printing time and the complexity of the attached structure being much higher in the case of the pedal built with transversally oriented filaments. The supports created are made of a material that can be chemically dissolved, resulting in parts of very high manufacturing quality.

2.2 Analysis of the mechanical behaviour of 3D printed pedals by electrical strain gauges

For the experimental analysis of the mechanical behaviour of the 3D printed pedals in relation to the injected model used on the Dacia Duster vehicle, a test stand has been designed to allow similar boundary and loading conditions of the pedals to those existing in reality. The experimental stand, shown in Figure 3 was also designed to be mounted in the universal testing machine (Instron 3366, 10 kN), so that the applied force, application point and the control of the displacement can be rigorously controlled. Several test were performed for the same load in order to check the measuremet duplicability. The stiffness of the stand is considered to be higher than the pedal but cannot be ignored. Tis is the reason why the displacement of the testing machine crosshead cannot be considered to be equal to the deflection of the pedal at that point. To avoid introducing of measuremnts errors strain gauges (type HBM 1-LY1x-6/120) were glued at the same distance from the point of force application to measure the strains and to evaluate the stiffness of the pedals. Figure 3 shows the pedals analysed as well as the mounting and positioning of the strain gauges.

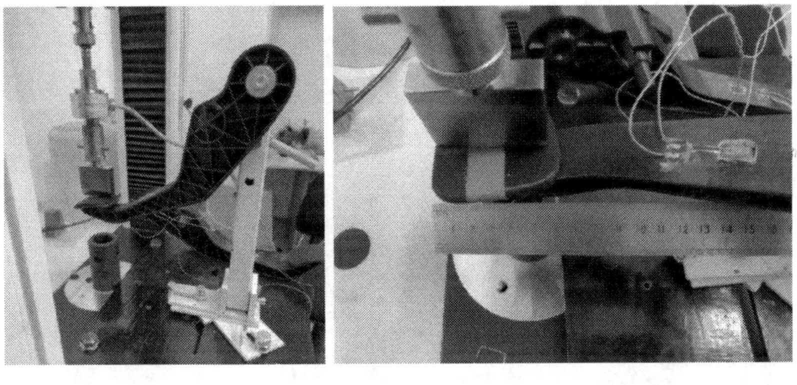

Figure 3. Measurement of strains in the investigated pedals by strain gauges: a) experimental set-up positioned in the tensile test machine, b) mounted strain gauge along longitudinal axis

3. Results and discussions

The pedals were subjected to bending by the same force, the used value was 1/10 from the maximum force that can be applied when using a car with ABS. Measurements of longitudinal strain for the same value of applied force for the three pedals (two 3D printed and the commercial ones) are summarised in Table 1.

In order to perform a comparative study of the experimentally obtained results, pedals with similar mechanical characteristics (Table 2) were analysed numerically analysed by the finite element method. For preprocessing of the model was used ANSA Software (BETA CAE Systems, Greece), the generated mesh size had a target lengt de 1mm with a minimum length 0.2mm.

IOP Conf. Series: Materials Science and Engineering **1319** (2024) 012006 doi:10.1088/1757-899X/1319/1/012006

Table 1. Strain values measured on different types of pedals.

Pedal type	Applied force (N)	Strain (μm/m)
Longitudinal filament printing	19.2	428
Transverse filament printing	19.2	453
Commercial model (injected)	19.2	99,8

The boundary conditions were similar to the experiment and the solution was done using EPYLISIS (BETA CAE Systems, Greece) solver. For the 3D printed pedals only one set of mechanical characteristics was used [11], without taking into account their change in relation to the orientation of the pedal (build orientation) in the 3D printer workspace. The mechanical constants for the commercial pedal are not available, the injected ABS was considered in the simulations [12].

Table 2. Mechanical constants used in the FE simulation

Material	Modulus of elasticity (MPa)	Poisson's ratio	Density (g/cm³)
Injected pedal	7375	0.35	1.57
Printed	1452	0.39	1.05

The results in terms of displacements in the force direction and longitudinal strain values are presented in Figure 4.

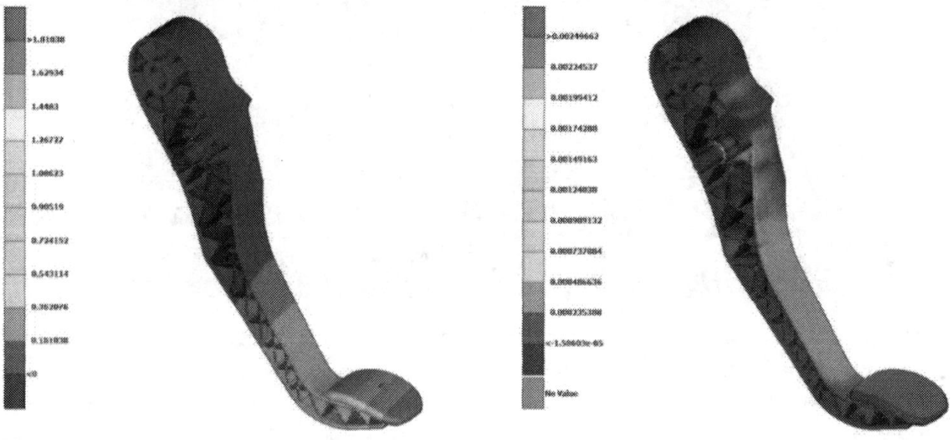

Figure 4. Numerical analysis of the 3D printed pedal: a) displacements (mm) along the applied force, b) maximum principal strain (mm/mm) distribution in the 3D printed pedal

The results obtained representing the strains in the longitudinal direction (identical to the mounting direction of the strain gauges) are depicted only for the 3D printed pedals, the state of strains being identical as long a linear elastic material models were used. The obtained strain values are 486 μm/m for the printed pedals and 102 μm/m for the injected (commercial) pedal, respectively. The maximum displacements in the application direction of the force were 1.8 mm for 3D printed pedals and max deflection 0.356 mm for injection moulded pedal. It is noticeable the significand difference in stiffness between AM pedals and the commercial ones.

For the average driver a common brake pedal force is about 200 N for a car with ABS. At this loading level the stress value in the printed pedal is about 25MPa, close to the tensile strength of printed ABS-M30 material of 31MPa. Using 3D printed materials with higher strength is a solution to overcome this drawback but it will increase the costs of the part. Stress level in the injection molded pedal is similar but the injected part has a higher tensile strength of about 68 MPa [13], so the strength condition is fulfilled.

4. Conclusions

It is observed that there is a small difference (5.5%) in the strain between the 3D printed pedals, the model where the filaments are oriented longitudinally being slightly stiffer compared to the one with transverse filaments. The commercial version of the plastic injection moulded pedal is much stiffer (more than 4 times), having a higher modulus of elasticity. The computer simulations closely matched the real-world tests, with deviations of only 13% for the pedal printed with outer filaments aligned lengthwise, 7.3% for the pedal printed with outer filaments aligned crosswise, and a very good agreement of 2.8% for the pedal made from a homogeneous material.

The results obtained for this case study demonstrate on the one hand the need to carry out tests or simulations of 3D printed materials in order to determine as close as possible to reality their mechanical characteristics, the idea of isotropy and uniform distribution of the material inside the 3D printed elements leading to a simplified model less accurate in comparison with the experimental results. On the other hand demonstrate the possibility of replacing in the future complex parts made today by conventional technologies with printed elements having mechanical characteristics allowing them to fulfil a functional role.

References

[1] A. Cassaignau. "3D Printing transforms the Automotive Industry." https://www.sculpteo.com/blog/2016/01/20/3d-printing-transforms-the-automotive-industry/ (accessed 02.03.2024).
[2] C. Clarke. "3D PRINTING SECRET TO SUCCESS IN FORMULA 1?" https://3dprintingindustry.com/news/3d-printing-secret-success-formula-1-99683/ (accessed 02.03.2024).
[3] K. K. Dhande, N. I. Jamadar, S. Ghatge, Design and Analysis of Composite Brake Pedal: An Ergonomic Approach, International J. Mech. Eng. Robotic Res., vol. 3, no. 3 (2014), 475-482.
[4] S. Sapuan, A Conceptual Design of the Concurrent Engineering Design System for Polymeric Based Composite Automotive Pedals, Am. J. Appl. Sci 2 (2) (2005) 514–515.
[5] M.N. Sudin, M. Mohd Tahir, F.R. Ramli, S.A. Shamsuddin, Topology Optimization in Automotive Brake Pedal Redesign, International Journal of, Eng. Technol. 6 (1) (2014) 398.
[6] J.G. Prada, A. Cazon, J. Carda, A. Aseguinolaza, Direct digital manufacturing of an accelerator pedal for a Formula Student racing car, Rapid Prototyp. J. 22 (2) (2016) 311–321.
[7] M.I.M. Sargini, S.H. Masood, Suresh Palanisamy, Elammaran Jayamani, Ajay Kapoor, Additive manufacturing of an automotive brake pedal by metal fused deposition modelling, Materials Today: Proceedings, 45(6) (2021), 4601-4605.
[8] M. H. Muntean and M. C. Dudescu, Development through reverse engineering of a fixture for automated 3D scanning, Academic Journal of Manufacturing Engineering, vol. 19, no. 4, 2021.
[9] Dudescu, C., & Racz, L. (2017). Effects of raster orientation, infill rate and infill pattern on the mechanical properties of 3D printed materials. Acta Universitatis Cibiniensis. Technical Series, *69*(1), 23-30.
[10] Racz, L., & Dudescu, M. C. (2022). Numerical investigation of the infill rate upon mechanical proprieties of 3D-printed materials. Polymers, *14*(10), 2022.
[11] Dudescu, M. C., Racz, L., & Popa, F. (2023). Effect of infill pattern on fatigue characteristics of 3D printed polymers. Materials Today: Proceedings, 78, 263-269.
[12] Shaik, Y.P., Naidu, N.K., Yadavalli, V.R. and Muthyala, M.R. (2023) The Comparison of the Mechanical Characteristics of ABS Using Three Different Plastic Production Techniques. Open Access Library Journal, 10, 1-18.

IOP Conf. Series: Materials Science and Engineering **1319** (2024) 012007 doi:10.1088/1757-899X/1319/1/012007

Unidirectional steel wire reinforced metal matrix syntactic foams

Benedek Szovák[1,2*], Alexandra Kemény[1,2] and Imre Norbert Orbulov[1,2]

[1] Department of Materials Science and Engineering, Faculty of Mechanical Engineering, Budapest University of Technology and Economics, Műegyetem rkp. 3., H-1111 Budapest, Hungary
[2] MTA–BME Lendület „Momentum" High-performance Composite Metal Foams Research Group, Műegyetem rkp. 3., H-1111 Budapest, Hungary
*E-mail: szovak.benedek@edu.bme.hu

Abstract. A double composite structure has been successfully produced using vacuum infiltration. In this structure, the matrix was unalloyed aluminium filled with Lightweight Expanded Clay Aggregate Particles to create metal matrix syntactic foam, and stainless steel wires were placed inside this matrix as unidirectional reinforcement. The wires created a shape-locking bond with the matrix owing to their wave-like form, which helped to prevent reinforcement pull-out. Based on the acquired tensile data, the tensile strength of the specimens calculated with the theoretical maximum area of reinforcement wires is $R_{m,\,theoretical} = 1063 \pm 192$ MPa, while the tensile strength of the individual wires is $R_{m,\,wires} = 882 \pm 2$, which leads to the conclusion that the matrix distributed some of the stress.

1. Introduction

One of the main objectives of materials science today is to reduce the weight of structural elements. In this regard, porous materials such as metal foams are an excellent choice due to their low density, approximately 25% to 50% of the density of the bulk material [1–4]. Owing to their low density, metal foams have high specific mechanical properties such as energy absorption capacity in case of compression [5,6]. However, due to their porous structures, they cannot withstand tensile loads [7–9].

One way to improve a material property is to use composites, where two or more material types are combined to take advantage of each other's properties while negating each other's disadvantages [10–13]. Composites consist of three components: (i) the matrix, which transfers the load to the reinforcement and protects it from the environment; (ii) the reinforcement, which handles the load; and (iii) the interface between them, allowing the load to be transferred. The reinforcement can be in the form of grains or fibres [14–16].

The goal of this study was to create a double composite structure from metal matrix syntactic foam (MMSF) combined with longitudinally placed wires as unidirectional fibre reinforcement. The result is a lightweight structure with advantageous compressive properties [5] of the metal matrix syntactic foam but capable of withstanding tensile loads without the drastic increase in density.

2. Materials and methods

2.1 Materials

Unalloyed aluminium (Al99.7) was chosen as the matrix material due to its high ductility. The chemical composition of the matrix was measured with a WAS PMI-MASTER SORT spectrometer (Worldwide Analytical Systems AG, Uedem, Germany) and is given in Table 1. Lightweight Expanded Clay Aggregate Particles (LECAPs) were used as the filler material due to their low density and low cost, as shown in Figure 1a.

Table 1. Chemical composition of Al99.7 matrix material

Element	Al	Fe	Si	Mn	Ti	Cu	Mg	Zn	Other
Weight %	99.8	0.08	<0.01	<0.01	0.01	<0.01	<0.001	<0.01	bal.

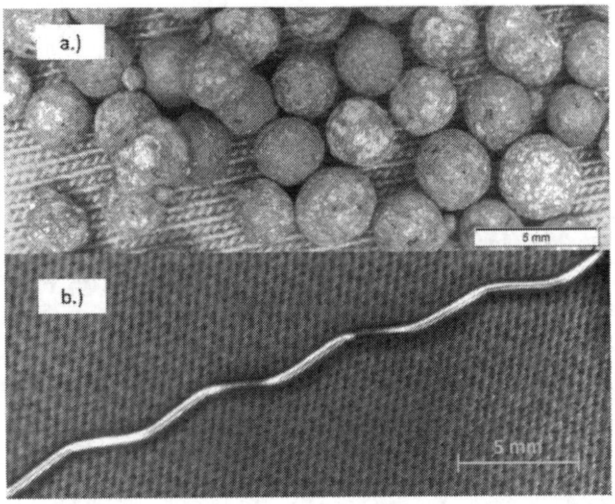

Figure 1. Macro image of the (a) LECAPs particles and (b) stainless steel wire

1.4307 (X2CrNi18-9) stainless steel wires with a diameter of Ø0.64±0.02 mm were used for reinforcement because of their high strength and easy accessibility. It was purchased as a steel mesh, which was broken down into individual fibres with a length of 150 mm. This resulted in a wavy structure that could act as a shape-locking joint between the matrix and the reinforcement. The wire is shown in Figure 1b.

2.2 Methods

The fibres were oriented using a unique frame, with both ends attached to a steel mesh of the same material. This method was used to place the individual fibres at a uniform distance from each other. This is essential to allow the filler particles to move freely between the reinforcing fibres and fill the space between them.

The frame was placed into a S235JR crucible with external dimensions of 60x60x300mm and a wall thickness of 2mm. During placement, the frame was filled with LECAPs. The particles were compacted by manually vibrating the crucible. A layer of alumina quilt was placed on top of the frame inside the crucible to filter out dirt and oil during infiltration. The solid matrix material was placed on top of the quilt, and the crucible was placed into the furnace.

The samples were prepared by vacuum infiltration in a Blücher vacuum furnace. The pressure was reduced to 10^{-4} Pa during preheating, while the temperature was raised to 550°C. After one hour of holding, the temperature was raised to 800°C, held for a half hour and then infiltrated with 0.4 MPa of 99.9% argon gas. The mould was air-cooled to room temperature. The parameters of the infiltration are shown in Figure 2.

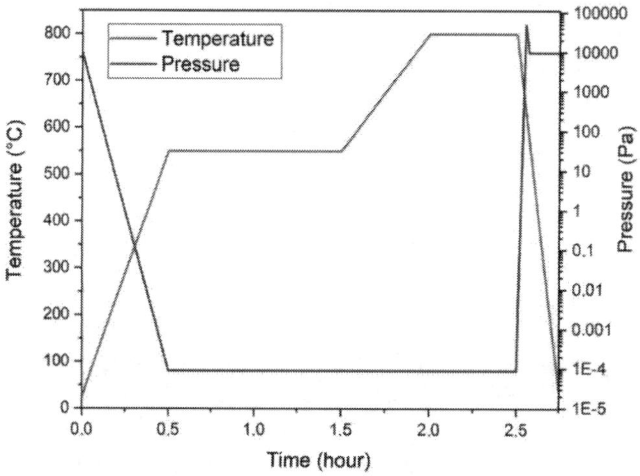

Figure 2. The parameters of the infiltration process

Specimens with a diameter of Ø20 mm and a tensile length of 100 mm were prepared by conventional machining. The density of the specimens was measured as bulk density, using a calliper to define the outer volume and a precision scale for measuring the mass of each specimen. Tensile tests were performed in accordance with the ISO 6892-1 standard [17], on a TiraTest 2300 universal material testing machine (TIRA GmbH, Schalkau, Germany) with a maximum load capacity of 100 kN and mechanical grips. The speed of the cross-head was 4 mm·min^{-1}. Tensile tests were also performed on individual wire pieces with an Instron 5965 universal testing machine (Instron, Norwood, MA, USA) with a maximum load capacity of 5 kN and a cross-head speed of 3 mm·min^{-1}.

The diameter of the wires was examined on images taken with an Olympus SZX16 stereomicroscope (Olympus Corporation, Tokyo, Japan) using JmicroVision software. The same microscope was used to examine the reinforcement position inside a specimen after the tensile test.

3. Results

3.1 Density

The density of the specimens is 1.69±0.02 g·cm^{-3}. The density of the Al99.5 MMSF filled with LECAPs is approximately 1.54 g·cm^{-3} while the density of the bulk Al99.5 is 2.7 g·cm^{-3} [5]. The relative density of the MMSF is 0.57, while the relative density of the reinforced MMSF is 0.62. Therefore, the reinforcement increases the relative density by 8.9 % compared to the MMSF.

3.2 Tensile tests

Figure 3 a shows the engineering stress–engineering strain curves of the tensile test of the individual wires. The slope decreases after an initial phase. The explanation is that during this initial phase, the wires are straightened, and the area where the waveform was formed is plastically deformed, increasing the strength of the wire.

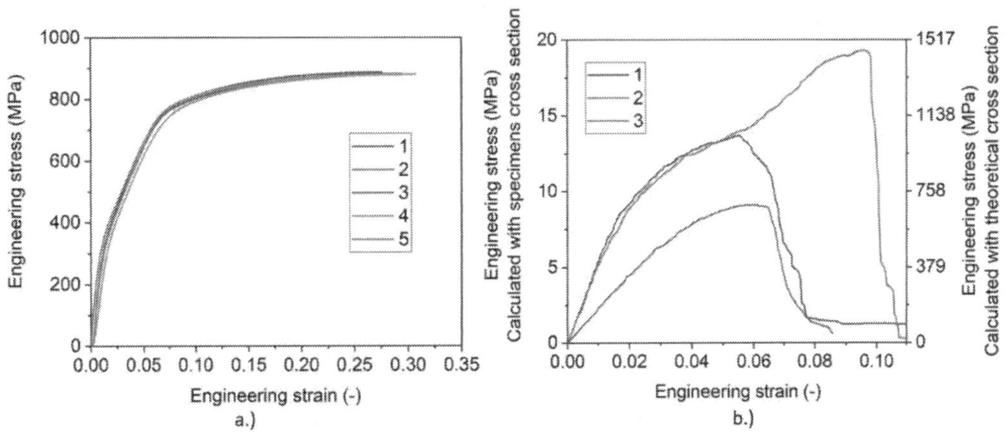

Figure 3. Engineering stress–engineering strain tensile curves of the individual wires (a) and the double composite structure (b) (the specimens were numbered for identification)

The engineering stress–engineering strain curves of the tensile tests are shown in Figure 3 b. On the left-hand side of the diagram are the engineering stress values calculated using the total cross-sectional area of the specimens, while on the right-hand side are the stress values calculated using the theoretical cross-sectional area of the 13 reinforcement wires within a specimen. The average tensile strength calculated from the cross-sectional area of the specimens was $R_{m,\,specimens} = 14 \pm 3$ MPa, while by estimating from the theoretical wire cross-section, the value was $R_{m,\,theoretical} = 1063 \pm 192$ MPa. This significantly exceeds the average tensile strength of the individual wires ($R_{m,\,wires} = 882 \pm 2$ MPa). This leads to the conclusion that the MMSF matrix can withstand significant tensile loading. This is further supported by the fact that the matrix did not begin to crack on the surface until 9 % engineering strain, as shown in Figure 4.

Figure 4. Characteristic behaviour of the tensile specimens (start of the cracking highlighted with red circle)

The stress values drop rapidly after the curves reach their maximum. A possible explanation is that there is no definitive chemical bond between the matrix and the reinforcement and that the wires were pulled out of the matrix within the crack zone after straightening. Although no significant bond was observed in the crack zone, the wires were not pulled out. This indicates that the corrugated structure of the wires forms a connection with the surrounding matrix.

3.3 Optical microscopy

After the tensile tests, the specimens were examined by optical microscopy to determine the functional cross-sectional area of the reinforcing fibres. An example is shown in Figure 5. The wires inside the specimens were randomly placed because, although they were attached to the net at both ends of the frame, they were not stretched. Due to this, they got misplaced inside the specimens. The average number of reinforcement wires was 7 ± 2 per specimen. Using the number of wires and the original diameter value, the tensile strength was calculated as $R_{m,\,measured}$ = 2031 ± 356 MPa. This further supports the conclusion that the MMSF matrix can withstand tensile loading.

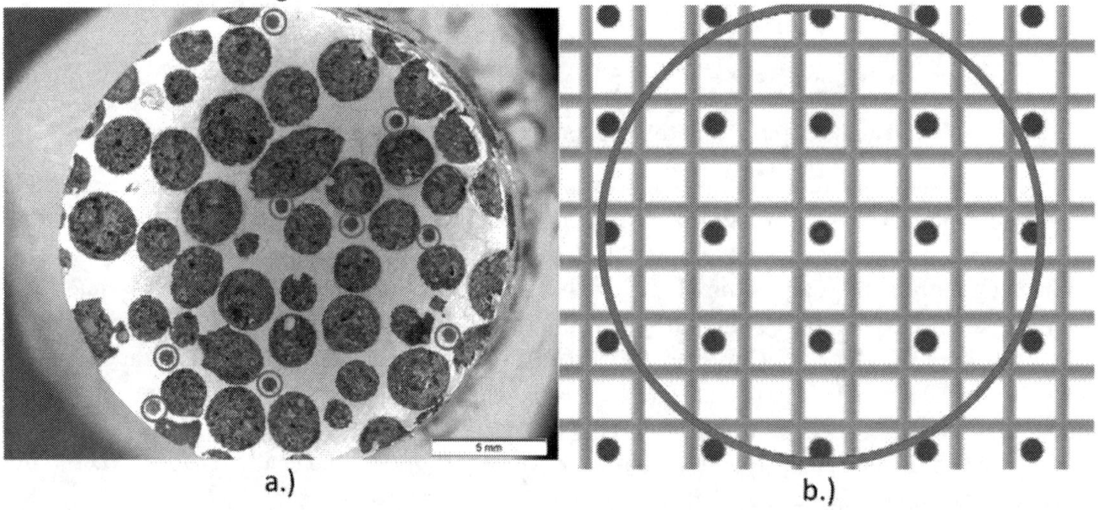

Figure 5. (a) An example of reinforcement wires inside a specimen (reinforcement wires marked with a red circle) and (b) the placement of the wires at both ends of the frame

4. Conclusion

MMSFs with unidirectional stainless steel wire reinforcement were successfully produced by vacuum infiltration. Tensile tests were conducted on this material and individual reinforcement wires. The data obtained during this study leads to the conclusion that the MMSF matrix can handle a significant amount of tensile load with the used Al99.5 matrix. The wires were found to form a shape-locking bond with the surrounding matrix, preventing them from being pulled out. This effect diminishes in the crack zone of the matrix.

5. Acknowledgements

This work was partially supported by the National Research, Development and Innovation Fund (grant number OTKA FK_21 138505). Project no. TKP-6-6/PALY-2021 has been implemented with the support provided by the Ministry of Culture and Innovation of Hungary from the National Research, Development and Innovation Fund, financed under the TKP2021-NVA funding scheme.

6. References

[1] Linul E and Khezrzadeh O 2021 Axial crashworthiness performance of foam-based composite structures under extreme temperature conditions *Compos. Struct.* **271** 114156
[2] Amin K F and Rashed H M M A 2019 Steel Used in Construction Industries *Ref. Modul. Mater. Sci. Mater. Eng.*
[3] Thalmaier G, Sechel N A, Csapai A, Popa C O, Batin G, Gábora A, Mankovits T and Vida-Simiti I 2022 Aluminum Perlite Syntactic Foams *Materials (Basel).* **15** 1–10
[4] Thalmaier G, Sechel N A and Vida-Simiti I 2020 Syntactic Aluminum Foam from Recycled Sawing Chips *Jom* **72** 3377–82
[5] Kincses D B, Károly D and Bukor C 2020 Production and testing of syntactic metal foams with graded filler volume *Mater. Today Proc.* **45** 4225–8
[6] Kádár C, Kubelka P and Szlancsik A 2023 On the compressive properties of aluminum and magnesium syntactic foams: Experiment and simulation *Mater. Today Commun.* **35**
[7] Kalpakoglou T and Yiatros S 2022 Metal foams: A review for mechanical properties under tensile and shear stress *Front. Mater.* **9**
[8] Mei Y, Fu C, Fu Y, Ding Y, Wang E and Yang Q 2022 Tensile Behavior and Performance of Syntactic Steel Foams Prepared by Infiltration Casting *Metals (Basel).* **12** 1–14
[9] Jenei P, Kádár C, Szabó Á, Hung S M, Kuo C J, Choe H and Gubicza J 2022 Mechanical behavior of freeze-cast Ti foams with varied porosity *Mater. Sci. Eng. A* **855**
[10] Sivananthan S, Ravi K and Samuel S J 2020 Effect of SiC particles reinforcement on mechanical properties of aluminium 6061 alloy processed using stir casting route *Mater. Today Proc.* **21** 968–70
[11] Messaoud B and Amrane M N 2024 Vibration Analysis of Damaged Viscoelastic Composite Sandwich Plate *Period. Polytech. Mech. Eng.* **68** 85–96
[12] Szabó G and Váradi K 2023 FE micromodel of a cord-rubber composite tube sample subjected to uniaxial tension *J. Reinf. Plast. Compos.* **42** 393–404
[13] Belhadj B, Abdelkader L and Chateauneuf A 2021 Weibull probabilistic model of moisture concentration build up in a fiber graphite/epoxy polymer composite under varying hydrothermal conditions *Period. Polytech. Mech. Eng.* **65** 27–38
[14] Kumar N, Chittappa H C and Ezhil Vannan S 2018 Development of Aluminium-Nickel Coated Short Carbon Fiber Metal Matrix Composites *Mater. Today Proc.* **5** 11336–45
[15] Ucsnik S, Großalber A, Reiter J, Scheerer M and Mozdzen G 2019 Advances in continuous carbon fibre reinforced Aluminium matrix composites *8th Eur. Conf. Aeronaut. Sp. Sci. EUCASS_2019* **1** 2
[16] Bhanuprakash L, Manikandan N, Varghese L and Joseph J 2023 Study on aluminium metal matrix composites reinforced with SiC and granite powder *Mater. Today Proc.*
[17] Iso M S Z E N 2020 Metallic materials - Tensile testing - Part 1: Method of test at room temperature (ISO 6892-1:2019)

Failure Analysis of Cu-DHP Joining Processes: A Comparative Study of FSW and SFSW Techniques

L N Boțilă[1*], I A Perianu[1], B Radu[2], C Codrean[2], M Nicolaescu[3] and E-F Binchiciu[1]

[1] National Research & Development Institute for Welding and Material Testing – ISIM, Mihai Viteazu Blvd. no.30, Timisoara 300222, Romania
[2] Politehnica University Timisoara, Piata Victoriei, no. 2, Timisoara 300006, Romania
[3] National Institute for Research and Development in Electrochemistry and Condensed Matter Timisoara, A.P. Podeanu Street 144, Timisoara, 300569, Romania

*E-mail: lbotila@isim.ro

Abstract. This article presents a failure analysis conducted on specimens extracted from 2.5 mm sheets of Cu-DHP (Cu99), which were joined using friction stir welding (FSW) and submerged friction stir welding (SFSW) processes. To evaluate the performance of the welded joints, destructive and non destructive tests, including tensile tests were performed. Additionally, morphologic analysis using scanning electron microscopy (SEM) combined with energy dispersive X-ray (EDX) and fractography investigations were carried out on the samples. The results of experimental research show refined microstructures both for FSW and SFSW welding. Improved mechanical properties have been obtained for SFSW welding, compared to FSW. SFSW specimens demonstrates superior tensile strength and a higher hardness compared to FSW specimens, by performing joining process underwater. The fracture surfaces of the tensile test specimens from the base material (BM), FSW and SFSW joints, revealed the ductile fracture mechanism of the joints. EDX analysis confirms compositional integrity of the base and welded metals. Results highlights suitability of the FSW and SFSW processes for joining of copper Cu-DHP.

1. Introduction

Copper, with a density of 8.96 g/cm^3, exhibits low hardness and has a high melting point of 1083°C. Copper and copper alloys are widely used in various industries, including in the field of electrical and thermal energy conduction, heat transmission, electronics, automotive industry, aeronautics, aerospace, chemical industry, etc. due to their excellent thermal and electrical conductivity [1-4], ranking on second position after silver in these terms [5], being resistant to corrosion and wear [2]. Considering that the thermal conductivity of copper exceeds that of steel and nickel alloys by 7-30 times, there are considerable difficulties to carry out welded joints of it using conventional fusion welding techniques [2, 3].

Among the different joining techniques, special attention is paid to friction stir welding (FSW), a solid-state welding process that could be a viable alternative to fusion welding processes, with promising results regarding the joining of different categories of metallic materials, such as aluminium, copper, magnesium, titanium alloys, various type of steels, etc. [2],

[6-23]. FSW offers benefits compared to fusion welding, such as solid state welded joints made through an environmental friendly process which does not require additional or filler materials, no noxes or radiations are emitted, the chemical composition and metallurgical characteristics of the base materials are preserved, no processing edges of the materials to be joined is necessary, the residual stresses and distortion are reduced, defects that may appear during fusion welding of metallic materials are avoided, obtaining good mechanical properties and increased joint strength, with less energy consumption [2, 3], [6], [12], [16], [23-28]. In the case of copper, considering its high conductivity, FSW welding can be an alternative to fusion welding [7].

In recent years, a new variant of FSW called submerged friction stir welding (SFSW) has been developed [12], [23], [29-35], attracting the interest of researchers due to its unique characteristics and potential benefits in terms of process control and weld quality.

To establish the foundation of the research, relevant references regarding FSW and SFSW welding, but also regarding the possibilities to apply of these to copper, have been consulted. A review of the current status and development of friction stir welding FSW [2, 3], [6-11], [13], [16-23], [25-27], [33], [36-41] and submerged friction stir welding SFSW [11], [23, 24], [31], [34], [36] was done; noting that issues related to processes and work techniques, applications, challenges and potential advancements of SFSW and FSW are addressed, consolidating the current knowledge and research progress in the field. Studies that focused on FSW welding of copper compared to other welding processes showed that welded joints with superior mechanical characteristics compared to conventional fusion welding processes, e.g. TIG, can be obtained [42], [21]. Some studies have been carried out on the influence of the geometry and dimensions of the tool, the process parameters and the amount of heat introduced in the process on the appearance and quality of the weld, structural analyses, the formation of defects in the joint and the mechanical properties of the welded joint [3], [8], [10], [17], [19-23], [27], [37- 41].

Other studies have focused on submerged friction stir welding in general, but also on copper and the quality of welded joints [23], [31-34], analysis of joint formation mechanisms and numerical simulations [19], [21], [34], [43, 44], showing that welded joints with better tensile strength compared to FSW welding can be obtained by using SFSW welding [31-33]. An analysis of the specialized literature in the field of friction stir welding shows that there are current researches aimed at optimizing the process parameters, the geometry and the dimensions of the welding tools, in order to increase the quality and improve the tensile strength and mechanical properties of the welded joints, to reduce the formation of defects for increasing the corrosion and fatigue resistance, respectively [9], [36], [45-49].

All of these [1-49] contribute to a better knowledge of FSW and SFSW welding processes and of potential to apply to the joining of metallic materials.

Friction stir welding process is an unconventional welding process, of interest both for researchers and for industry, due to advantages compared to conventional welding processes. Both FSW and SFSW are solid state welding processes which are recommended to be used for joining metallic materials which are difficult to be joined using fusion welding processes.

Submerged friction stir welding has advantages compared to FSW and a few of them are mentioned below. Using SFSW welding process, the overheating of the welding tool and welding device are limited/avoided, so the service life of them could be increased. Excessively plasticization of welding materials is also limited/avoided. The process temperature can be maintained in a defined interval, by monitoring and control of the water temperature inside the

IOP Conf. Series: Materials Science and Engineering **1319** (2024) 012008 doi:10.1088/1757-899X/1319/1/012008

water enclosure. SFSW can be used for joining materials which are sensitive to overheating. The peak of temperature in SFSW welding is also lower than in FSW welding process, contributing to obtain a more refined grains microstructure and better mechanical properties compared to FSW welding process. Another reason to use SFSW welding is that prevents the oxidation of the surface of materials to be joined, resulting a better aspect of the weld compared to FSW.

The above mentioned are reasons for which experimental research regarding use of these welding processes to join copper plates were performed.

Results from previous scientific programs performed by ISIM on FSW copper alloys were a starting point regarding SFSW of copper alloys, contributing to a better understanding of FSW processes and the properties and potential applications of Cu-DHP (Deoxidized High Phosphorus Copper) alloys in various industries [48].

To align with the current state of the art development in the FSW and SFSW processes, the authors developed FSW and SFSW joints, made from 2.5 mm sheets of Cu-DHP, then analysed the joint characteristics by conducting destructive tests on extracted specimens, including tensile and bending tests, morphologic analysis using scanning electron microscopy (SEM) with energy dispersive X-ray (EDX) and fractography investigations.

2. Materials and Methods

2.1 Materials

The experimental program specifically targets 2.5 mm Cu-DHP sheets, joined using FSW and SFSW with the aim to provide insights into their performance and behaviour in joining processes, and improving the understanding and optimization of joining techniques for this material. Tables 1 and 2 present the chemical composition and mechanical properties of Cu-DHP.

Table 1. Chemical composition of Cu-DHP alloys.

Material	Cu mass %	Zn mass %	Al mass %	Si mass %	P mass %	Mg mass %	Other mass %
EN Cu-DHP	99.8	0.1	0.03	0.02	0.02	0.01	0.02

Table 2. Mechanical properties of Cu-DHP.

Mechanical properties	Value
Tensile strength, R_m	267 MPa
Vickers hardness, HV5	71 HV1
Elongation at break, A	11.7%
Bending angle	1800

2.2 Equipment and water supply module used for welding

The friction stir welding machine FSW 4-10 belongs to ISIM Timisoara and was used to carry out the FSW experiments (Figure 1a). Figure 1b shows a module designed and built by ISIM, that was integrated on the FSW welding machine to allow the underwater friction stir welding.

A water enclosure of an assembled metal frame from special aluminum profiles with transparent plastic material (as walls) was made, sealed mounted on the table of the welding machine and equipped with controlled mechanism for water supply and evacuation. It also incorporates faucets and the necessary connections for water inlet/outlet in/from the water enclosure. For submerged friction stir welding experiments, ordinary water from water supply network was used. A water temperature monitoring and control system has also been mounted to ensure process control by detect and adjust temperature variations that can lead to process failures.

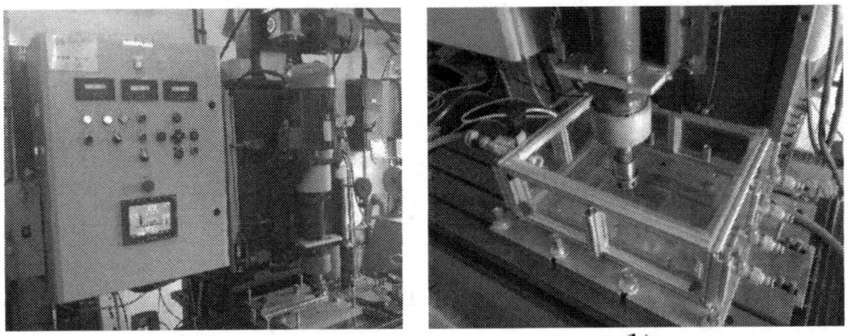

(a) (b)

Figure 1. Constructive solutions of the FSW/SFSW welding system: (a) friction stir welding machine FSW 4-10; (b) module for underwater friction stir welding, integrated on the FSW welding machine.

The materials to be welded are positioned and fixed on a steel base plate, inside the water enclosure and completely immersed underwater.

2.3 Joining process parameters

The process parameters used to develop Cu-DHP similar butt joints through the conventional friction stir welding (FSW) technique and our proprietary submerged friction stir welding (SFSW) equipment concept are described in Table 3.

The process parameters were established by previous experimental research programs of FSW and SFSW welding of copper sheets. In order to obtain a good plasticization of the materials to be joined by SFSW process, a higher rotation speed and a lower welding speed were used compared to FSW process.

Figure 2. Welding tool used for experimental research.

Table 3. FSW and SFSW butt weld experiment process parameters for Cu-DHP 2.5 mm sheets.

Welding process	Tool			Welding parameters			Rotation direction	**Experiment No.**
	Tool material	Pin tool	Pin length L_{pin} (mm)	Type/shoulder diameter $Ø_{tool\ shoulder}$ (mm)	Rotation speed n (rpm)	Welding speed v (mm/min)		
FSW	P20S	smooth conical	2.05	Flat, 22	950	60	Counter-clockwise	Exp.3.1
SFSW	P20S	smooth conical	2.05	Flat, 22	1250	40	Counter-clockwise	Exp.3.1A

A welding tool made of sintered tungsten carbide was used for experiments (Figure 2). The welding tool has a flat shoulder and a conical shape pin of 2.05 mm length.

IOP Conf. Series: Materials Science and Engineering **1319** (2024) 012008 doi:10.1088/1757-899X/1319/1/012008

2.4 Characterization

The surface morphology of weld samples was examined by scanning electron microscopy (SEM). A FEI Inspect S model coupled with an energy dispersive X-ray analysis detector (EDX) was used for the inspecting of the chemical composition. Tensile test specimens were prepared according to the EN ISO 6892-1/2020 standard and tested on a universal machine 400 KN, type EDZ 40; digital calliper, mark KS Tools 300.0520; digital thermo-hygrometer CEM, in 23 0C, ambient environment. The surface characteristics of the joints were thoroughly examined and analysed to evaluate their visual appearance, according to SR EN ISO 25239-5:2020, SR EN ISO 9712:2022 and ISIM LIEA testing laboratory internal specifications. The size of the tensile specimens taken from the FSW and SFSW welded joints is presented in Table 4.

Table 4. The size of tensile test specimens FSW and SFSW process.

Material Cu 99	Specimen no.	Specimen thickness a (mm)	Specimen width b (mm)	Section a x b (mm²)
3.1	T1-3.1	2.55	20.74	52.89
3.1	T2-3.1	2.55	20.79	53.01
3.1 A	T1-3.2	2.56	20.83	53.32
3.1 A	T2-3.2	2.57	20.75	53.33
Base material	T-BM	2.55	20.34	51.87

3. Results

3.1 Visual analysis

The surface appearance of the Cu-DHP joints, welded using the process parameters from Table 3, is depicted in Figure 3. In the case of FSW welding, the strong thermal imprint of the material is visible due to a large amount of heat released in the process, necessary to plasticize the materials to be welded. In the thermally imprint area of the materials to be joined, a change in their color due to surface oxidation can be observed. In the case of SFSW welding, due to the liquid working environment, the process temperature is lower compared to FSW welding, the thermal imprint and oxidation of the materials to be joined being not visible. This aspect could be an advantage for various application where are strongly demands related to quality of the weld surface.

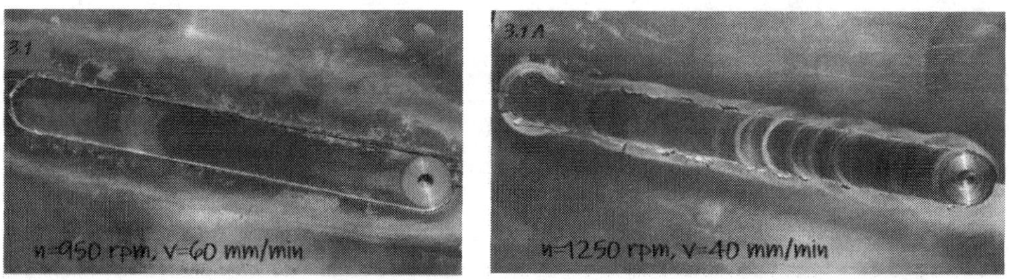

(a) (b)

Figure 3. Surface appearance of Cu-DHP joints: (a) FSW sample, (b) SFSW sample.

3.2 Hardness measurements

The hardness measurements were performed on the thickness midline of each welded joint sample and also for the base material. The hardness measurement was 71 HV5 for the base material. For FSW and SFSW welded joints, hardness values in the middle of the weld were 64 HV5 and 91 HV5, respectively. In the adjacent areas of the welded joint, measured hardnesses have value close to that the base material for FSW weld and 99 HV5 for SFSW weld.

3.3 Tensile tests

The results of tensile tests and elongation at break are visible in Table 5.

Table 5. Tensile test results Cu-DHP butt welded using FSW and SFSW process.

Material Cu 99	Specimen no.	Maximum force F_{max} (N)	Tensile strength R_m (N/mm²)	Breaking area	Elongation at break (%)
3.1	T1-3.1	10939	207	weld	6.38
3.1	T2-3.1	10627	200	HAZ	5.05
3.1 A	T1-3.2	12927	242	weld	2.66
3.1 A	T2-3.2	11399	214	weld	2.13
Base material	T-BM	13830	267	BM	11.7

It was observed that for the SFSW welded joints the tensile strength is higher and the elongation at break is lower compared to the FSW welding. In the case of FSW welding, the higher process temperature can lead to an effect equivalent to a recrystallisation annealing, which reduce the tensile strength and contribute to the increase in the elongation at break. In case of SFSW welding, this effect does not occur due to the water cooling which leads to a lower process temperature than in the case of FSW welding and to less heat affected materials.

These findings demonstrate the influence of the welding process on the mechanical strength of the joints and indicate that both FSW and SFSW welding techniques exhibit favorable strength characteristics relative to the base material.

3.4 Structural Characteristics

First, the authors extracted a base metal sample (sample 1), made of Cu-DHP, which was subjected to scanning electron microscopy (SEM) analysis to investigate its microstructural characteristics. SEM analysis is important for understanding the properties of the material, optimizing the processing parameters, and ensuring the desired performance of Cu-DHP in various industrial applications. Results are presented in Figure 4.

Microstructural examination of the alloy reveals the presence of a coarse grain structure, characterized by larger grain sizes of the copper.

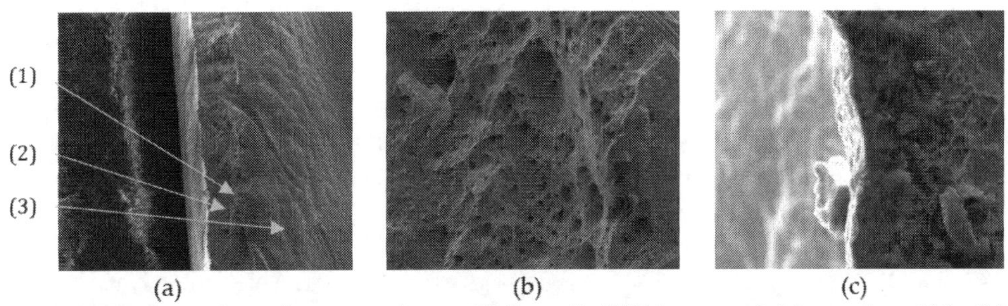

Figure 4. SEM images of sample no. 1 made out of 2.5 mm Cu-DHP base metal: (a) sample at 106x; (b) zone (1) at 800x; (c) zone (2) at 3000x, (1.3) dimpled fracture surfaces.

Analyzing the fracture surfaces of the tensile test, using SEM (Figure 4.a, detail 1 and 2, Figure 4.b) revealed the ductile aspect, characterized by micro-voids coalescence which develop and grow during plastic deformation of the material. On those surfaces, can be observed the dimpled structure of the material, also characteristic for a ductile fracture (Figure 4.a, detail 3).

Continuing with the investigation, the authors conducted a detailed SEM analysis on Cu-DHP similar joints (sample 2) made using the friction stir welding (FSW) process (Figure 5). The FSW process induces dynamic recrystallization and grain refinement in the welded region due to the severe plastic deformation and the high strain rates involved. This leads to the formation of finer grains in the weld zone compared to the base metal. The refined grain structure of the FSW joints offers several advantages, including improved mechanical properties such as higher strength and enhanced resistance to fatigue and cracking. Furthermore, the FSW process helps in breaking down and redistributing any inclusions or precipitates present in the base metal, resulting in a more uniform distribution throughout the weld zone. This homogenization of the microstructure contributes to the improved integrity and reliability of the FSW joints.

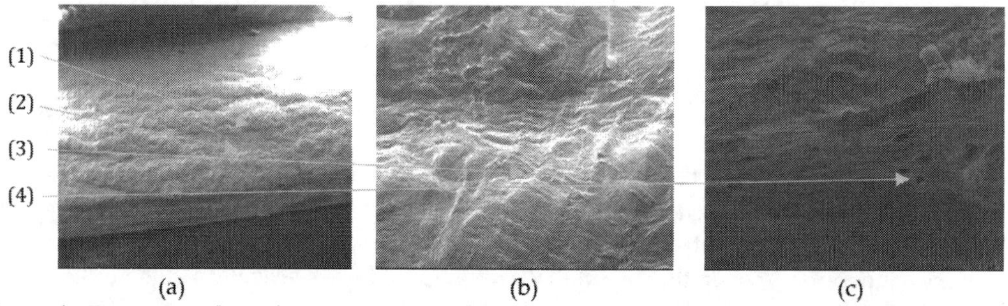

Figure 5. SEM images of sample no. 2 made out of 2.5 mm Cu-DHP similar joint obtained by FSW: (a) sample at 104x; (b) zone (1) at 800x; (c) zone (2) at 3000x, (a.3) fracture step lines, (c.4) micro-voids.

The SEM micrographs of the 2.5 mm Cu-DHP similar joints exhibit a refined micro-structure with significantly smaller grain sizes compared to the base metal. It is noteworthy that the grain size dimensions of the welded material correspond to an ultrafine grain structure. The FSW process induces significant grain refinement and microstructural homogenization in the Cu-DHP base metal, leading to improved properties and performance in the welded joints compared to the base material.

Images of the fracture surfaces of the specimens subject to tensile tests revealed that the material is ductile, making it very easy to observe the advancement of the fracture steps/lines (Figure 5.b, detail 3). Using a higher magnification (3.000x), the same micro-voids are observed, but their dimension is smaller than those observed in base material, due to the plastic deformation at which the material in this region was subjected during the FSW process (Figure 5.c, detail 4).

The SEM analysis conducted for similar Cu-DHP butt joints welded using submerged friction stir welding (SFSW) is presented in Figure 6. The SEM micrographs reveal the detailed morphology and structural features of the welded region. The analysis demonstrates a well-defined and continuous bond line between the joint surfaces, indicating successful joining and metallurgical bonding during the SFSW process.

The grain structure of the SFSW joints shows similarities to the base metal and the FSW samples, with comparable grain sizes and orientations. This observation suggests that the SFSW process has effectively preserved the microstructural integrity of the Cu-DHP alloy, minimizing any significant distortion during the joining operation. The micro-graphs indicate the absence of prominent defects such as cracks or inclusions, but evidence of porosity can be seen.

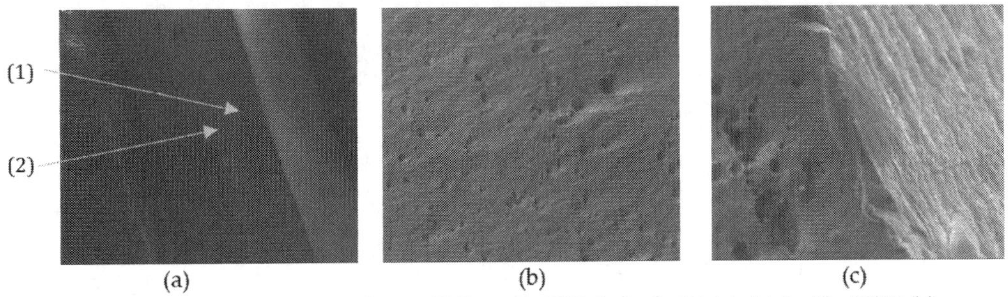

Figure 6. SEM images of sample no. 3 made out of 2.5 mm Cu-DHP similar butt joint obtained by SFSW: (a) sample at 100x; (b) zone (1) at 800x; (c) zone (2) at 3000x.

Micrographs reveal the presence of small voids or pores distributed within the welded zone. The occurrence of porosity can be attributed to various factors, including gas entrapment which may occur due to the turbulence caused by the swirl rotation movement of the water around the welding tool during the welding process. While the porosity may not be extensively prevalent, its presence can have implications on the mechanical properties and structural integrity of the joints. Aspect of the fracture surfaces, characterized by voids and micro-voids (Figure 6.b) as well as steps (terraced appearance) (Figure 6.c in the right-side half) resulting from the slip mechanism under which was produced the failure proves once more than the material was ductile.

The presence of fracture surfaces with voids together with other fracture surfaces with steps proves that two mechanisms are involved in the development of the fracture: one mechanism driven by the cavity growth and the other one driven by the shear respectively [49].

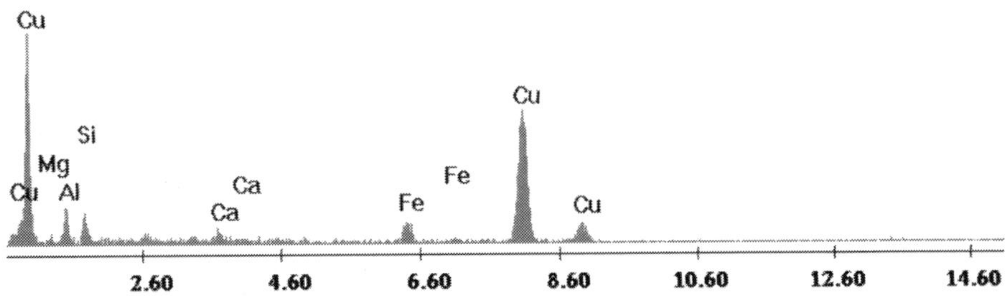

Figure 7. EDX spectrum of sample no.1, made out of 2.5 mm Cu-DHP base metal.

By analyzing the spectral lines (Figure 7) corresponding to different elements, such as copper (Cu), impurities or alloying elements, valuable information was obtained on the composition and purity of the Cu-DHP base metal. This spectral analysis serves as a tool for quality control, material characterization and ensuring the desired properties of Cu-DHP for various industrial applications [12]. The Energy Dispersive X-ray analysis EDX of the Cu-DHP base metal samples revealed spectra that are characteristic of the alloys under investigation, indicating a consistent composition. However, minor deviations were observed in the chemical composition, namely Fe and Ca. These slight variations in the iron and calcium content can be attributed to factors such as impurities from the manufacturing process, or external contamination (e.g. from the welding process, tool, previous processing and so on).

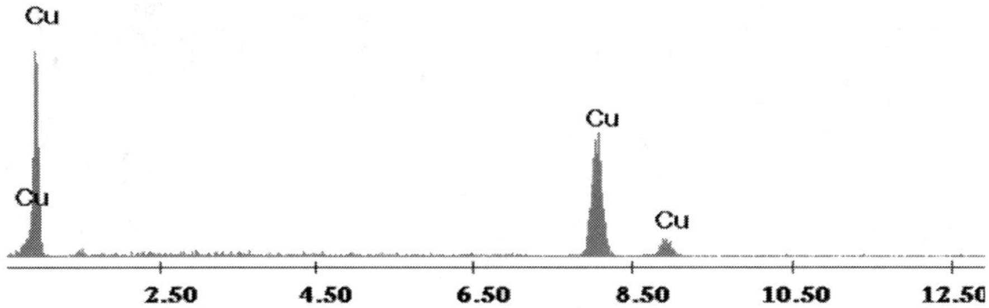

Figure 8. EDX spectrum of sample no. 2 made out of 2.5 mm Cu-DHP alloy similar butt joints obtained by FSW.

Further, the authors performed the spectral analysis on Cu-DHP similar joints produced through the friction stir welding (FSW) process, analyzing the middle of the weld (Figure 8).

The EDX spectra acquired from the analysis of Cu-DHP samples joined using friction stir welding (FSW) exhibit characteristic patterns that are consistent with the alloys under investigation. No visible variations in the chemical composition of the material were observed. This observation suggests a high degree of compositional integrity and uniformity in the welded Cu-DHP joints, confirming the efficacy of the FSW process in preserving the desired chemical characteristics of the base material.

The authors performed the spectral analysis on Cu-DHP similar joints produced through the submerged friction stir welding (SFSW) process (Figure 9), analyzing the middle of the weld.

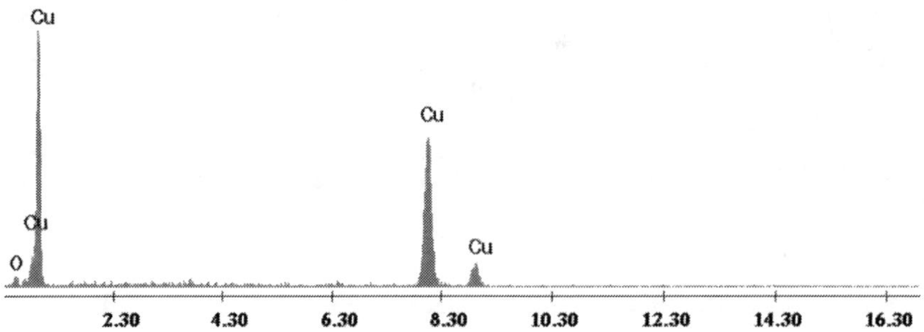

Figure 9. EDX spectrum of sample no. 3 made out of 2.5 mm Cu-DHP similar butt joints obtained by SFSW.

The EDX spectrum obtained reveals a prominent peak corresponding to copper (Cu), indicating the dominance of this element within the weld composition.

4. Discussions

Drawing upon the experimental results stemming from the joining of Cu-DHP alloys through both friction stir welding (FSW) and submerged friction stir welding (SFSW) techniques, carried out under comparable operating conditions and technological parameters, a set of significant discussions can be derived. To start, it is evident that both welding methods effectively facilitate the successful joining of Cu-DHP sheets.

In the case of FSW a certain degree of material weakening is observed, it could be attributed to the intricate interplay of mechanical forces and heat generation inherent in the FSW process, leading to localized changes in material properties. Conversely, the SFSW technique demonstrates a comparatively higher average tensile strength, it signifies a more efficient and robust joining of the material through the SFSW process. The additional submerged environment leads to better heat dissipation in the welding area, avoiding the overheating of the welding tool and welding device. The process temperature can be maintained in a defined interval, by monitoring and control of the water temperature inside the water enclosure. All of these contribute to carry out stable conditions for the welding procesees, that ensure a uniform plasticization and material flow during the welding process, which can lead to a enhanced weld quality. The observed disparity in tensile strengths thus accentuates the importance of choosing the appropriate welding technique based on the desired mechanical properties for the specific application. Further insight reveals that both procedures induce dynamic recrystallization and concurrent grain refinement within the welded area, leading to finer grained structures when compared to the base metal. This shift towards a refined grain arrangement presents a distinct advantage, resulting in improved mechanical properties, enhanced resistance to fatigue induced deterioration and cracking, and heightened microstructural uniformity.

The alignment in composition between the FSW, SFSW joints and Cu-DHP alloys is evident. The SFSW welding provides good protection of the materials to be joined. In case of other welding processes applied to join copper sheets, the presence of brittle inclusions emerges as a critical consideration. These inclusions, interfacing with the softer copper matrix, possess the potential to induce defects and compromise the overall structural integrity of the welded components.

5. Conclusions

In summary, the analysis suggests that the FSW and SFSW processes are effective in producing joints with refined microstructures and improved properties. FSW induces slight material weakening owing to mechanical forces and heat effects. Conversely, SFSW demonstrates superior tensile strength and a higher hardness by enhancing the joining process in a submerged environment, meant underwater. Images of the fracture surfaces of the specimens subject to tensile tests revealed the ductile fracture mechanism of the joints. EDX analysis confirms compositional integrity of the base and welded metals. The SFSW welding provides good protection of the materials to be joined so that brittle precipitates do not occur. Both methods induce grain refinement, enhancing mechanical properties and uniformity.

SFSW process provide an environment that extracting the heat from the welded zone, prevents the recrystallization annealing effect of the heat produced during the friction process. In this way the tensile strength increases and elongaion at break decreases, compared with those obtained with FSW process. In the same time, in FSW samples, fracture will take place immediately in the vicinity of the base material, in the Thermo Mechanical Affected Zone – TMAZ, where the grains increased and effect of the cold hardening of the base material plates is erased.

Due to the dynamic recrystallization that took place during the FSW process, as a result of plstic deformation under hgh forces, the grains are smaller in welded zone compared with the base material, providing better properties of the joint despite the fact that temoerature in this zone increased during welding process. SFSW process has even better effect in this aspect.

Acknowledgments: This paper has been developed within the project PN 23 37 01 02 entitled "Research on the modification of metallic materials properties using the innovative and environmentally friendly method of friction stir processing in liquid environment" (funded within the Program Nucleu of ISIM Timisoara - contract 16N/2023), in collaboration with UPT-Faculty of Mechanics, Material Science Department.

References

[1] Farrokhi H, Heidarzadeh A, Saeid T 2013 *Sci. Technol. Weld. Join* **18 (8)** pp 697–702
[2] Sahlot P, Singh A K, Badheka V J, Arora A 2019 *Trans. Indian Inst. Met.* **72** pp 1339–1347
[3] An T, Li X, Xu H, Lan C, Song L, Hui H 2024 *Mater. Res. Express* ad3168
[4] Fattah-alhosseini A, Taheri A H, Keshavarz M K 2016 *Trans. Indian Inst. Met.* **69** pp 1423–1434
[5] Binchiciu E F, Codrean C, Boțilă L N, Nicolaescu M, Iacobici R I 2022 *Defect Diffus. Forum* **416** pp 55–65
[6] Nyang'au M G, Muumbo A M, Mutua F N 2024 *Afr. J. Eng. Res.* **12(1)** pp 1–11
[7] Constantin M A, Nițu E L, Bădulescu C 2019 *IOP Conf. Ser.: Mater. Sci. Eng.* **564** 012031
[8] Vukčević M, Šibalić N, Janjić M 2014 *J. trends dev. mach. assoc. technol.* **18(1)** pp 77-80
[9] Kilic S, Ozturk F, Demirdogen MF 2023 *J. Eng. Res.* in press, https://doi.org/10.1016/j.jer.2023.09.005
[10] Jayabalakrishnan D, Naga Muruga D B 2019 *IJTRET* **3(4)** pp 1-5
[11] Yeshitla G G, Singh B 2020 *HTL* **26(11)** pp 682–705
[12] Gyasi, E.A. Welding processes of metals for offshore environment: underwater welding, Research reports., Lahti University of Technology LUT, Finland, 2019, Series 103, https://urn.fi/URN:ISBN:978-952-335-433-3.
[13] Meng X, Huang Y, Cao J, Shen J, dos Santos J F 2021 *Prog. Mater. Sci.* **115** 100706
[14] Singh G, Thakur A, Singh S, Sharma N 2020 *Indian J. Eng. Mater. Sci. (IJEMS)*, **27** pp 709–716
[15] Xu A 2020 *J. Phys. Conf. Ser.* **1676**, 012019
[16] Singh R P, Dubey S, Singh A, Kumar S 2020 *Mater. Today Proc.* **38** pp 6–11
[17] Khan N, Rathee S, Srivastava M 2021 *Mater. Today Proc.* **47(5)** pp 7196–7202
[18] Sharma A, Dwivedi V K 2022 *Lecture Notes in Mechanical Engineering (LNME);* Dubey, A.K.; Sachdeva, A.; Mehta, M., Eds., Springer, Singapore, pp 183–191
[19] Bhardwaj N, Narayanan R G, Dixit U S, Hashmi M S J 2019 *Adv. Mater. Technol.* **5(3)** pp 461–496
[20] Mahananda C, Jeet S, Kar S 2018 *IJSER* **9(4)** pp 51–54
[21] Soori M, Asmael M, Solyalı D 2021 *SAE Int. J. Mater. Manuf.* **14(1)** pp 63–80
[22] Naik S, Panda S, Padhye S, Lobo G, Joshi G, Deshmukh S, Ingle A 2021 *Mater. Today: Proc.* **47** pp 3117–3122
[23] El-Sayed M M, Shash A Y, Abd-Rabou M, ElSherbiny M G 2021 *J. Adv. Join. Process.* **3** 100059
[24] Jayaseelan P, Christy T V, Gowtham S 2020 Friction Stir Welding; Principles and Applications. *In Friction Stir Welding; Principles and Applications,* 1st ed.; McWilliams, M., Ed.; Nova Science Publishers: Hauppauge, NY
[25] Heidarzadeh A, Mironov S, Kaibyshev R, Çam G, Simar A, Gerlich A, Khodabakhshi F, Mostafaei A, Field D P, Robson J D, Deschamps A, Withers P 2021 *Prog. Mater. Sci.* **117** 100752
[26] Nagabharam P, Srikanth Rao D, Manoj Kumar J, Gopikrishna N 2018 *Mater. Today Proc.* **5(1)**, pp 1264–1270
[27] Liu X C, Sun Y F, Nagira T, Ushioda K, Fujii H 2019 *Sci. Technol. Weld. Join.* **24(4)** pp 352–359
[28] Ambrosio D, Morisada Y, Ushioda K, Fujii H 2023 *J. Mater. Process. Technol.* 2023 **320**, 118116
[29] Rathinasuriyan C, Pavithra E, et. al. 2021 *Int. J. Precis. Eng. Manuf.-Green Tech.* **8**, pp 687–701
[30] Xue P, Xiao B L, Zhang Q, Ma Z Y 2011 *Scr. Mater.* **64(11)** pp 1051–1054
[31] Wahid M A, Siddiquee A N 2018 *Trans. Nonferrous Met. Soc. China* **28(2)**, pp 193–219,
[32] Dumpala L R, Salavaravu 2016 *Int. J. Adv. Res. Sci. Eng.* **5(10)** pp 5-9
[33] Moaref A, Rabiezadeh A 2020 *Trans. Nonferrous Met. Soc. China* **30(4)** pp 972–981
[34] Paramaguru D, Pedapati, S R, Awang M 2019 *Lecture Notes in Mechanical Engineering (LNME)*, Awang M., Ed., Springer, Singapore, pp 71–83.
[35] Ethiraj N, Sivabalan T, Meikeerthy S, Kumar K L V R, Chaithanya G, Reddy G P K 2019 *In Proceedings of the AIP Conference*, Tamilnadu, India, 8-9 March 2019, **2128 (1)**, 030003

[36] Bharti S, Kumar S, Singh I, Kumar D, Bhurat S S, Abdullah M R, Rahimian Koloor S 2024 *J. Mar. Sci. Eng* **12**, 71
[37] Teimurnezhad J, Pashazadeh H, Masumi A 2016 *J. Manuf. Process.* **22** pp 254–259
[38] Barenji R V 2016 *Trans. Indian Inst. Met.* **69(5)** pp 1077–1085
[39] Xu N, Song Q, Bao Y 2017 *Mater. Lett.* **205** pp 219–222
[40] Dialami N, Cervera M, Chiumenti M 2020 *Eur. J. Mech. A/Solids* **80** 103912
[41] Chandana R, Saraswathamma K 2023 *Mater. Today Proc.* **76** pp 602–606
[42] Lin J W, Chang H C, Wu M H 2014 *J. Manuf. Process.* **16(2)** pp 296–304
[43] Akbari M, Asadi P, Sadowski T 2023 *Materials,* **16(17)** 5890
[44] Azizi A, Hajinejhad M 2018 *J. Mech. Eng.* **48 3(84)** pp 187–195
[45] Bhardwaj N, Narayanan R G, Dixit U S, Hashmi M S J 2019 *Adv. Mater. Process. Technol.* **5(3)** pp 461–496
[46] Sezhian M V, Ramadoss R, Giridharan K, Chakravarthi G, Stalin B 2020 *Mater. Today Proc.* **33** pp 4842–4847
[47] Zhang Y M, Yang Y P, Zhang W, Na S J 2020 *J. Manuf. Sci. Eng.* **142 (11)**110816
[48] Cojocaru R, Boţilă L N, Daşcău H F, Duma I 2020 *BID-ISIM* **4** pp 3–8.
[49] Ghahremaninezhad A, Ravi-Chandar K 2011*Int. J. Solid. Struct.* **48** pp 3299–3311

IOP Conf. Series: Materials Science and Engineering **1319** (2024) 012009 doi:10.1088/1757-899X/1319/1/012009

Optimizing Milling Processes through FEA: An In-depth Analysis of Cutting Tool Geometric Parameters and Their Impact on Performance of milling.

Martin Necpal[1*], Jozef Peterka[1] and Marek Vozár [1]

[1] Faculty of Material Science and Technology in Trnava, Slovak University of Technology in Bratislava , Slovakia

*E-mail: nartin.necpal@stuba.sk

Abstract. In the rapidly advancing world of manufacturing, the optimization of cutting processes is of paramount importance for enhancing efficiency and product quality. This work underscores the necessity of employing Finite Element Analysis (FEM) for a comprehensive understanding and improvement of cutting processes. By leveraging DEFORM we create a detailed 2D simulation of the milling process, allowing for an unprecedented inspection of the procedure in a virtual environment. Central to our investigation is the variation of geometric parameters of the cutting tool within the FEM simulation model, offering insights into how minute alterations can significantly affect the outcome of the milling process. The study systematically analyzes the impact of the milling tool's geometric parameters on various process parameters, including chip formation and cutting forces. This analysis is critical for identifying optimization opportunities in tool design and process settings. Furthermore, we delve into the stress-strain state of the cutting zone, a factor crucial for understanding the material behavior under operational conditions and ensuring the structural integrity of the final product. Through our findings, we demonstrate that strategic modifications in the geometric parameters of the cutting tool can lead to substantial improvements in process performance. This research not only contributes to the body of knowledge on milling processes but also provides practical recommendations for engineers and manufacturers striving to optimize their cutting operations through the application of FEM.

1. Introduction

In the field of manufacturing by machinning processe , the optimization of cutting processes is a key factor in improving efficiency and product quality (Zhang & Wang, 2015) (Melkote et al., 2022). In general, the effort of modeling and improvement of simulation techniques is to relieve imagination from a lot of experiments and costly tests when deploying new tools for new machining materials. (Arrazola et al., 2013) present the state-of-the-art in predictive performance models for machining operations and to discuss the strengths, weaknesses, and applicability of various modelling techniques used in the prediction of complex machining operations industry. It emphasizes the importance of linking fundamental process variables to industry-relevant performance measures such as product quality, tool-wear, surface and subsurface integrity, chip-form/breakability, and part accuracy in order to make the models useful for industry applications. A number of studies focus on verifying the simulation model and setting boundary conditions by comparing experiments and measuring results of experiments. (Hussain et al., 2022) focuses on investigating the effects of chamfering angle and chamfering width on polycrystalline cubic boron nitride tools during the hard turning of GCr15 hardened steel. It aims to study the cutting force, cutting temperature, and the formation of saw-tooth chips under constant cutting conditions, utilizing both simulation analysis and experimental methods. Within the framework of comparing experiments with simulation techniques, the problem of tool wear and durability is one of the key issues. (Shi et al., 2021) investigates the effects of tool wear on cutting performance, considering factors such as stress and temperature distribution on wear rate. Additionally, it aims to validate the model's accuracy through cutting experiments and to use the validated model to study how a worn tool affects cutting performance in terms of rake angle, stress distribution, cutting force, and temperature. As the milling process belongs to a complex problem with a large representation of milling operations in industry, it is possible to meet many studies focused on milling. The study (Wojciechowski et al., 2019) introduces a novel approach for predicting cutting forces in micro end milling, accounting for chip thickness accumulation. By considering various factors such as micro end milling kinematics, machine tool errors, and workpiece deformations, the proposed model demonstrates improved accuracy compared to traditional rigid micro end milling models, particularly in scenarios with close-to-minimum uncut chip thicknesses. (Wu et al., 2013) presents an enhanced cutting force model for circular end milling processes, addressing the limitations of traditional linear milling force models. By accounting for curvature effects on chip thickness and entry/exit angles, the proposed model accurately predicts instantaneous cutting forces during circular end milling operations, validated through experiments with both constant and variable radial depth. By incorporating these advancements in the modeling and simulation of cutting processes, we aimed to provide a more accurate understanding of the behavior of materials during milling (Kundrák et al., 2021). In addition to standard FEM methods, special simulation techniques also find their application in machining processes. Research (Afrasiabi et al., 2021) utilized both Smoothed Particle Hydrodynamics and Finite Element Method. The study introduced an innovative approach for measuring rake face temperature without impeding chip flow and proposed novel, empirically-derived friction coefficients tailored to AISI 1045 steel and Ti6Al4V alloy under specific cutting conditions.

This work is focused on the use of 2D FEM orthogonal cutting technique of commercial software. With the use of 2D simplification, it is possible to set up a fine enough mesh so that it is possible to examine the rounding of the cutting edge and thereby assess the design of the tool. The main purpose is the investigation of chip formation, cutting forces and the stress-deformation state at the cutting zone. In the end, there is also a demonstration of the effect on various machined materials in terms of their strength and machinability.

2. FEM Setting and Boundary Condition

The FEM simulations were conducted using the commercial DEFORM 2D software version 13, renowned for its capabilities in analyzing complex material deformation phenomena. The implicit solver employed within DEFORM utilizes a Lagrangian formulation to address the nonlinear time incremental problem inherent in the cutting process. The time increment is determined based on the size of the element situated at the root of the section. To ensure comparability across different cutting edge roundings, it was imperative to employ a small element size of 0.003 mm. Consequently, the corresponding time step size is established at 0.0005 s. To achieve continuous frictional behavior, the simulation necessitates approximately 1000 steps to be executed. This formulation of the problem enables a comprehensive analysis spanning from the initiation of cutting to the critical moment when chip separation commences, typically occurring within a few microseconds. Such meticulous setup ensures that the simulation captures the intricacies of the cutting process with sufficient temporal resolution, facilitating detailed insights into machining dynamics.

DEFORM's sophisticated chip separation methods are facilitated through its re-mesh functionality, which allows for precise control over parameters such as increment depth between objects, maximum stroke increment, maximum time increment, or maximum step increment. Automatic re-meshing is triggered if the Jacobi determinant of resulting elements is negative ensuring numerical stability throughout the simulation. For the milling operation under investigation, the re-mesh criterion was conservatively set to an interference depth of 0.01 mm.

The parallel solver within DEFORM offers flexibility in the choice of iteration method, with options including the Newton-Raphson or direct iteration method. To expedite computation while maintaining accuracy, the implicit setting employs the Newton-Raphson method by default, as it typically yields favorable convergence rates. In cases where convergence is not achieved within prescribed tolerances, the solver seamlessly transitions to the direct iteration method. This adaptive approach ensures efficient computation while safeguarding against convergence issues.

It is worth noting that while the computational demands of 2D simulations are generally lower compared to their 3D counterparts, careful consideration was given to element discretization to optimize computational efficiency. By judiciously allocating mesh density, finer elements are concentrated in regions of interest, such as the cutting zone, while coarser elements are utilized in less critical areas. This strategy minimizes computational overhead while preserving solution accuracy, thereby maximizing computational efficiency.

Furthermore, aside from mesh density, the computational time is influenced by various factors including friction models, material models, and other solver settings. Balancing these parameters is crucial to achieving accurate and efficient simulations, ensuring that computational resources are utilized effectively.

2.1 Tool and Workpiece material

The selection of workpiece materials plays a pivotal role in shaping the chip formation process and influencing the mechanical properties at the cutting zone. In this study, three distinct materials—aluminum alloy, steel, and titanium alloy—were chosen due to their disparate properties, thereby affording a comprehensive exploration of machining behavior. According to ISO standards, aluminum alloy stands out as one of the lighter materials conducive to machining processes. Steel, on the other hand, serves as a ubiquitous benchmark in machining operations, given its widespread utilization across various industrial sectors. Titanium alloys present a formidable challenge in machining, characterized by their inherent hardness, propensity for early tool wear, and unfavorable machining conditions such as elevated temperatures and considerable cutting forces.

The materials under investigation were sourced from the standard DEFORM database, ensuring a consistent and validated framework for simulation. Each material exhibits plastic properties spanning from ambient to elevated temperatures, mirroring the thermal conditions encountered during machining operations. It is imperative to note that the stress-strain behavior of these materials is intricately tied to the strain rate, a critical consideration for accurately simulating the cutting process.

By incorporating these diverse materials into our simulation framework, we aim to elucidate the nuanced interplay between material properties, cutting parameters, and machining performance. This comprehensive approach not only enhances our understanding of individual material behaviors but also facilitates comparative analyses essential for optimizing machining strategies across different workpiece materials.

In the simulation, the cutting tool material is represented as an ideally rigid entity, a simplification that does not significantly impact the outcome. Nonetheless, given the rapid generation of heat during the machining process, it becomes imperative to incorporate heat balance considerations into the simulation model. To this end, the tool model undergoes meshing to facilitate thermal calculations. The tool material is characterized by properties including thermal conductivity, specific heat, and boundary conditions governing heat transfer. Consistency across simulations is ensured by maintaining uniform conditions, with the material properties aligned with those of carbide. By incorporating thermal considerations and aligning material properties with real-world counterparts, the simulation framework is enhanced to better capture the thermal dynamics inherent in the cutting process. This refinement facilitates a more accurate representation of machining phenomena, thereby enriching the insights gleaned from the simulations.

2.2 Tool and workpiece geometry

The Finite Element Method (FEM) simulation employed in this study is confined to a 2D framework, specifically orthogonal cutting. While orthogonal cutting is an idealization, particularly from a milling perspective where maintaining strict orthogonality may pose challenges, it suffices for preliminary analyses. The foundational aspects of the simulation, including boundary conditions, tool, and workpiece motion, are illustrated in Figure 1.

Upon conducting the basic simulation and examining the stress distribution, it becomes evident that the influenced area is relatively compact. Consequently, simplifying the geometries of both the tool and workpiece becomes feasible, as depicted in Figure 2. This simplification strategy involves focusing on areas where variables undergo significant changes while retaining accuracy in critical regions.

By streamlining the objects and concentrating on regions of variable alteration, it becomes feasible to augment the volumes of elements along the cutting edge. This optimization enables the execution of simulations incorporating diverse cutting edge roundings.

Through these methodological refinements, we aim to enhance computational efficiency while retaining fidelity to the underlying physical processes. This approach allows for a nuanced exploration of the effects of cutting edge roundings on machining performance, shedding light on critical parameters for optimizing machining strategies.

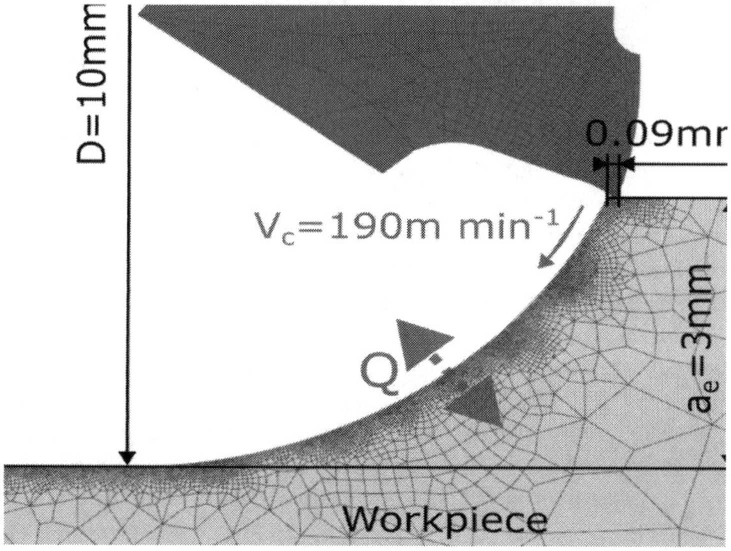

Figure 1. illustration of the setting of the boundary conditions of the tool and workpiece objects.

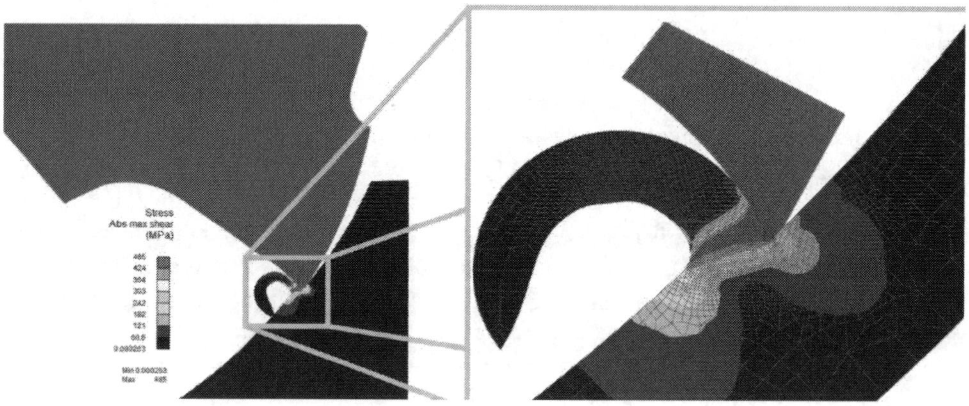

Figure 2. Representation of the simulation model simplification following the base Simulation.

3 Result and discussion

The simulation and analysis of the cutting process are primarily directed towards investigating cutting forces, tensile stresses, deformation at the cutting zone, and the formation of chips. The outcomes of this analysis were examined across various types of workpiece materials. This comprehensive approach enables a thorough examination of the interplay between cutting parameters and material properties, shedding light on critical aspects of the machining process. By scrutinizing these key variables across different materials, valuable insights are gained into the mechanisms governing machining behavior and the factors influencing machining performance.

IOP Conf. Series: Materials Science and Engineering **1319** (2024) 012009 doi:10.1088/1757-899X/1319/1/012009

3.1 Analysis of chip shape and cutting forces

The initial simulations focused on steel material, incorporating various cutting edge roundings: 10 µm, 20 µm, and 30 µm. Figure 3 illustrates the impact of cutting edge roundness on chip formation.

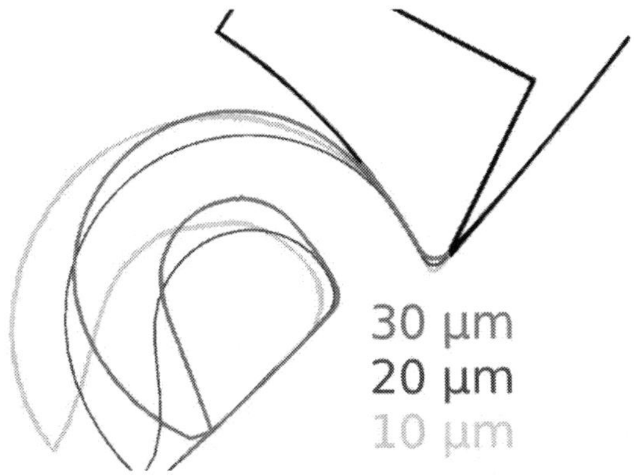

Figure 3. Impact of cutting edge roundness on chip formation.

Notably, a larger radius of cutting edge rounding results in a more pronounced curvature of the chip, as depicted by the 30 µm curve highlighted in red. For steel material, the evolution of forces from initial penetration into the material to the formation of a continuous chip was documented. Figure 4 presents the progression of forces in both the X-direction and Y-direction since the inception of the simulation.

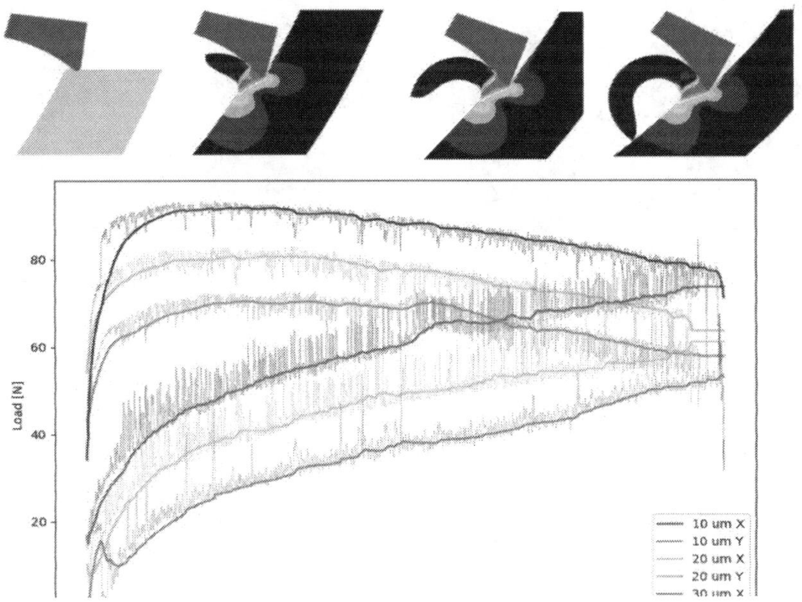

Figure 4. Progression of forces at different cutting edge roundings, from tool penetration into the workpiece to chip formation.

Each waveform corresponds to a specific radius of roundings of the cutting edge, with the dotted line representing the force waveform generated by the DEFORM software. The solid line denotes the smoothed waveform of cutting forces achieved through moving averages. Additionally, the upper portion of the figure provides an approximation of the chip's state and the cutting zone.

3.2 Shear stress distribution at the cutting zone

The simulation offers the advantage of observing any variable included in the calculation, providing insights into critical aspects of the machining process. For instance, analyzing the distribution of shear stresses at the cutting zone offers valuable information regarding chip formation and stress-strain states. Figure 5 illustrates the distribution of shear stresses across all three cases of cutting edge radius. Upon examination of the figure, it becomes apparent that the maximum shear stress values and the overall stress field distribution remain relatively consistent across all scenarios. This consistency is further corroborated by the distribution of cutting forces, wherein the magnitude variations are minimal. Consequently, it can be inferred that while there may be fluctuations in the forces' magnitudes, their resultant effects remain comparable. Moreover, an observation can be made regarding the relationship between forces in the X and Y directions. A higher force magnitude in the X direction corresponds to a comparatively lower force magnitude in the Y direction. This observation suggests that the overall magnitude of the forces can be regarded as consistent, despite variations in their directional components. These findings underscore the utility of simulation in elucidating complex machining dynamics and highlight the potential for further insights into optimizing machining processes.

Figure 5. Distribution of shear stresses across all three cases of cutting edge radius.

3. 3 Influence of the material to be cut on chip formation and the cutting process

A simulation was conducted with a cutting edge featuring a rounding radius of 30 μm, focusing on two distinct workpiece materials: aluminum alloy and titanium alloy. For the aluminum alloy, renowned for its ease of machinability, the chip morphology and shear stress distribution mirrored those observed in previous simulations, albeit at lower magnitudes. Figure 6 illustrates the distribution of shear stresses, revealing a maximum shear stress value within the shear plane of approximately 398 MPa. In contrast, machining of the challenging titanium alloy, known for its hardness and brittleness, yielded distinct outcomes. Despite its hardness, the material exhibited a tendency towards fragmentation, resulting in the formation of segmented chips characterized by high shear stresses, reaching up to 895 MPa, as depicted in Figure 6. In such scenarios involving the machining of hard and brittle materials like titanium alloy, striking a balance in the design of the cutting edge becomes imperative. This necessitates careful consideration of factors such as tool geometry and material properties to ensure that the tool can withstand the elevated forces and stresses encountered during machining operations.

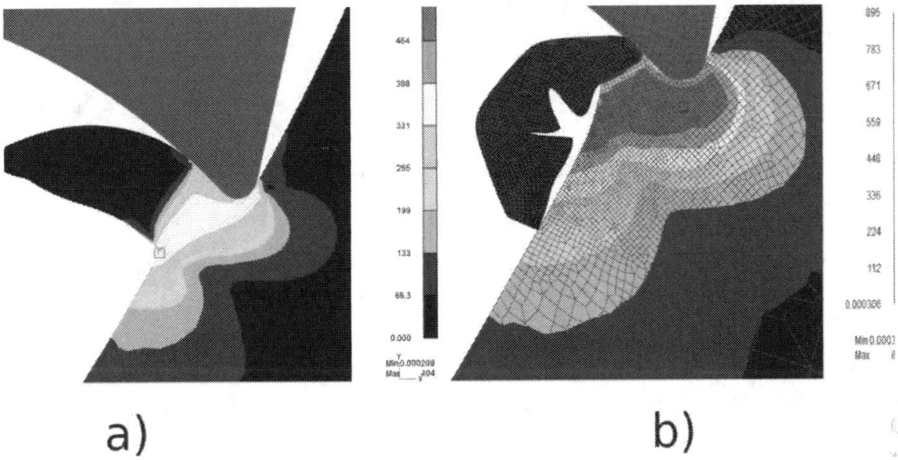

Figure 6. Chip shape and shear stress distribution for hard titanium alloy material and lightweight aluminum alloy material.

Conclusion

In conclusion, the findings of this study underscore the critical influence of cutting edge geometry and workpiece material properties on machining performance. Through comprehensive finite element simulations, we have examined the intricate dynamics of orthogonal cutting processes, elucidating the behavior of cutting forces, shear stresses, and chip formation across different materials and cutting edge radius size. The simulations revealed that while easy-to-machine materials like aluminum alloy exhibit predictable chip morphology and relatively low shear stresses, challenging materials such as titanium alloy present unique machining challenges due to their hardness and brittleness. In such cases, achieving a balance in cutting edge design becomes essential to withstand the high forces and stresses encountered during machining.

These insights contribute to a deeper understanding of machining processes, guiding the development of optimized machining strategies tailored to specific materials and applications. Moving forward, further research in this field holds the potential to refine machining techniques, enhance tooling designs, and ultimately improve manufacturing efficiency and product quality.

Acknowledgement

This work was supported by the Slovak Research and Development Agency under the contract No. APVV-21-0071. This work was supported by the Science Grant Agency - project VEGA 1/0266/23.

References

Afrasiabi, M., Saelzer, J., Berger, S., Iovkov, I., Klippel, H., Röthlin, M., Zabel, A., Biermann, D., & Wegener, K. (2021, October 22). A Numerical-Experimental Study on Orthogonal Cutting of AISI 1045 Steel and Ti6Al4V Alloy: SPH and FEM Modeling with Newly Identified Friction Coefficients. Multidisciplinary Digital Publishing Institute. https://doi.org/10.3390/met11111683

Arrazola, P., Özel, T., Umbrello, D., Davies, M., & Jawahir, I. (2013, January 1). Recent advances in modelling of metal machining processes. Elsevier BV. https://doi.org/10.1016/j.cirp.2013.05.006

Hussain, G., Alkahtani, M., Alsultan, M., Buhl, J., & Gupta, M K. (2022, November 1). Chip formation, cutting temperature and forces measurements in hard turning of Gcr15 under the influence of PcBN chamfering parameters. Elsevier BV. https://doi.org/10.1016/j.measurement.2022.112130

Kundrák, J., Karpuschewski, B., Pálmai, Z., Felhő, C., Makkai, T., & Borysenko, D. (2021, January 1). The energetic characteristics of milling with changing cross-section in the definition of specific cutting force by FEM method. https://doi.org/10.1016/j.cirpj.2020.11.006

Melkote, S N., Liang, S Y., Özel, T., Jawahir, I., Stephenson, D A., & Wang, Q. (2022, June 10). <i>100th Anniversary Issue of the Manufacturing Engineering Division Paper</i> A Review of Advances in Modeling of Conventional Machining Processes: From Merchant to the Present. https://doi.org/10.1115/1.4053522

Shi, Z., Li, X., Duan, N., & Yang, Q. (2021, April 1). Evaluation of tool wear and cutting performance considering effects of dynamic nodes movement based on FEM simulation. Elsevier BV. https://doi.org/10.1016/j.cja.2020.08.003

Wojciechowski, S., Matuszak, M., Powałka, B., Madajewski, M., Maruda, R W., & Królczyk, G. (2019, December 1). Prediction of cutting forces during micro end milling considering chip thickness accumulation. Elsevier BV. https://doi.org/10.1016/j.ijmachtools.2019.103466

Wu, B., Xue, Y., Luo, M., & Gao, G. (2013, August 1). Cutting force prediction for circular end milling process. Elsevier BV. https://doi.org/10.1016/j.cja.2013.04.003

Zhang, X., & Wang, H. (2015, June 1). Finite Element Analysis of Milling Parameters for Metallurgy Saw Blade Cutting Process. https://doi.org/10.1109/icicta.2015.278

IOP Conf. Series: Materials Science and Engineering **1319** (2024) 012010 doi:10.1088/1757-899X/1319/1/012010

Microstructure and surface properties of Co-based brazed alloys coatings

Iasmina-Mădălina Anghel (Petculescu)[1], Ion-Dragoș Uțu [1], Diana Uțu [2] and Albert Titus Constantin [3]

[1] Department of Materials and Fabrication Engineering, Politehnica University Timisoara, Bd. Mihai Viteazul nr. 1, 300222 Timisoara, Romania
[2] Department of Pharmacology-Pharmacotherapy, Faculty of Pharmacy, Victor Babes University of Medicine and Pharmacy Timisoara, Eftimie Murgu Sq 2, 300041 Timisoara, Romania;
[3] Department of Hydrotechnical Engineering, Civil Engineering Faculty, Politehnica University Timisoara, Splaiul Spiru Haret 1A, 300022 Timisoara, Romania

*E-mail: diana.utu@umft.ro

Abstract. The paper investigates the characteristics of two different manufactured high temperature brazed coatings using a cobalt based powder applied onto the surface of a stainless-steel substrate (AISI 904L). The coatings are intended to shield components exposed to high temperatures in an atmosphere where wear is expected. Energy dispersive analysis (EDX) together with scanning electron microscopy (SEM) was used to characterise the coating microstructure and morphology, and X-ray diffraction was used to identify the phase composition. Properties evaluations were carried out on the coatings' surface characteristics, including hardness, erosion behaviour and hot corrosion resistance. The hot corrosion test was conducted at 650°C in a salt mixture that contained 75 wt. % sodium sulphate (Na_2SO_4) and 25 wt.% sodium chloride (NaCl). Alumina particles were used in the erosion tests, which hit the samples surface with an impact angle of 90°.The degree of corrosion attack, erosion behavior and hardness depended on the microstructure of the manufactured vacuum brazed coatings.

Keywords: corrosion, wear, brazed coating, cobalt alloys

1. Introduction

Cobalt-based coatings have emerged as a crucial solution for enhancing the durability and performance of materials in demanding industrial environments [1]. These coatings, known for their remarkable hardness, wear resistance, and thermal stability, are extensively applied in aerospace, automotive, and energy sectors [1-3]. The primary methods for applying cobalt-based coatings include electroplating, thermal spraying, and chemical vapor deposition, each contributing to the coatings' unique microstructural characteristics and performance attributes [4-6]. Recent advancements in the development of cobalt coatings have focused on improving

their resistance to corrosion and oxidation, which are vital for extending the service life of components exposed to harsh conditions. Furthermore, innovative approaches such as nanostructuring and the incorporation of secondary phases have shown significant enhancements in the mechanical properties and functionality of cobalt-based coatings [4-8].
Co-based alloys have been the subject of various studies, each focusing on different aspects of their properties and applications. Zhang et. al explored the microstructural characteristics of Co-based ODS alloys, highlighting the role of mechanical alloying in inducing the formation of a supersaturated solid solution [9]. Schwarz investigated the formation ranges of Co-based amorphous alloys and their application as cladding materials, with a focus on wear protection [10]. Wu and collab. developed a weldable, break-resistant Co-based alloy with wear and corrosion resistance [11]. Panda et.al. discussed the development of nanostructured CoFe-based alloys for high-temperature magnetic applications, emphasizing the role of Fe substitution by Co in enhancing their magnetic properties [12].
Each mentioned deposition technique provides certain advantages, and it is selected depending on the required application, including desired coating properties, substrate material, and operational environment.
Among these, the brazing of cobalt-based alloys as coating techniques has been explored in various studies. Lu and collab. investigated the microstructure and bonding strength of a WC reinforced Ni-based alloy brazed composite coating, finding that the cobalt coated on the surface of WC particles dissolved into the liquid nickel-base alloy [13]. Rass and team discussed the production of functional coatings through brazing, highlighting the use of cobalt as a filler matrix material [14]. Otterloo et. al. focused on the microstructure and abrasive wear of cobalt-based laser coatings, emphasizing the low dilution and controllability of the laser cladding process [15]. Criss and collab. developed a new method of joining cobalt by braze-welding it with a silver-copper filler, which could potentially increase the strength of the weldments [16].
The present work aims to use brazing process as a new alternative for surface protection. Although brazing is typically used as a joining technique, the goal of this study is to assess if it may also be used as a coating technology, offering novel and creative approaches to the deposition of functional coatings on various surfaces. It would be anticipated that this coating process would offer strong cladding cohesion and adherence to the substrate material [17].
The experimental investigations were designed to evaluate the microstructure, erosion and hot corrosion properties of a cobalt based powder alloys applied on an AISI 904L substrate by high temperature vacuum brazing method as an alternative to improve the durability of industrial elements which are operating at high temperatures.

2. Materials. Experimental procedure

Commercially Amdry MM509B-C powder which is a cobalt activated diffusion braze alloy (Oerlikon Metco) has been used for experiments. This powder should be used as mix material for braze repairs, such as repairing cracked surfaces, worn or damaged areas, or surface imperfections on gas turbine components used in industrial and aerospace applications. The material contains boron, which acts as a melt-depressant [18].

The cobalt based powder with a particle granulation between–106 +45 μm, and the chemical composition by weight percent balance Co, Cr (22.5 – 24.25), Ni (9.0 – 11.0), W (6.5 – 7.5), Ta (3.0 – 4.0), B (2.0 – 3.0), Ti (0.15 – 0.30), Zr(0.30 – 0.60) and C (0.55 – 0.65) was combined and fixed using a water-based adhesive (Aleene's Tack-It Over and Over) to create flexible tape coatings. By incorporating 5 weight percent and 6 weight percent adhesive into the brazing powder, two tape compositions were produced. The powder combination was then formed into flexible tapes by a constant thickness of 2 mm using a rolling mill.

After obtaining the appropriate dimensions, the tapes were cut and placed onto the used tainless steel substrate. The high temperature vacuum furnace (HITERM 80-200) was used to manufacture the vacuum brazing as a step of the coating process. The AISI 904L substrate was machined to remove surface oxides and impurities and cleaned with ethanol prior coating deposition.

The parameters of the brazing process were set up according to the Amdry brazing powder supplier recommendations (1175-1260 °C). The heating was made in stages (200 °C, 500°C, 870°C, 1000°C, 1200°C) with maintaining stages in order to allow the organic binder to completely evaporate and the powder alloy to melt and partially fill the spaces left by the binder, as well as to braze the coating to the substrate [19].

Using energy dispersive X-ray spectroscopy (EDX)analysis and scanning electron microscopy SEM (Quanta FEG 250 microscope), the morphology and microstructure of the coatings were investigated.

The coatings were placed in an oven set at 650 °C for 100 hours, in a salt mixture that contained 75% wt. Na_2SO_4 and 25% wt. NaCl, in order to assess the hot corrosion. Using a SEM microscope, the degree of corrosion attack was examined in cross-section.

The ASTM G76 guidelines were followed when conducting the erosion tests. After being cleaned with acetone, the samples were dried and weighed. The Al_2O_3 particles (d50 of 50μm), the impingement angle of 90°, the nozzle to sample distance of 10 mm, the compressed air at 2 bar, the particle mass flow rate of 7 g/min, and the testing time of 20 minutes (split into 4 segments of 5 minutes each) were the settings for the test. At the start of the test and after every five minutes of testing, the samples were weighed.

The Zwick/Roell YHVμ-S equipment was used to evaluate the microhardness (HV0.3) of the brazed coatings.

3. Experimental results

3.1 Microstructural analysis

Figure 1 shows the morphologies at different magnifications of the coated layers obtained after vacuum heat treatments using 5 wt% (Fig. 1a) and 6 wt% organic binder (Fig. 1b). Both covering layers are dense, with a homogeneous distribution of phases and a small amount of pores. In the case of the sample obtained with 6 wt.% organic binder, it can be observed that the number of pores is slightly higher compared to the other sample (5 wt.%). While the tapes being laminated to a 2 mm thickness, the densification of the deposited material during the brazing process was materialised in a reduction in coating layer thickness of 1.3-1.5 mm. The sample with a higher amount of organic binder had lower values measured. A good adhesion of the coating layer to the substrate is evidenced by the interdiffusion line (marked by arrows). The microstructure comprises dendrites and Laves eutectic phases dispersed in a cobalt solid solution.

Microscopic images in Fig. 1 (a–d) highlight a hypoeutectic dendritic microstructure which has primary dendrites of the Co-based solid solution encircled by an interdendritic lamellar eutectic comprising carbides of Cr, W, or even Laves phases. It is discontinuous and randomly oriented. Depending on the local chemical composition, these zones (rich in Cr or rich in W) show in different shades of grey.

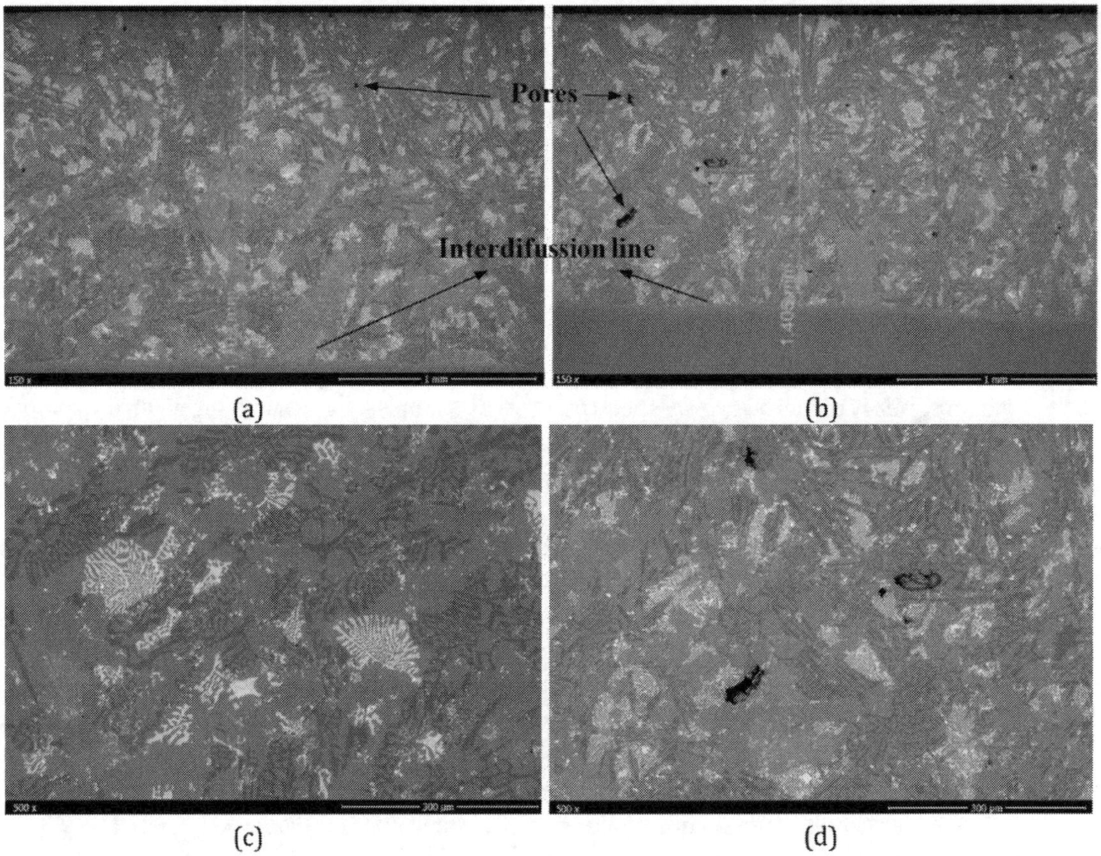

Figure 1. SEM micrographs of the coatings after the brazing process: (a,c) 5 wt. % organic binder, (b, d) 6wt%. organic binder.

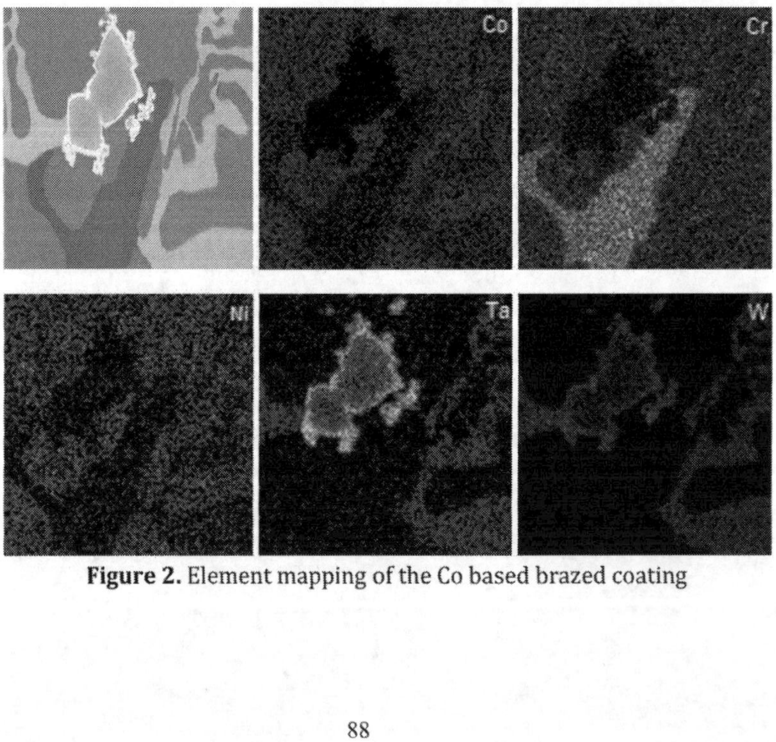

Figure 2. Element mapping of the Co based brazed coating

The performed EDX mapping analysis (Fig. 2) has revealed that the lighter dendrites zone contain higher amounts of W in opposite with dark and darker grey dendrites which have more chromium. The cobalt solid solution contains a random distribution of undissolved tantalum inclusions.

3.2 Hot corrosion evaluation

Because no significant microstructure modifications were observed regarding the hot corrosion of the coatings, the results which are presented below are for the sample with 5 wt.% organic binder.

After the immersing the coatings into the salt mixture at 650 °C a material degradation in forms of preferential (localized) metal dissolution has been noticed.

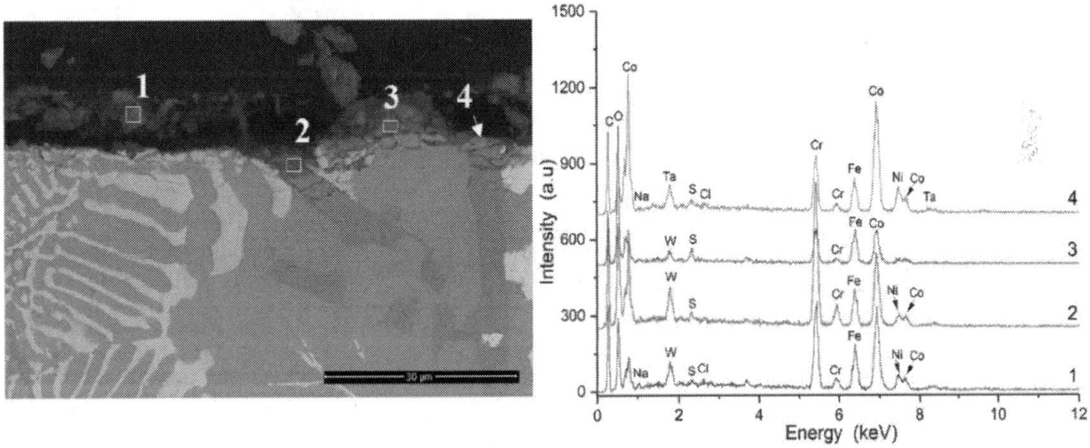

Figure 3. SEM micrographs of the corroded coating at 650 °C, 5000x magnification, and EDXanalysis

According to the SEM micrograph from Figure 3a, the corrosion attack of the coating exposed at 650 °C is mainly concentrated in the regions with higher Cr content. The EDX chemical analysis (Fig. 3b) demonstrate that the predominant corrosion products found on the sample surface were chromium, oxygen and sulf.

According to published reports, a phase segregation may develop because of the temperatures and susceptibility to hot corrosion [19,20]. In order to guarantee a good phase distribution and avoid separation of the tungsten, tantalum, or chromium-rich phases, the processing temperature throughout the coating production process is critical. Because internal stresses can quickly cause damage to heterogeneous materials, the coating's cooling procedure needs to be done carefully.

3.3 Hardness and solid particle erosion properties

The erosion test was made both the deposited coatings marked with C1 (5wt.% organic binder) and C2 (6wt.% organic binder). The mass loss recorded during the erosion test at an impact angle of 90 ° of the Al_2O_3 abrasive material is shown in Figure 4.

To ensure the reproducibility of the measurements, 3 samples from manufactured coating (C1 and C2) were initially weighed and after every 5 minutes testing time. Following a 20-minute testing period, the coatings mass losses were approximately similar 4.8±0.05 mg (C1) and 5.0±0.03 mg (C2).

IOP Conf. Series: Materials Science and Engineering **1319** (2024) 012010 doi:10.1088/1757-899X/1319/1/012010

Generally, harder materials tend to have better resistance to erosion, as they are less susceptible to surface deformation and wear.

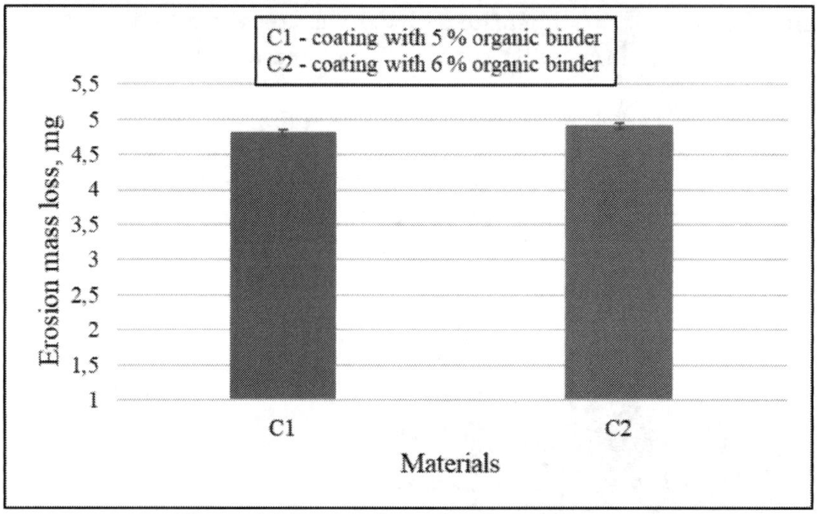

Figure 4. Materials loss after the erosion test

For this purpose microhardness measurements were performed on the cross section of the both coatings starting from top to the near zone of the interface with the substrate. The obtained data recorded are displayed in Figure 5.

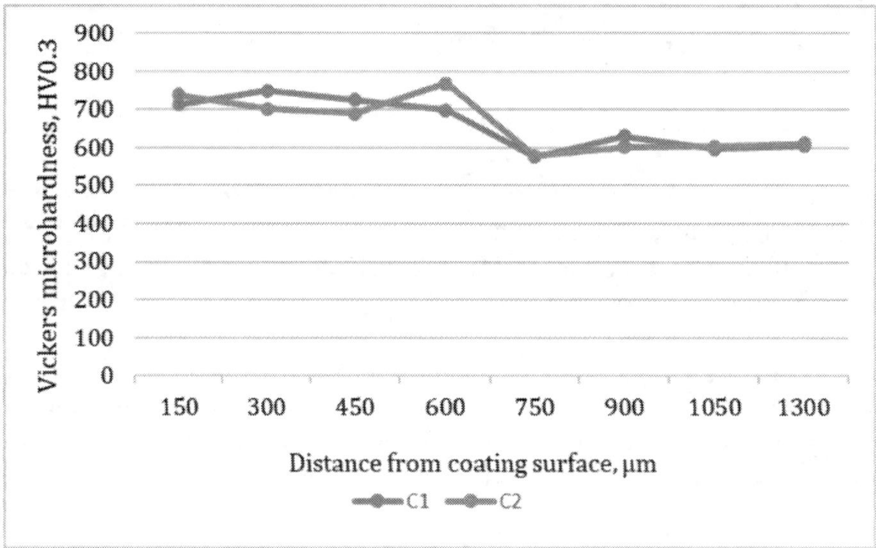

Figure 5. Recorded values after the microhardness measurements

Regarding the susceptibility to fracture development, the hardness values of both coatings ranged between two limits, 570 and 780, based on the region where the micro-indentation was conducted. The microhardness measurements showed almost similar values. The variations depended on the microstructure of the coatings.

Conclusion

Two different cobalt-based powders were deposited by vacuum brazing onto the surface of a AISI 904L substrate and studied in terms of microstructure, hot corrosion, and solid particles erosion resistance. The results showed that both manufactured coatings were dense, with a homogeneous distribution of phases and a small number of pores.

The material degradation during exposure of the coatings at 650 °C into a salt mixture of sodium sulphate and sodium chloride was materialised from corrosion products primary composed of chromium oxide. The corrosion was localized in the vicinity of the phase's rich in chromium.

Both coatings showed almost similar mass losses recorded during the erosion test. The hardness values of the ranged between two limits, 570 and 780 HV0.3.

References

[1]. Liu, X., Wang, L., & Zhang, G. (2016). "Cobalt-based coatings for high-temperature applications: A review." Journal of Alloys and Compounds, 663, 170-185.
[2]. Sharma, A., & Kumar, R. (2018). "Development and characterization of cobalt-based coatings: A review." Surface and Coatings Technology, 344, 149-164.
[3]. Pang, J., Liu, Z., & Yan, F. (2020). "Recent advances in cobalt-based coatings: Fabrication, characterization, and applications." Materials Science and Engineering: A, 776, 139072.
[4]. Xu, W., Li, X., & Zhang, Y. (2019). "Microstructural evolution and properties of cobalt-based coatings under thermal exposure." Surface and Interface Analysis, 51(4), 301-311.
[5]. Zhang, X., & Chen, L. (2017). "Electroplating and properties of cobalt-based nanocomposite coatings." Journal of Materials Science, 52, 12345-12356.
[6]. Huang, Y., & Ma, Z. (2019). "Enhancement of wear resistance and corrosion protection of cobalt-based coatings by adding secondary phases." Corrosion Science, 150, 38-49.
[7]. Sharma, A., & Kumar, R. (2018). "Development and characterization of cobalt-based coatings: A review." Surface and Coatings Technology, 344, 149-164.
[8]. Chen, Q., & Scharf, T. W. (2013). "Cobalt-based hard coatings: Tribological performance and applications." Wear, 297(1-2), 1010-1016.
[9]. Zhang, L., Qu, X., He, X., Din, R., Liu, H., Qin, M., & Zhu, H. (2012). "Microstructural Characterization of Co-Based ODS Alloys." Journal of Materials Engineering and Performance, 21(11), 2487–2494.
[10]. Schwarz, W., & Warlimont, H. (1997). "A new series of Co-based amorphous alloys and their application as cladding materials." Materials Science and Engineering: A, 226–228, 1098-1101.
[11]. Wu, J. B. C., & Yao, M. (2008). "Weldable, break-resistant Co-based alloy, application methods and components." Materials Science and Engineering.
[12]. Panda, A. K., Mohanta, O., Ghosh, M., & Mitra, A. (2009). "Development of Nanostructured CoFe-Based Alloys for High Temperature Magnetic Applications." Journal of Nanoscience and Nanotechnology, 9(9), 5600-5603.
[13]. Lu, S.-P., & Kwon, O.-Y. (2002). "Microstructure and bonding strength of WC reinforced Ni-base alloy brazed composite coating." Surface and Coatings Technology, 153(1), 40-48.
[14]. Rass, I. (2022). "Functional Coatings Made by Brazing." In Thermal Spray 2022: Proceedings from the International Thermal Spray Conference (pp. 601-606). Paper No: itsc2022p0601.
[15]. De Mol van Otterloo, J. L., & De Hosson, J. Th. M. (1997). "Microstructure and abrasive wear of cobalt-based laser coatings." Scripta Materialia, 36(2), 239-245.
[16]. Criss, E. M., & Meyers, M. A. (2015). "Braze welding of cobalt with a silver–copper filler." Journal of Materials Research and Technology, 4(1), 44-59.
[17]. Pascal, D. T. (2017). Development of high temperature vacuum brazed WC-Co-NiP functional composite coatings. PhD thesis, Politehnica Publishing, Timişoara.
[18]. https://www.metcojoiningcladding.com/ecoma/files/DSM-0292.2_Amdry_MM509.pdf
[19]. Esmaeili, N., & Ojo, O. A. (2018). "Analysis of Brazing Effect on Hot Corrosion Behavior of a Nickel-Based Aerospace Superalloy." Metallurgical and Materials Transactions B, 49, 912–918.
[20]. Fang, J.-H., Ma, H.-W., Xie, M., Chen, Y.-T., Yang, Y.-C., Hu, J.-Q., & Wang, S. (2019). "Effect of Joining Temperature and Bonding Time on Evolution of Interfacial Microstructure and Brazing Properties for 4J29/Ag–27Cu–4Ga/4J29 Brazed Joint." Vacuum, 167, 459–470.

Compression behavior of *ex-situ* PVC foam-filled tubes

Emanoil Linul[1*], Cristina Vălean[2] and Andreea-Iustina Rusu[1]

[1] Politehnica University Timisoara, Department of Mechanics and Strength of Materials, Mihai Viteazu Blvd. 1, 300222-Timisoara, Romania
[2] Politehnica University Timisoara, Research Institute for Renewable Energies, Musicescu Gavril Str. 138, 300774-Timisoara, Romania

*E-mail: emanoil.linul@upt.ro

Abstract. The present work investigates the mechanical characteristics of tubes filled with Polyvinyl chloride (PVC) foam. The tubes used are made of aluminum and were filled ex-situ. Static compression tests were performed on both axially (AL) and laterally (LL) loaded tubes. Comparisons between foam-filled (FFT) and empty (ET) tubes are presented, highlighting the foam-tube interaction effect. The emphasis is on elastic, strength and strain properties, but energy absorption performances are not neglected. Discussions regarding the failure mechanisms of ETs and FFTs are also presented. It was obtained that, regardless of the loading direction, FFTs show clearly superior mechanical properties to ETs. At the same weight, the specimens tested axially support higher loads than those tested laterally. This aspect is due to deformation mechanisms that take place in the samples during the tests. It was noted that the compressive strength is more affected by the filling than by the compressive modulus. Under lateral loads, the ETs samples fail quasi-brittle through complete failure of the tube, while in the case of FFTs, a ductile fracture with stable deformation of the sample is obtained.

1. Introduction

The structures that absorbs energy have become very important in numerous engineering fields (e.g. aircraft, ship and vehicle) [1-3]. The most of engineering structures are made to absorb the impact energy by using thin-walled (TW) tubes or cellular materials [4-6]. For instance, to protect certain contents or to avoid injuries, the TW tubes are piled up lateral as an energy absorption device on the highway, which increase the crashworthiness for the vehicle body structure during impact [7]. Other researches propose the use of composite structures based on foam for protective system against failure of railway axles on impact with flying ballast [8], or for protective systems for offshore structures [9]. Because both TW tubes and foams have a high energy absorption (EA) potential, the combination of them is anticipated to intensify the strength and EA properties of the structures [10-12].

There are numerous studies on Empty Tubes (ETs) and foam-filled tubes (FFTs) [13, 14]. The most used filling cellular materials are polyurethane foam (PUF), metallic foam (MF) and honeycombs (HCs) [15-17]. However, due to the high costs and difficult manufacturing of MFs and HCs, PUF remain the main candidates for this purpose [18, 19]. Liu et al. [20] studied theoretically and experimentally the mechanical response and EA of ETs and aluminum FFTs with different geometries. The authors observed that the analytical model respects the experimental results of FFTs under lateral loading (LL). Also, under LL, the specimens show three well defined phases, called the elastic, stable plastic and densification phases. They reported that the densification modes of FFTs compared to ETs are modified by aluminum foam core. The behavior of square and circular FFTs under axial load (AL) was investigated by Onsalung et al. [21]. They found that as the foam density increases, the FFTs change their deformation mode by increasing the number of folds and circular FFTs tend to change their concertina mode to mixed (concertina+asymmetric) or asymmetric. The authors concluded that the EA could be improved by adding foam to the structure, but there is a possibility that the specific EA would not be improved. Su et al. [22] studied the axial and lateral compressive characteristics of the alumina-aluminum matrix syntax a TW tubes filled with foam. They noted that for axial compression, TW tubes and FFTs exhibit the concertina mode. For radial compression, the ETs and FFTs form four plastic hinge lines. Radial EA is less than axial EA for FFTs. Similar behaviors were also reported by Refs. [23, 24] for ETs and FFTs loaded axially and laterally and tested at high temperatures. Xianhui Li et al. [25] present an investigation on the axial impact strength of TW Al-alloy ETs and TW Al FFTs. The authors found that the most efficient option to increase the impact strength for both cases (ETs and FFTs), under the same impact energy, is to increase the thickness of the tube wall. They identified that the largest amount of impact energy of FFTs is taken up by TW Al-alloy. The authors also obtained that if the thickness of the wall increases, the energy absorbed by the foam decreases, but as the length of the tube increases, the energy absorbed by the Al-alloy foam increases. Partially FFTs, with and without adhesive, were tested in compression by Guden et al. [26]. The authors reported that in partially adhesive-free FFTs, the deformation mode was diamond, changing to concertina for adhesive-bonded FFTs. The reduction in fold length and the increase in average tube crush loads resulted from the influence of partial foam filling. Yao et al. [27] investigated the quasi-static axial compression of four different hybrid FFTs and their independent components. They noted that the collapse mode of the ETs changes by adding hybrid structures filled with PUF. This helps increase EA capacity and stability over time in compression tests.

The present work comes as a complement to the literature and investigates the mechanical behavior of tubes filled with Polyvinyl chloride (PVC) foam. Compression tests are performed through axial and lateral loads on ETs and FFTs. The failure mechanisms that take place in the samples depending on their configurations (AL or LL and ETs or FFTs) are presented.

2. Materials and methods

Closed-cell PVC polymer foams and square aluminum tubes were used to make the samples. The foams were obtained in the form of plates (20 mm × 40 mm × 8 mm), and the tubes in the form of bars (length of one meter, sides of 10 mm, and 1 mm thickness). The aluminum bars were cut into samples with a length of 10 mm. Part of the samples were tested empty (named Empty Tubes-ETs), and part were filled with PVC foams (named Foam Filled Tubes-FFTs). The ETs were filled by pressing the PVC foam inside the Al tube, process called *"ex-situ"* manufacturing. Thus, the effect of the filling on the compression characteristics of the manufactured samples was investigated. In addition to the filling effect, to find out the effect of the loading direction, some samples were loaded axially, and others laterally. Figure 1 shows an overview of the obtained samples, respectively the specification of the loading directions.

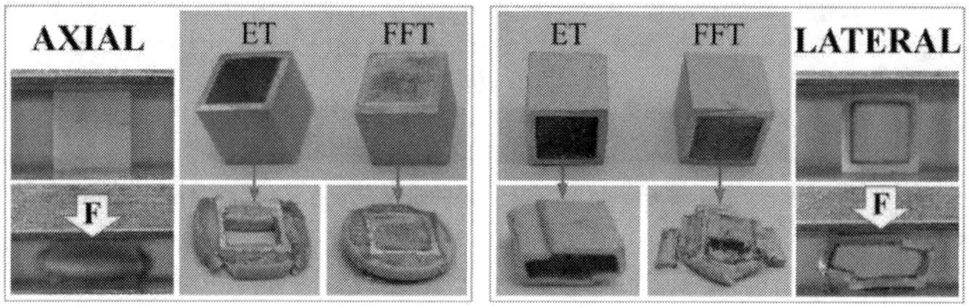

Figure 1. The analyzed specimens and the used loading directions.

The mechanical tests were carried out on the LBG TC100 machine. It has a load-cell of 100 kN and provided with a specialized data collection software. The tests were performed at 25 °C with a test speed of 10 mm/min.

3. Results and discussions

Figure 2 presents the compressive stress-strain curves for each type of specimen. Regardless of the specimen configuration (PVC foam, ET or FFT), the behavior is strongly influenced by the loading direction. For all specimens, the axial loading direction shows superior properties to the lateral one.

Under axial loads, the PVC foam shows an extended linear elastic zone followed by an obvious yield point (Figure 2a). On the other hand, lateral-loaded foam specimens have a short linear area with a hard-to-identify yield point. The plateau area corresponding to the two foam loading directions differs both in amplitude and shape. Axial loading gives a long and almost constant plateau with some oscillations in stress, while under lateral loading an increasing and almost linear plateau was obtained. For the axial direction a compressive strength value of 12.03 MPa was found, 811.4% higher than the lateral direction. Similar differences were obtained for the energy absorption, where the lateral direction (7.12 MJ/m^3) shows values up to about 511.2% higher than the lateral direction (1.17 MJ/m^3). Between the two loading directions equal differences (162.1%) are obtained for plateau stress and densification stress, both in favor of the axial direction. It was noted that the densification strain is over 61% for the axial direction and under 40% for the lateral direction. Double values were recorded for compressive modulus with respect to the loading direction, but this time the lateral direction showing higher values (22.09 MPa).

The compression behavior of the empty tube overlaps most of the elastic zone (only 3.3% differences). However, beyond the elastic zone, the properties are significantly different, all in favor of the axial direction. Under axial loads, ET presents a plateau zone, followed by a densification zone (>60% strain), while under lateral loads, ET collapse under the imposed loads and presents a sudden failure, through the complete fracture of the walls of the specimen. The yield point of the sample material is clearly higher (76.5%) in the case of axial loading.

FFTs show a behavior similar to ETs with the observation that under lateral loads it does not yield suddenly. It seems that the PVC foam contributes to the deformation mechanism of the specimen. Axial loaded FFT absorbs up to 396.5% more energy than ET. This attribute is given both by the length of the plateau and by the stress amplitude.

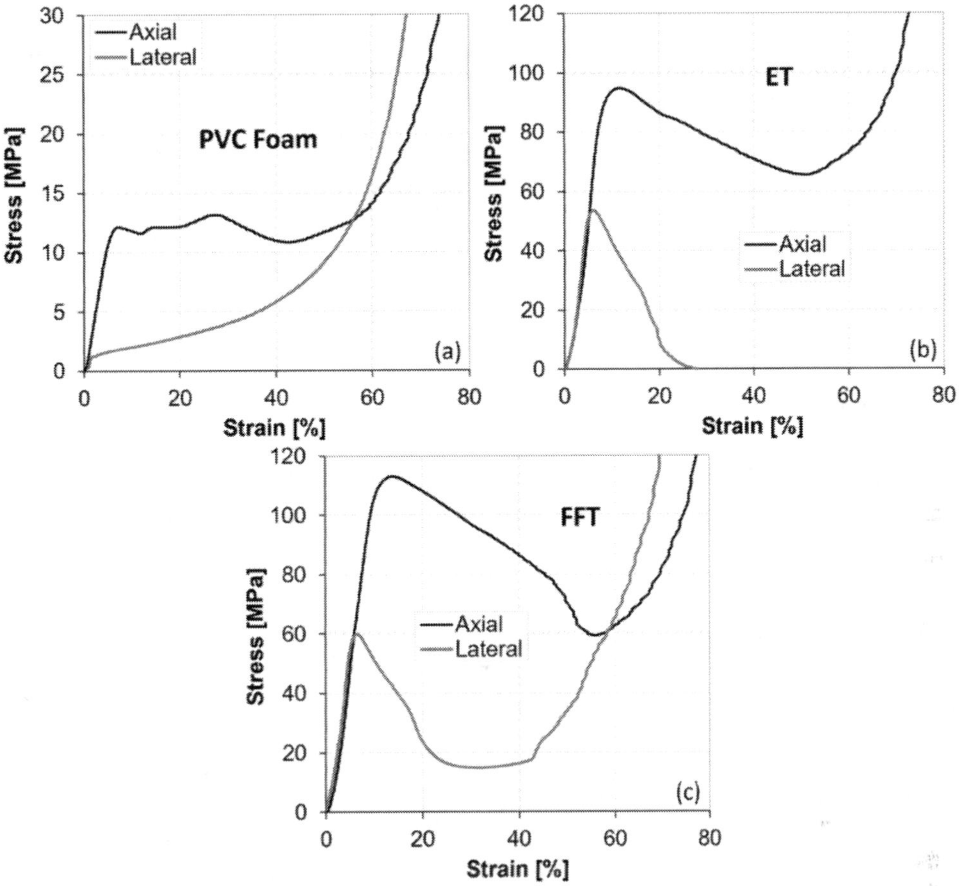

Figure 2. Stress-strain curves for PVC foam (a), ET (b) and FFTs (c) specimens.

A comparison of the energy absorption-strain and stress-strain curves for the two sample configurations (ET and FFT) are presented in Figures 3 and 4. It was observed that, regardless of the direction of loading (axial or lateral), the FFTs samples show a behavior superior to ETs. The only similarity is identified for the linear elastic zone, where the behaviors almost overlap. This improvement in the behavior of the FFTs samples is given by the effect of the foam-tube interaction. Under compression loads, on the one hand, the core of the samples deforms together with the tube, and on the other hand, the friction between them contributes significantly to the increase of the mechanical characteristics. In this sense, it can be easily seen that even under lateral loads (Figure 3), the tube does not collapse catastrophically, but shows a continuous deformation, this being supported by the PVC foam inside.

IOP Conf. Series: Materials Science and Engineering **1319** (2024) 012011 doi:10.1088/1757-899X/1319/1/012011

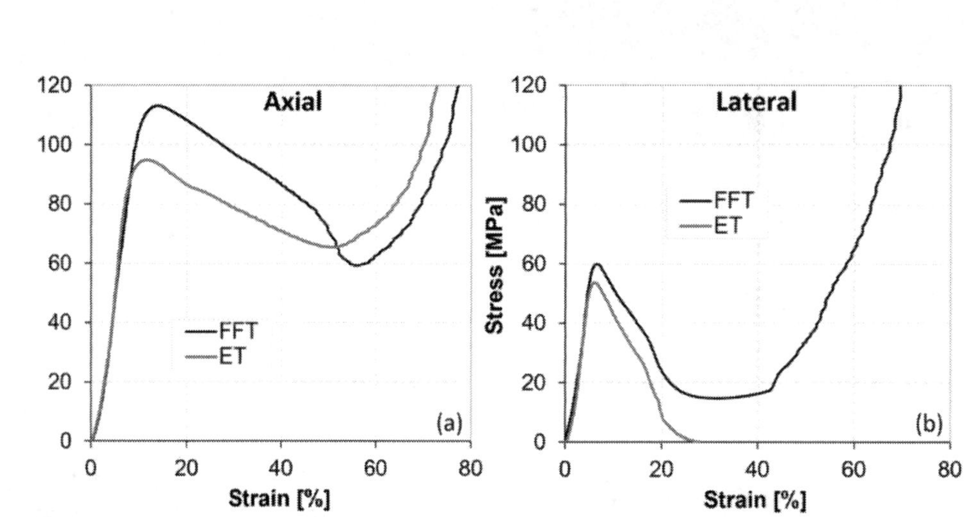

Figure 3. FFT-ET comparison of axial (a) and lateral (b) stress-strain curves.

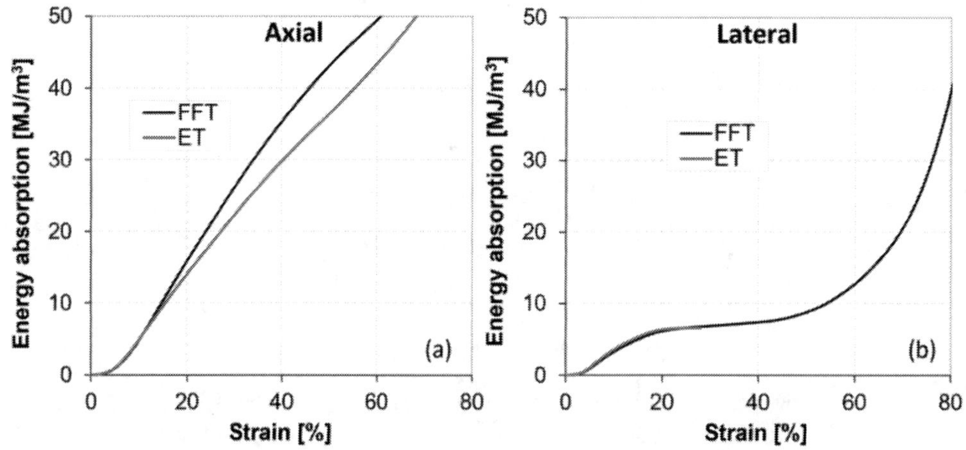

Figure 4. FFT-ET comparison of axial (a) and lateral (b) energy absorption-strain curves.

Figure 5 presents the variation of the studied compression properties for ETs and FFTs depending on the direction of loading. The column graphs highlight very well the difference between the ETs and FFTs samples for strength, modulus, strain at strength and EA. Under lateral loads, ETs show a compressive strength of 53.68 MPa, and FFTs of 59.92 MPa, the slightly larger difference (17.7%) is recorded for the axial direction (94.75 MPa for ETs and 111.48 MPa for FFTs). The smallest differences between ETs and FFTs (below 0.5%), for both loading directions, are recorded for compressive modulus. Thus, modulus values of 1405.4 MPa (ET) and 1412.6 MPa (FFT) were obtained for the lateral loading direction, respectively of 1451.3 MPa (ET) and 1456.5 MPa (FFT) for the axial direction. The obtained values are almost identical due to the overlap of the obtained stress-strain curves presented previously.

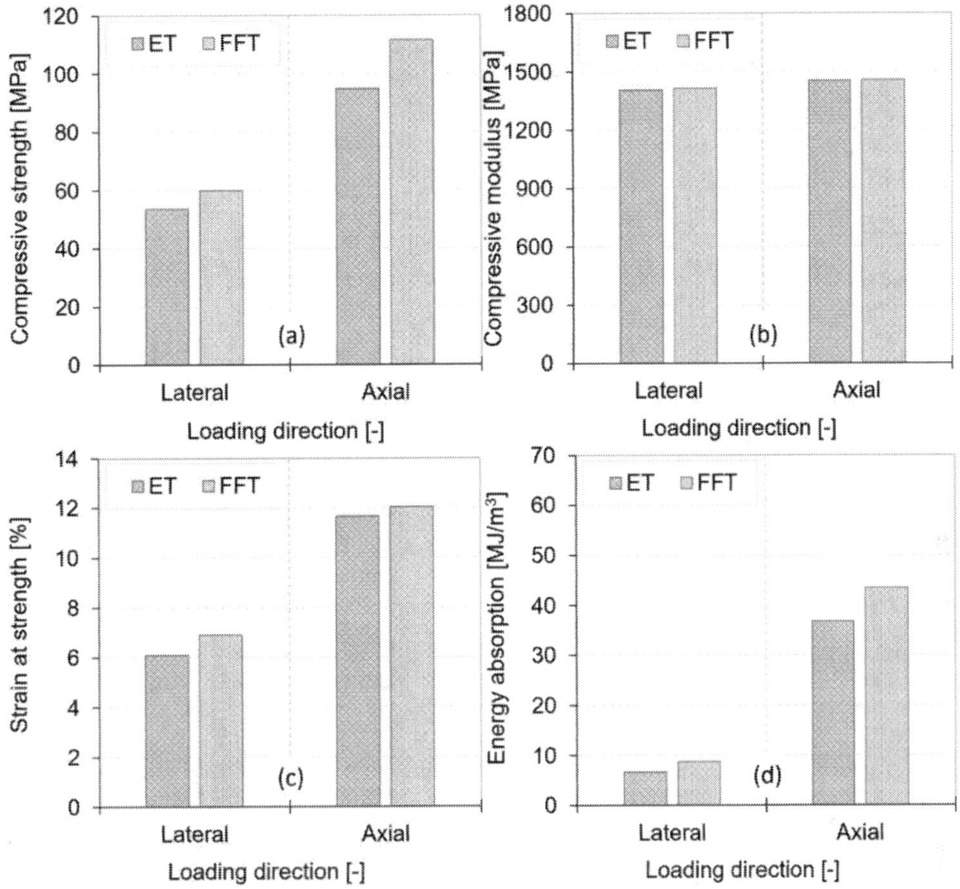

Figure 5. The variation of the compressive properties depending on the loading direction. ETs-FFTs comparison.

The samples filled with PVC foam show values of strain at strength slightly higher, with 31.1% for the lateral direction and 3.4% for the axial direction, than the empty samples. The largest percentage increases are found for the energy absorption. FFTs absorb up to 33.0% (lateral) and 18.2% (axial) more energy than ETs. This aspect is due to densification strain which is superior for FFTs compared to ETs for both loading directions.

4. Conclusions

The paper investigates the static compression behavior of empty tubes (ETs) and PVC foam-filled tubes (FFTs). Characteristics such as strength, modulus, strain at strength and energy absorption are considered. The failure mechanisms in the ETs and FFTs specimens are also presented. The following can be concluded:

- PVC foam shows a strong anisotropic behavior. Compared to LL, AL highlights higher values for compressive strength (811.4%), strain at strength (207.7%) and energy absorption (511.2%), respectively lower values for compressive modulus (52.1%).

- Regardless of the configuration of the samples, AL exhibited higher values than LL for compressive strength (76.5% for ETs and 86.1% for FFTs), compressive modulus (3.2% for ETs and 3.1% for FFTs) ,strain at strength (91.1% for ETs and 74.7% for FFTs) and energy absorption (458.5% for ETs and 396.5% for FFTs).
- FFTs show higher property values than ETs for both loading directions as follows: 11.6% (LL) and 17.7% (AL) for compressive strength, 13.1% (LL) and 3.4% (AL) for strain at strength and 33.0% (LL) and 18.2% (AL) for energy absorption. Almost identical values are obtained for compressive modulus (0.45% difference), but slightly in favor of AL.
- It was observed that ET samples do not withstand lateral loads, their walls breaking at a strain of 28%.
- FFTs show superior properties to ETs due to the interaction between the PVC foam and the ET.

References

[1] J.E. Maróti, I.N. Orbulov, 2023, Materials Science and Engineering: A 869, 144817.
[2] A. Kemény, B. Leveles, T. Bubonyi, et al., 2021, Composites Part A: Applied Science and Manufacturing 140, 106152.
[3] I. Duarte, L. Krstulović-Opara, J. Dias-de-Oliveira, et al., 2019, Thin-Walled Structures 138, 124-136.
[4] E. Linul, L. Marsavina, 2011, Iranian Polymer Journal (English) 20 (9135), 735-746.
[5] A. Niknejad, S.A. Elahi, G.H. Liaghat, 2012, Mater Des, 36 (2012), pp. 24-34.
[6] L. Marsavina, D.M. Constantinescu, E. Linul, et al., 2014, Procedia Materials Science 3, 1342-1352.
[7] G. Epasto, F. Distefano, H. Mozafari, et al., 2021, Applied Sciences 11 (3), 1148.
[8] G. Epasto, F. Distefano, L. Gu, et al., 2020, Materials & Design 196, 109120.
[9] H. Mozafari, F. Distefano, G. Epasto, et al., 2022, Journal of Marine Science and Engineering 10 (10), 1385.
[10] I. Duarte, L. Krstulović-Opara, M. Vesenjak, 2018, Composite Structures 192, 184-192.
[11] C. Kádár, R. Gorejová, P. Kubelka, et al., 2024, Advanced Engineering Materials, 2301496.
[12] E. Linul, D. Pietras, T. Sadowski, et al., 2021, Composites Part A: Applied Science and Manufacturing 149, 106516.
[13] A. Kemény, N. Movahedi, T. Fiedler, et al., 2022, Materials Science and Engineering: A 852, 143706.
[14] M. Vesenjak, I. Duarte, J. Baumeister, et al., 2020, Composite Structures 234, 111748.
[15] E. Linul, L. Marsavina, J. Kovacik, et al., 2017, Proceedings of the Romanian Academy - Series A 18 (4), 361-369.
[16] J. Jerz, N. Mináriková, L. Marsavina, et al., 2016, Frattura ed Integrita Strutturale 10 (36), 55-62.
[17] D. Pietras, E. Linul, T. Sadowski, et al., 2020, Composite Structures 249, 112548.
[18] S.A. Elahi, J. Rouzegar, A. Niknejad, et al., 2017, Thin Wall Struct, 114 (2017), 1-10.
[19] I. Duarte, L. Krstulović-Opara, J. Dias-de-Oliveira, et al., 2019, Thin-Walled Structures, 138 (2019), 124-136.
[20] Z. Liu, Z. Huang, Q. Qin, 2017, Composite Structures 175, 19-27.
[21] N. Onsalung, C. Thinvongpituk, K. Painthong, 2009, Proceedings of SPIE – The International Society for Optical Engineering, Vol. 7522.
[22] M. Su, H. Wang, H. Hao, 2019, Composite Structures 226, 111197.
[23] N. Movahedi, E Linul, 2021, Composite Structures 277, 114634.
[24] E. Linul, O. Khezrzadeh, 2021, Composite Structures 271, 114156.
[25] X. Li, X. Zhu, R. Wang, et al., 2023, International Journal of Crashworthiness, 29, 95-114.
[26] M. Guden, A.K. Toksoy. H. Kavi, 2006, Journal of Materials Science, 41, 6417-6424.
[27] S. Yao, Z. Chen, P. Xu, et al., 2021, Metals, 11, 118.

The possibility of recycling iron sludge in the steel industry

Adriana BOBORA[1*], Ana SOCALICI[1], Corneliu BIRTOK BANEASA[1] and Doru SAPTA[1]

[1] Politehnica University of Timişoara, Faculty Engineering of Hunedoara, Hunedoara, Romania

*E-mail: gulasadriana@yahoo.com

Abstract. The steel industry strives to achieve quality steel with improved mechanical properties. In order to meet these market requirements, steel manufacturers need to streamline technological operations, increase process efficiency and productivity, and improve product quality control. Old iron and scrap metal used in the production of steel come from various sources and special attention should be paid to their quality, in particular, to the content of elements harmful to steel. The paper presents the results of the research undertaken on the use as raw material of the steel-making aggregates, in addition to the classic assortments of old iron of by-products obtained from the processing of iron slats in the form of briquettes. The influence of the quality of waste used was assessed by expanding the raw material base in the context of sustainable development, on steel blanks by investigating the effect on the chemical composition as well as the possibility of recovery of alloying element.

1. Introduction

Old iron, ferrous waste in the form of by-products and iron sponge are used in metal load when developing steel in electric arc furnaces. Some small ferrous and/or powder waste resulting from current production flows are reintroduced directly into the circuit and the waste already stored is recycled in the steel industry. Waste that does not correspond dimensionally is converted into by-products. Also, waste containing elements harmful to the development of ferrous alloys is transferred and recycled to other industrial sectors (for example, high zinc electrofilter dust is used to produce zinc) [1,2].

The old iron produced from the steel industry is of superior quality and more easily usable than the old iron sorts collected and resulting from sources other than the steel ones. These old iron sorts resulting from various industries or by scrapping equipment, or, machinery often contains higher amounts of impurities and is recycled after significant time gaps between production and return to industry [3-5]. The quality of waste is a limiting factor for the use of scrap and ferrous waste. Certain types of waste contain large amounts of unwanted elements in the load of steel-making aggregates.

In the long run, the decrease in the quality of scrap used in the load and the increase in the quality of the developed steel pose serious challenges for recycling. The lower the quality of scrap, the lower the yield and therefore the higher the energy demand per tonne of production [5-8].

The expansion of the use of small and powder ferrous waste on current or landfilled flows leads to a reduction in energy needs to the extent that by-products from such waste replace the basic raw material when developing steel [9-12]. To increase productivity when using waste and replace high-cost electricity with lower-cost fuels, some plants preheat waste.

The aim of this study is to assess the influence of the quality of waste briquettes on the recycling process and the final product by investigating the effect on the chemical composition and specific energy consumption.

2. Industrial researches

The paper presents tests carried out at the industrial level on the use in the metal load of some by-products obtained by briquetting from steel ferrous slats. Given that the chemical compositions of scrap iron are highly variable and often impure with undesirable elements in steel, the tests carried out highlight the influence of these by-products on steel quality.

The industrial research was carried out in a steel plant where 9 C1010M steel chargers were cast according to ASTM 512. The steel chargers were created in an electric arc furnace, the secondary treatment of steel was carried out in the LF plant and the casting of steel was carried out continuously in the form of billets with a diameter of 177mm. The chemical composition of the elaborated steel: 99.18-99.62%Fe, 0.08-0.13%C, 0.3-0.6%Mn, 0.17-0.37%Si, under 0.05%S, under 0.04%P.

The analysis of the process was carried out throughout the technological flow of elaboration, secondary treatment and steel casting. When developing steel in the electric arc furnace, quality metal load was used. Technological parameters were observed in elaboration, secondary treatment and casting. Additions without moisture were used.

The metal load of the experimental chargers was made up of several old iron sorts and of briquette-type by-products obtained from ferrous slurry, as follows: E1 - Old iron easily processed; E3 - Hard scrap; E5 – Steel bar; E6 – Metal waste with small, thin, compressed/baloted residues; F – cast iron waste; B – Briquetting by-products. Figure 1 shows the distribution of the metal load to the elaborate steel. The loading of the melting furnace was carried out with 3-4 bene depending on the aprons used in the load (figure 2).

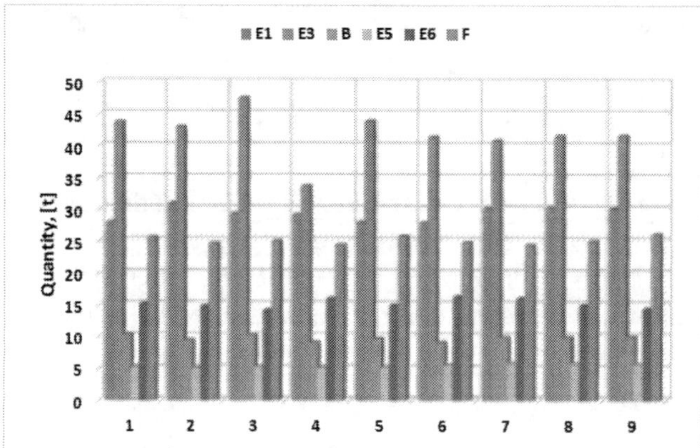

Figure 1. Distribution of metal load

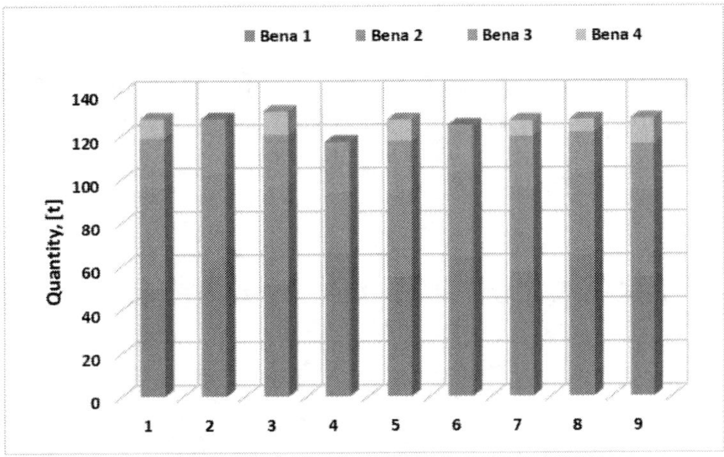

Figure 2. Loading the electric arc furnace

In the metal load besides the old iron assortments used in the current way, about 9.5 tons/large by-product briquettes were used for each batch, as follows: 9.05-10.35tons 7.3-8.1% (figure 3).

During the steel development and casting process, it was observed the influence of the use of the briquetting by-product in the metal load. The analysis was carried out on the sequences of elaboration and casting of steel. It was determined: the degree of metallization, dilution of elements, fuel consumption, methane gas consumption and oxygen consumption.

Also, on each sequence was determined the electricity consumption. Figure 4 shows the balance sheet of the metal load of the developed loads and the consumptions made are shown in figures 5 and 6.

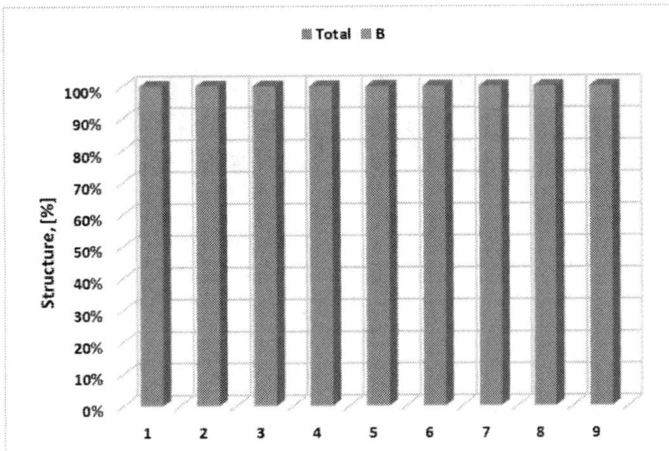

Figure 3. By-product of briquettes reported on total metal load

The elaborated steel was subjected to secondary treatment in the LF plant.

The casting of steel was carried out as follows: a sequence of 6 charge/sequence for the charge 1 - 6 and a sequence of 3 charge/sequence for the charge 7 - 9.

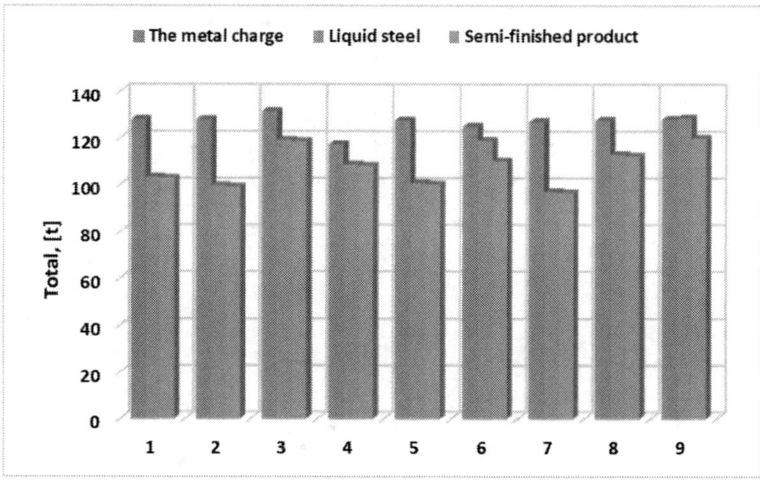

Figure 4. Balance sheet metal load

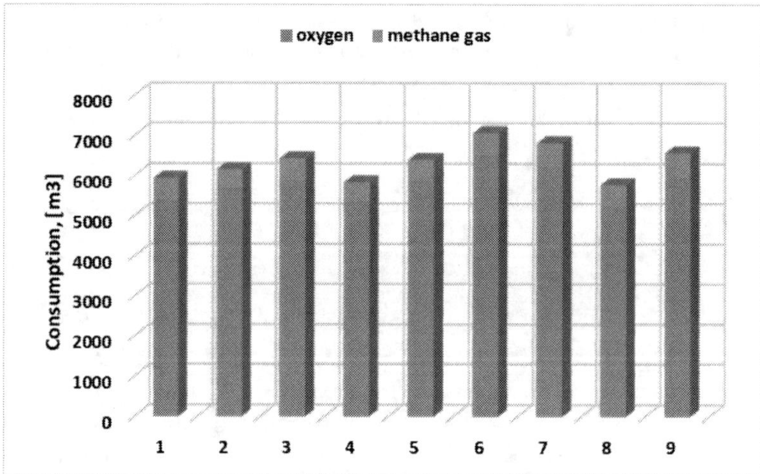

Figure 5. Consumption of oxygen and methane gas

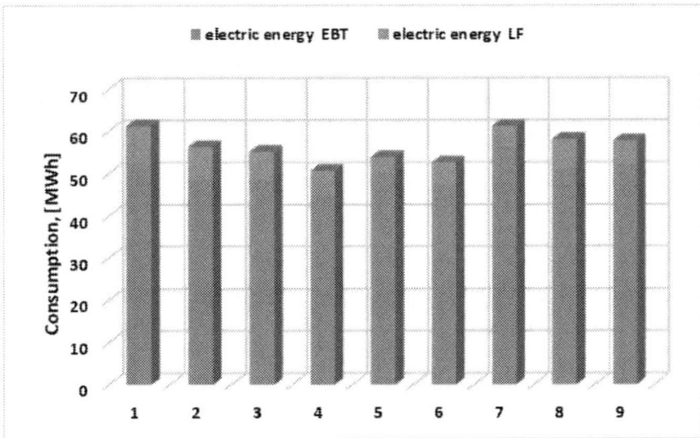

Figure 6. Consumption of electrical energy in the development of steel

The materials used in the production of steel had the characteristics determined in advance to obtain the experimental results and the results obtained were compared with the control data obtained in the elaboration of the same steel brand, under the same conditions of elaboration, secondary treatment and casting, with the specification that in case of control salts, no briquettes by-products were used in the load, but only the old iron and cast iron assortments. Quantitatively and qualitatively all technological parameters, specific consumptions, energy consumption, oxygen and fuels were detected. There were taken samples throughout the development and casting of experimental barges for steel and slag. The resulting semi-finished quantities as well as the resulting quantities of waste were determined.

3. Conclusions

From the analysis of the experimental data, the following results:
- The specific consumption of metal load was 1152.38 kg/t steel;
- The degree of metallization 87%;
- Clay 172 kg/t;
- Consumption of oxygen 52,57 m^3/t;
- Methane gas consumption 5,07 m^3/t;
- Electricity consumption 455kWh/t;
- Electrode consumption 2.49 kg/t.

By comparison with the standard consumptions for the control charge, an exceedance of 12kWh/t is observed for the specific electricity consumption, of 10 m^3/t for the specific oxygen consumption and 0.11kg/t/t to the electrode consumption, fact explained by the carbon content at higher melting resulting from the component of the cast iron load (loaded on the barges ca, 25t/charge). The chemical composition of steel and slag were also analysed. Emphasis was placed on the copper content. The by-product in the form of briquettes does not influence the copper content, which is directly influenced by the old iron E1 assortment. It must be limited to a maximum of 30t/charge instead the E2 and E6 assortments may vary within 40-50t/sarge limits.

Currently, there is an increase in the use of iron-containing waste in the form of by-products in the metal load of steel-making aggregates. The use of iron by-products in compilation aggregates generates substantial reductions in raw materials, fuels and energy. The main objective of the steel industry is zero waste, so the recovery and use of by-products within and outside the steel industry, combined with the responsible management of natural resources, contribute to material efficiency and waste prevention.

References

[1] M. Hauput, C. Vadenbo, C. Zeltner, S. Hellweg, Influence of Input-Scrap Quality on the Environmental Impact of Secondary Steel Production: Influence of Scrap Quality on EAF Steel Production, Journal of Industrial Ecology, 21(2), 2016

[2] EC (European Commission), The circular economy. Connecting, creating and conserving values. Brussels: European Commission, 2024

[3] https://worldsteel.org/steel-topics/sustainability/

[4] T. Heput, A. Socalici, E. Ardelean, M. Ardelean, N. Constantin, R. Buzduga, Recovery of small and powdery ferrous waste, Politehnica Publishing House Timisoara, 2011

[5] C. P. Mititelu, M. Hritac and N. Constantin, Laboratory experiments for determination of optimal characteristics of ultrafine ferrous waste briquettes to be used in cupola furnace, Scientific Bulletin Series B-Chemistry and Materials Science, vol. 77, nr. 1, 2015, pp. 157-164

[6] Project nr. 31-098/2007, "Prevention and fighting pollution in the steelmaking, energetic and mining industrial areas through the recycling of small-size and powdery wastes", Program 4 Partnerships in priority areas, 2007-2010

[7] C. Pandelescu, N. Constantin, D. Gheorghe, E.F. Plopeanu, Experimental Research on the Metallic Material Defects Appeared at the Operation of Pumping Aggregates, Scientific Bulletin Series B-Chemistry and Materials Science, 82, 2020, pp. 295-306

[8] I. Farcean, G. Prostean, E Ardelean, Life cycle analysis of steel made in the electric arc furnace with eccentric bottom tapping, Scientific Bulletin Series B-Chemistry and Materials Science, vol. 85, Iss. 3, 2023, pp.219-230

[9] A. Zamfir, A. Meghea, M. Mihaly, Physical – chemical characterization of some mining waste for capitalization, Scientific Bulletin Series B-Chemistry and Materials Science, vol. 85, Iss. 3, 2023, pp.113-122

[10] O. Lupu, The recovery of small waste resulted from the process of steel manufacturing and processing, PhD Thesis, University Politehnica of Timisoara, 2023

[11] M. Moskal, P. Migas, M. Karbowniczek, Multi-Parameter Characteristics of Electric Arc Furnace Melting, Materials, 15, 2022, p.1601

[12] Carlsson, L.S.; Samuelsson, P.B.; Jönsson, P.G. Predicting the Electrical Energy Consumption of Electric Arc Furnaces Using Statistical Modeling. Metals, 9, 2019, p.959

Quality Assessment of Aluminum Alloys for Automotive Rims

Doru SAPTA[1*], Ana SOCALICI[1], Corneliu BIRTOK BANEASA[1] and Vasile PUTAN[1]

[1] Politehnica University of Timişoara, Faculty Engineering of Hunedoara, Hunedoara, Romania

*E-mail: saptad94@gmail.com

Abstract. This paper presents experimental research conducted at a factory specializing in the development of aluminum alloys and the semi-continuous production of bars intended for rim manufacturing. The semi-finished products are made of 6082 aluminum alloy and undergo forging processes to obtain rims. The charge load consists of various types of aluminum waste and primary aluminum in different proportions, depending on the quality of the waste in the experimental recipes. The elaboration of the aluminum alloy takes place in melting furnaces with a capacity of 80 tons, equipped with two chambers. The casting is performed using a semi-continuous casting installation, and the cast semi-finished products undergo further heat treatment. The casting process involves the addition of alloying elements, in the form of wire, which serve to refine the structure. Experimental data were processed, and correlations between the qualitative characteristics of the alloy and the elements in the chemical composition, as well as grain size, were analyzed. From the obtained correlation analysis, it is observed that the alloying elements refine the grain size, while the grain size and quality of the resulting product are of interest for industrial practice.

1. Introduction

Aluminum and its alloys have good weldability, which is determined by the processing and casting technology, the method of processing, the heat treatment that needs to be applied to the welded parts, and the design and conception of the welded parts made from these alloys [1].

Aluminum alloy wheels have the advantage of being lightweight and aesthetically pleasing. The disadvantage is that, following an uncontrolled impact, cracks or breaks can occur. These cracks can be repaired by welding, with the possibility of retaining the initial technical characteristics [2]. Alloys from the 6xxx series are widely used as deformed products in the automotive or aerospace industries. The strengthening phase in these alloys is Mg_2Si, with the size of the phase being determined by the presence of iron and silicon. Alloys in this group use magnesium and silicon, in different proportions, to form magnesium silicide, can be heat-treated, and have good forgeability and high corrosion resistance [3].

The most used aluminum alloys for manufacturing wheels are the deformable ones from group 6, namely 6082 (ISO: AlSi1MgMn) and 6061 (ISO: AlMg1SiCu), as well as the casting ones

AlSi11 (EN AC-44000) and A356.0 (ISO: Al-7Si-0.3Mg), whose chemical composition and mechanical, thermal, and physical properties comply with SR EN 573-1: 2005 [1,4].

Alloy 6082 is widely used in machine construction, as well as in welded constructions. It is used for producing oscillating, rotating component parts, the base plates of these subassemblies, as well as for making housings for hydraulic pumps or component parts of hydraulic systems. Corrosion resistance in both marine environments and the atmosphere is good. Alloy 6082 is used for high-load structures, truck or trolleybus frames, boilers, or bicycles, flanges, poles, towers, motorboats, masts, and sideboards in boat construction, and metal scaffolding [1,5].

Aluminum alloy 6082 has medium strength and is known as a structural alloy. Alloy 6082 has started to replace alloy 6061 in many applications due to its higher strength. Increasing the titanium content allows for the control of the grain structure, resulting in a stronger alloy. Aluminum alloy 6082 has similar, but not equivalent, physical characteristics compared to alloy 6061 [1,6,7,8].

Aluminum alloy 6082, like other aluminum alloys, easily forms a layer of aluminum oxide (Al_2O_3) in contact with air, which acts to prevent further corrosion. Because the volume occupied by aluminum oxide is greater than that of the oxidized metal, the oxide has a compact structure and prevents oxygen from diffusing to the aluminum-oxide interfacial surface [9,10].

For the manufacturing of automotive wheels, worldwide, technologies for manufacturing from aluminum alloys by casting (e.g., alloy AlSi11) and plastic deformation (alloys 6062 and 6082) are being developed.

2. Industrial research

The paper presents the industrial experiments conducted at a plant producing aluminum for manufacturing automotive wheels. The aluminum alloy is produced in two streams within the plant. The main technological process carried out in the plant is the production of secondary aluminum by recycling aluminum waste in melting furnaces. The waste melting is done on two different lines. The first line uses waste with a high aluminum content in the charge, while the second line uses aluminum waste with a lower aluminum content, specifically aluminum recovered from the slag resulting from the technological process of the first production line.

The technological flow for producing aluminum semifinished products consists of:

- Three side-loading melting furnaces with two chambers MF 1-2-3 (figure 1), with a capacity of 80 tons;
- Two casting furnaces HF1&2, with a capacity of 50 tons;
- One induction melting furnace, with a capacity of 7.5 tons and an installed power of 2500kW;
- Two gravitational casting installations for bars and profiles VC1&2 (figure 2);
- Three homogenization furnaces HO-Batch 1-Batch 2;
- An installation for ultrasonic quality inspection of semifinished products (figure 3).

The main phases of the technological process are:

- Procurement, inspection, sorting, and storage of raw materials;
- Melting phase of the raw materials;
- Casting phase of the molten aluminum;
- Homogenization phase of the aluminum profiles and bars resulting from casting;
- Packaging and storage phase of finished products.

Figure 1. The melting furnaces with two chambers are called MF furnaces

Figure 2. The casting of the alloy

Figure 3. Quality control of the semifinished products

The raw materials used in the technological flow are: waste with aluminum content of 76.3% - 90%, aluminum with purity of 99%, and alloying metals (Figure 4).

Figure 4. Raw materials used in the elaboration of the alloy

The melting furnace has two chambers. The melting chamber is directly heated by two 4 MW gas burners, up to a temperature of 1050°C, while the waste preheating chamber is indirectly heated by hot gas recovered from the melting chamber, up to a temperature of 750-800°C. The liquid alloy is conducted using an electromagnetic pump through the Charge Well module, with the pump having a capacity of 8 tons per minute. During the alloy production and casting process, samples are taken to determine the chemical composition of the melt. For casting, the liquid alloy is transferred through a gutter system to the degassing installation (impurities such as hydrogen, magnesium, or other metals are removed using chlorine, nitrogen, and argon). Then, the alloy passes through a ceramic filter, which retains unwanted oxides and suspended particles. The round profiles resulting from casting go through the homogenization phase. Each semifinished product is introduced into the inspection chamber for potential nonconformities, and the inspection is carried out using ultrasound.

For the industrial experiments conducted at the plant, 30 batches of aluminum alloy 6082 were produced for the automotive industry to manufacture automotive wheels. Bars with diameters ranging between 305 - 400 mm and a length of 7500 mm were cast, these being delivered to the final customer, the plant where the automotive wheels are forged. The raw material used for producing the aluminum alloy consists of aluminum waste: casting remnants, defective cast parts, defective semifinished products, remnants from cutting semifinished products, defective wheels, shavings from mechanical processing, and waste from processing other products. Figure 5 shows the composition of the charge for the produced aluminum alloy batches.

The variation in the content of elements in the chemical composition of the produced aluminum alloy is shown in Figure 6. In the casting flow, between the SNIF filter and the casting installation, $AlTi_5B_1$ / $AlTi_3B_1$ is added in the form of 10 mm diameter wire to refine the structure. Then, the aluminum alloy passes through a ceramic filter that retains existing impurities in the melt. To obtain cast bars approximately 7500 mm in length, the casting duration is 1 hour and 45 minutes, and the bars are cooled in water, with the cooling parameters being: air pressure = 3.5 bar, water flow = 120 m^3/h, and water temperature = 22°C. The temperature of the bars when removed from the installation is 50-60°C, after which they are prepared for homogenization heat treatment.

The homogenization heat treatment involves introducing the bars into the furnace at 30°C and heating them to 532°C over 160 minutes, maintaining this temperature for 6 hours, with an average cooling rate of 430°C/h. Subsequently, they are cooled in air to ambient temperature.

To correlate the refining and determine the grain sizes in the production flow, two samples are taken—one from the furnace and one from the semifinished product. The surfaces of these samples are processed and polished, and the determination is carried out using the OPTICAST AB measurement method (Figure 7).

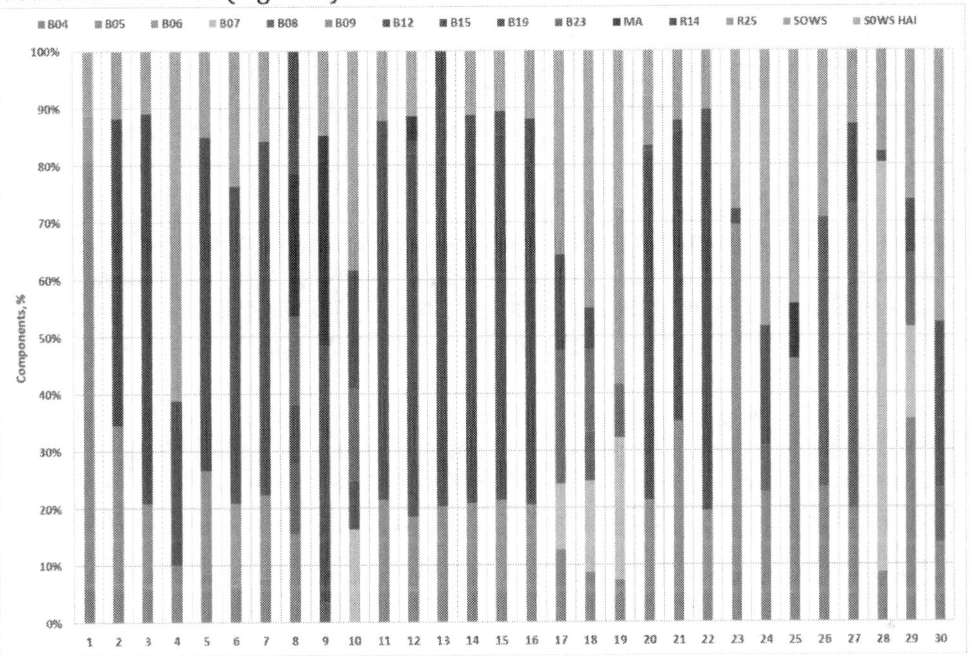

Figure 5. The composition of the charge

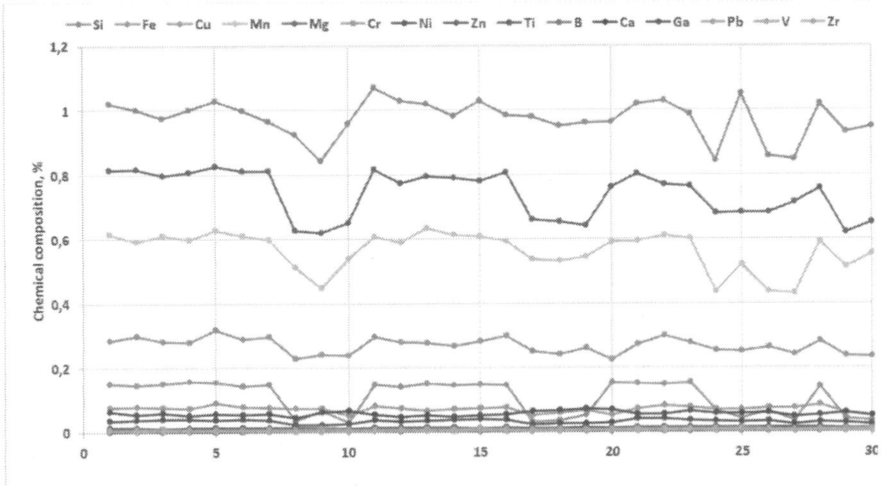

Figure 6. The chemical composition

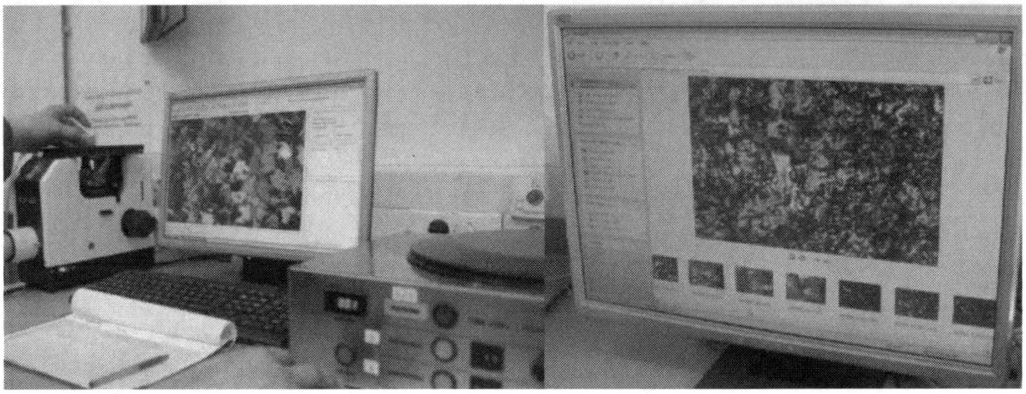

Figure 7. The XJP-6A microscope for determining grain sizes

The alloying elements refine the grain structure, and the grain size in the semifinished product is of practical interest. A titanium content of 0.022 – 0.041% ensures a grain size between 160 – 120 μm, while a magnesium content of 0.80 – 0.90% ensures the same limits for grain size. The same limits for grain size are also obtained for the 2Mg/Si ratio between 1.35 – 1.68. A simultaneous increase in magnesium content and the 2Mg/Si ratio results in finer grain structure, reaching up to 120 μm for magnesium values between 0.8 – 0.9%. A simultaneous increase in the values for titanium and magnesium content, or titanium and silicon, also leads to a refined grain structure. In conclusion, it is found that with a Ti content of 0.022 - 0.031%, Mg content of 0.82%, Si content of 1.01%, and 2Mg/Si ratio of 1.35 – 1.68, the smallest grain sizes are obtained.

3. Conclusions

The analysis of the experimental data reveals the following:

- the charge consisted of 15 components, 14 of which were waste and 1 was primary aluminum, with the number of components in a batch ranging from a maximum of 8 to a minimum of 4.

- the main component was waste from alloy 6082, of the same grade as the produced alloy, which represents an advantage both technologically and economically. The proportion of this type ranged between 35% - 80%.

- all the waste used in the charge was of very good quality, without soil, slag residues, or plastic waste. The high quality of the charge and the large proportion of 6082 waste resulted in the use of only 15% primary aluminum.

- the good quality of the charge is also reflected in the chemical composition of the alloy at the end of melting. In this regard, it can be observed that, in terms of the main alloying elements, Si, Mn, and Mg, the maximum limits were not exceeded. Moreover, the Si content falls within the limits specified in the standard.

- at the furnace exit, the temperature varied between 714 – 721°C, the technological range being 710 – 725°C, and the casting temperature was within the imposed limits of 690 – 710°C.

- the specific consumption of $AlTi_5B_1$ ranged between 600 – 750 g/t, and for each batch, the grain size was determined, falling within the norms imposed by the customer.

- the casting speed was maintained constant at 1.16 mm/s (68 mm/min), and the cooling parameters were respected, so there were no issues regarding the surface quality of the bars and internal defects.

References

[1] N. Bularda, Metal alloys used for the manufacture of rims intended for road vehicles, PhD Thesis, University Politehnica of Timisoara, 2015

[2] www.weldingtipsandtricks.com/tig-welding-aluminum.html

[3] Puțan V. - Bazele elaborării și turnării aliajelor neferoase, Editura Politehnica Timișoara,2010

[4] www.engineersedge.com

[5] https://metal-automotive.ro/category/aluminiu/

[6] E. Xia, T. Ye, S. Qiu, J. Liu, J. Luo, L. Sun and Y. Wu, Mechanical Properties and Microstructural Evolution of 6082 Aluminum Alloy with Different Heat Treatment Methods, Coatings, 14, 2024, p.602.

[7] Xu, X., Zhu, W., Yuan, M., Liang, C., Deng, Y., The effect of Zn content on the microstructure and mechanical properties of the Al-mg-Si alloy, Mater. Charact., 2023, 198, 112714

[8] Mròvka-Nowotnik G., Influence of chemical composition variation and heat treatment on microstructure and mechanical properties of 6xxx alloys, Archives of Materials Science and Engineering, 46(2), 2010, pp. 98-107

[9] Mrówka-Nowotnik G., Sieniawski J., Wierzbińska M., Intermetallic phase particles in 6082 aluminum alloy, Archives of Materials Science and Engineering, 28(2), 2007, pp. 69-76

[10] Fujda M., Kvačkaj T. - Microstructure and Mechanical Properties of EN AW 6082 Aluminum Alloy Prepared by Equal-Channel Angular Pressing, Journal of Metals, Materials and Minerals, 17(2), 2007, pp.23-27

IOP Conf. Series: Materials Science and Engineering **1319** (2024) 012014 doi:10.1088/1757-899X/1319/1/012014

Behaviour of AZ91 magnesium alloy in tension and torsion: experimental investigations and numerical modelling

Mihai-Alin MURARIU[1*], Lucian RUSU[2], George BELGIU[2] and Dan-Andrei ȘERBAN[1]

[1] Department of Mechatronics, Politehnica University Timisoara, Romania
[2] Department of Mechanics and Strength of Materials, Politehnica University Timisoara, Romania
[2] Department of Management, Politehnica University Timisoara, Romania

*E-mail: mihai.murariu@student.upt.ro

Abstract. This study is concerned with the evaluation of the most appropriate plasticity model for simulating the behaviour of the AZ91 magnesium alloy in terms of accuracy and analysis times, based on experimental data obtained from tensile and torsion tests. The material was supplied in ingot form, was subsequently cut using and automated saw and ISO 6892 round specimens were obtained through lathing. The test data was used to calibrate three hardening functions, namely the multi-linear hardening, Johnson-Cook power law hardening and the Ramberg Osgood plasticity model, with the former determining the best results.

1. Introduction

The market for magnesium is a significant component of the larger field of lightweight materials and metallurgy. Magnesium alloys, that primarily include aluminium, zinc, and manganese, provide a desirable strength-to-weight ratio which recommends them for a number of industries, such as aerospace, electronics, medical devices, and automotive [1, 2, 3].

In contrast to the aforementioned benefits, the use of magnesium alloys presents some challenges, the main issues being their vulnerability to corrosion (as certain surface treatments are necessary) and the scarcity of high-quality magnesium resources, which implies advancements in the recycling methods, in order to ensure sustainability and reduced costs [4].

The market for magnesium alloys is expected to grow as businesses place a greater emphasis on weight reduction [5]. The development of new alloys and manufacturing processes is a high priority, with the goal of improving these materials' performance and overcoming current constraints. As a result, it is anticipated that the market trends will be crucial to the migration of numerous industrial applications toward lightweight materials.

Because of their high castability and the recommended manufacturing technology (die casting), magnesium alloys are widely used in high-performance car bodywork, engine blocks, steering wheels, cylinder covers, and gearbox housings in the automotive industry [1, 6], which represent parts with complex geometries that exhibit many stress concentrators. In order to prevent early product failure, companies are employing virtual testing and validations as a crucial phase in the product design process. [7,8]

According to the EN 1753 standard [9], AZ91 or AZ91D is a magnesium alloy for die-casting applications with its main alloying elements being aluminium (9%) and zinc (1%) [10,11]. The benefits of magnesium die casting are the possibility of manufacturing components with complex shapes (low viscosity in molten form), weight reduction (magnesium is the lightest structural metal, with the density of around 1810 kg/m3), strength-to-weight ratio (tensile strengths comparable to some aluminium alloys at two thirds of the density), thermal stability, corrosion resistance and recyclability [1,2]

2. Materials and methods

The AZ91 Magnesium alloy investigated in this study was provided by TCG Unitech in cast form through 1150 mm x 65 mm x 65 mm ingots (Figure 1).

Figure 1. AZ91 magnesium alloy ingot

For this study, round ISO 6892 specimens were considered [12], the calibrated region having a 5.5 mm diameter and 60 mm length. In cosnequence, the magnesium ingots were initially cut into 150 mm x 15 mm x 15 mm prisms using an automatic saw and subsequently lathed into the final shapes (Figure 2) on a Doosan Lynx 220 L.

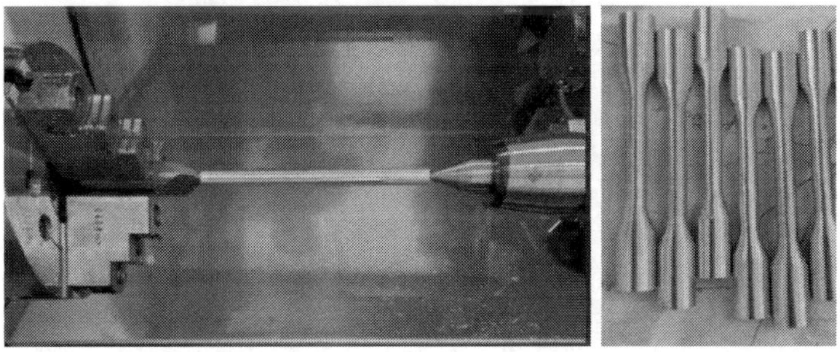

a) b)

Figure 2. Setup for the lathing of the semi-finished parts (a) and finished specimens (b)

3. Experimental procedures

The experimental procedures consisted of tensile and torsional loadings and were performed at ambient temperature on a servo-hydraulic Instron 8874 bi-axial testing machine, with a force capacity of 25 kN and a torque capacity of 100 Nm. The tensile tests were performed at a corsshead travel speed of 5 mm/min while the torsion tests used a 30°/min rotation speed. Three specimens were tested for each configuration.

Based on the data recorded during the tests, the tensile stress and strain was calculated using relations (1) and the shear stress and strain (at the outer limit of the specimen) were evaluated using relations (2)

$$\begin{cases} s(t) = \dfrac{4 \cdot F(t)}{\pi \cdot D^2} \\ e(t) = \dfrac{l(t) - L}{L} \end{cases} \quad (1)$$

$$\begin{cases} t(t) = \dfrac{16 \cdot T(t)}{\pi \cdot D^3} \\ g(t) = \dfrac{d \cdot \varphi(t)}{2L} \end{cases} \quad (2)$$

where $s(t)$ is the nominal tensile stress variation with time, $F(t)$ is the recorded load variation with time, D is the specimen initial diameter, $e(t)$ is the variation of the nominal strain with time, $l(t)$ is the variation of the extensometer travel with time, L is the initial gauge opening of the extensometer, $t(t)$ is the nominal sear stress variation with time, $T(t)$ is the recorded torque variation with time, $g(t)$ is the nominal shear strain variation with time and $\varphi(t)$ is the variation of the rotation with time.

The average resulting mechanical properties are presented in Table 1 while representative stress-strain curves for each type of test are presented in Figure 3.

Table 1. Mechanical properties of AZ91 magnesium alloy

	Stiffness [GPa]	Yield stress $R_{p0.2}$ [MPa]	Ultimate strength [MPa]
Tension	44.38	88.28	194.68
Torsion	15.82	60.51	175.29

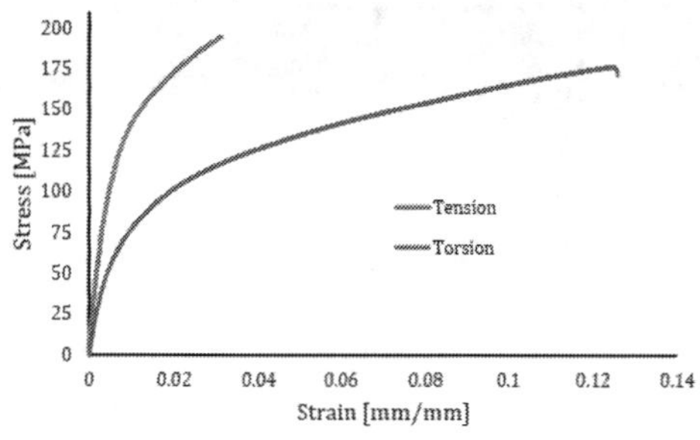

Figure 3. Stress-strain curves for the tensile and torsional tests

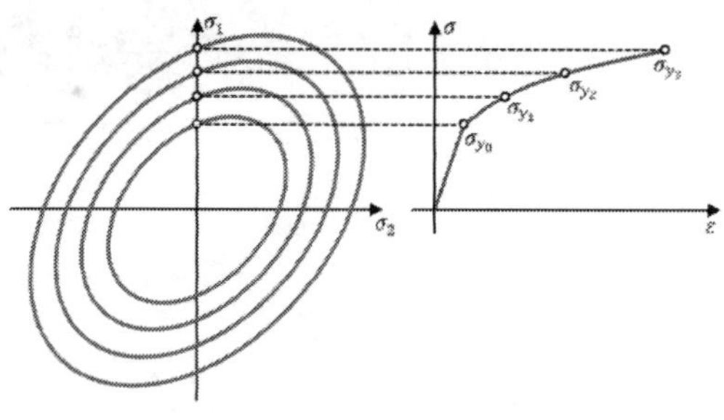

Figure 4. Isotropic hardening model for uniaxial loading

4. Numerical modelling

Considering that the experimental procedures were performed in monotone loading scenarios (no cyclic data is available), the investigated plasticity formulations will use the isotropic hardening model (the yield surface expands uniformly as the loading increases). For a given stress state, the isotropic plasticity model implies the consistency condition, namely that, during loading, the load point must remain on the yield surface (Figure 4) [13,14].

$$f(\boldsymbol{\sigma},\bar{\varepsilon}^p)=\sigma^e-\sigma^y(\bar{\varepsilon}^p)=0 \tag{3}$$

where $\boldsymbol{\sigma}$ is the Kirchhoff stress tenson, $\bar{\varepsilon}^p$ is the accumulated equivalent logarithmic plastic strain, σ^e is the equivalent Kirchhoff stress and $\sigma^y(\bar{\varepsilon}^p)$ is the equivalent logarithmic plastic strain dependend Kirchhoff yield stress.

In general, the plastic strain dependent Kirchhoff yield stress has the form [13]

$$\sigma^y(\varepsilon^p)=\sigma_0^y+r(\varepsilon^p) \tag{4}$$

where $r(\varepsilon^p)$ is called the isotropic hardening function

The Kirchhoff stress tensor $\boldsymbol{\sigma}$ is defined as the Cauchy stress tensor $\boldsymbol{C}$ multiplied by the Jacobian $\boldsymbol{J}$ of the deformation gradient $\boldsymbol{F}$. Since the Cauchy stress tensor components cannot be directly determined from the nominal stress tensor $\boldsymbol{S}$ data, some manipulations are required, leading to the relation [8, 9]

$$\boldsymbol{\sigma}=\boldsymbol{J}\cdot\boldsymbol{C}=\boldsymbol{F}\cdot\boldsymbol{S} \tag{5}$$

where

$$\boldsymbol{F}=\begin{pmatrix}\frac{\partial x}{\partial X} & \frac{\partial x}{\partial Y} & \frac{\partial x}{\partial Z}\\ \frac{\partial y}{\partial X} & \frac{\partial y}{\partial Y} & \frac{\partial y}{\partial Z}\\ \frac{\partial z}{\partial X} & \frac{\partial z}{\partial Y} & \frac{\partial z}{\partial Z}\end{pmatrix}\Longrightarrow \boldsymbol{F}(t)=\begin{pmatrix}\frac{d(t)}{D} & 0 & 0\\ 0 & \frac{l(t)}{L} & 0\\ 0 & 0 & \frac{d(t)}{D}\end{pmatrix} \tag{6}$$

$$\boldsymbol{S}(t)=\begin{pmatrix}0 & 0 & 0\\ 0 & s(t) & 0\\ 0 & 0 & 0\end{pmatrix} \tag{7}$$

Considering the Poisson effect, the variation of the diameter $d(t)$ is

$$d(t)=D\left[1-\nu\frac{l(t)-L}{L}\right] \tag{8}$$

After some algebraic manipulations, it results that, for uniaxial loadings, the Kirchhoff stress $\sigma(t)$ can be expressed as a function of the nominal stress $s(t)$ and the nominal strain $e(t)$ [17]

$$\sigma(t)=s(t)[1-e(t)] \tag{9}$$

The logarithmic strain $\varepsilon(t)$ is defined as the sum of the infinitesimal stretches λ [8, 9]

$$\lambda(t)=\frac{dl(t)}{L} \tag{10}$$

$$\varepsilon(t)=\int_L^l\frac{dl(t)}{L}=ln\left[\frac{l(t)}{L}\right] \tag{11}$$

4.1 Multi-linear hardening model

The multi-linear hardening model (MLH) considers the isotropic hardening function of the form [18]

$$r(\varepsilon_k^p) = \begin{cases} h_1 \cdot \varepsilon_k^p, & \varepsilon_k^p \in [0,\varepsilon_1^p] \\ \sum_{i=0}^{n} h_{i+1} \cdot (\varepsilon_{i+1}^p - \varepsilon_i^p) + h_{n+1} \cdot (\varepsilon_k^p - \varepsilon_n^p), & \varepsilon_k^p \in (\varepsilon_n^p, \varepsilon_{n+1}^p], \ \varepsilon_0^p = 0 \end{cases} \tag{12}$$

where

$$h = \begin{cases} h_1, & \bar{\varepsilon}^p \in [0,\varepsilon_1^p] \\ h_2, & \bar{\varepsilon}^p \in (\varepsilon_1^p,\varepsilon_2^p] \\ & \vdots \\ h_k, & \bar{\varepsilon}^p \in (\varepsilon_{k-1}^p,\varepsilon_k^p] \end{cases} \tag{13}$$

The required input data consists of Kirchhoff yield stress – logarithmic plastic strain data pairs determined from experimental data (Figure 5). In equations (4) and (5), h_i represent the slope of the lines defined by the points $(\varepsilon_{i-1}^p, \sigma_{i-1}^y)$ and $(\varepsilon_i^p, \sigma_i^y)$.

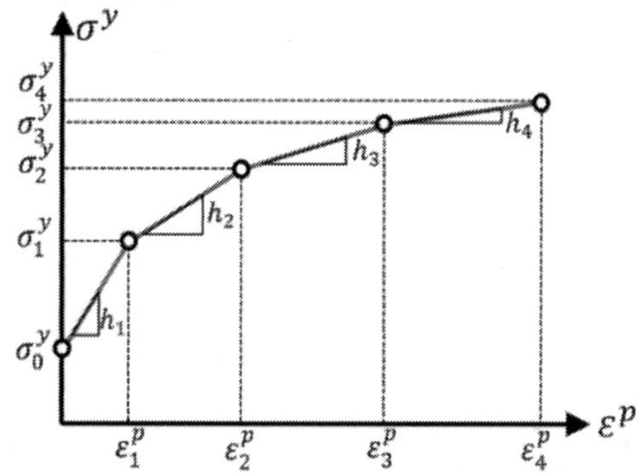

Figure 5. Multi-linear hardening input data

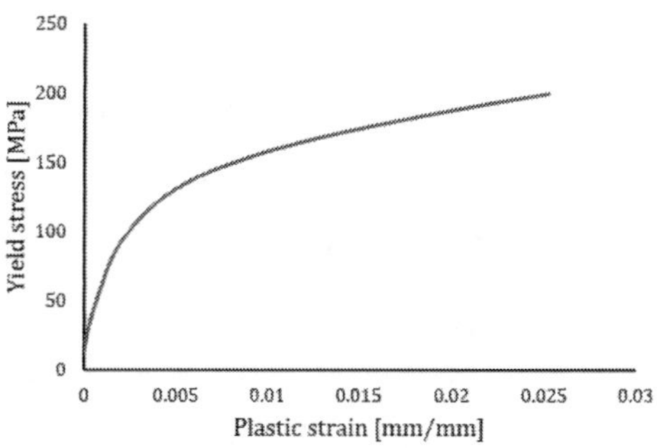

Figure 6. Yield stress variation with plastic strain for AZ91

In order to determine the points $(\varepsilon_i^p, \sigma_i^y)$, the strain decomposition principle was considered [13,17]

$$\varepsilon^t(t) = \varepsilon^e(t) + \varepsilon^p(t) \tag{14}$$

where $\varepsilon^t(t)$ is the total strain and $\varepsilon^e(t)$ is the elastic strain.

The experimental stress-strain curve (converted to Kirchhoff stress-logarithmic strain values) was used to extract the plasticity data by setting the plastic strain values and determining the intersection of the stress-strain curve with a line with the slope equal to the Young's modulus [17]. The resulting data is presented in graphical form in Figure 6. The first data point corresponds to the elastic limit (rather than the yield point $R_{p0.2}$), having the coordinates (0,13).

4.2 Johnson-Cook model

The Johnson-Cook model (J-C) assumes an exponential plasticity form, that can account for strain rate and temperature dependency [19]

$$\sigma^y(\bar{\varepsilon}^p) = [A + B \cdot (\bar{\varepsilon}^p)^n]\left[1 + C \cdot \ln\left(\frac{\dot{\bar{\varepsilon}}^p}{\dot{\bar{\varepsilon}}_0^p}\right)\right](1 - \widehat{\Theta}^m) \tag{15}$$

where A, B, C, n and m are material constants, $\dot{\varepsilon}^p$ is the current plastic strain rate, $\dot{\varepsilon}_0^p$ is the reference plastic strain rate and $\widehat{\Theta}$ is a nondimensional temperature parameter. Having no strain rate and temperature dependent data, the part of equation (12) dealing with these factors will be omitted.

The material constant A represents the elastic limit σ_0^y while the hardening function is $r(\varepsilon_k^p) = B \cdot (\tilde{\varepsilon}_p)^n$. However, for increased accuracy, in the calibration of the Johnson-Cook material model, all three parameters (A, B, and n) were fitted. For the parameter determination, a least square algorithm was employed, determining the values and the coefficient of determination presented in Table 2 and the graphic variation presented in Figure 7.

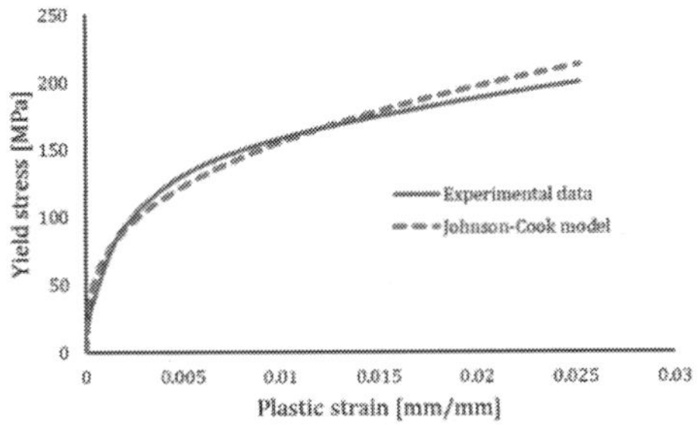

Figure 7. Experimental plasticity data and calibrated Johnson-Cook model

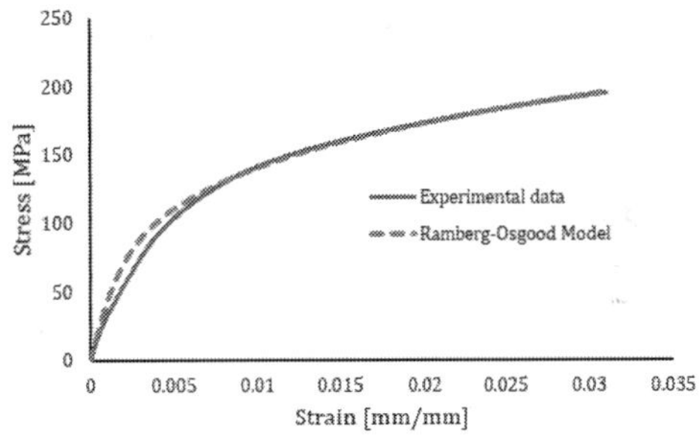

Figure 8. Experimental stress-strain data and calibrated Ramberg-Osgood model

Table 2. Parameters of the Johnson-Cook model for AZ91

A [MPa]	B [MPa]	n [-]	R^2 [-]
1	741.057	0.341	0.981

4.3 Ramberg-Osgood model

The Ramberg-Osgood (R-O) model provides a single equation that accounts for the entire stress-strain response of the material [20]. This fact implies that, even though the curvature is insignificant, there is no linear-elastic region for the material below the yield point. In addition it ensures a smooth transition from the apparent elastic region to the plastic flow region. Another difference is that the R-O model expresses the strain variation with stress. These factors determine the main drawback of the R-O model, that consists in the lack of information regarding the plasticity of the material. Its advantages lie in the fast computational times (no implicit algorithm for convergence is required for low increments and strains) and a very facile evaluation of the strain energy (which has an analytical expression, rather than requiring numerical integration with each increment). The Ramberg-Osgood formulation is

$$\tilde{\varepsilon} = \frac{\sigma^e}{E} + \alpha \frac{\sigma^e}{E} \left(\frac{\sigma^e}{\sigma_0^y} \right)^{(n-1)} \quad (16)$$

where E is the Young's modulus, σ_0^y is the elastic limit, n and α are material constants.

IOP Conf. Series: Materials Science and Engineering **1319** (2024) 012014 doi:10.1088/1757-899X/1319/1/012014

The same least square algorithm was applied to the R-O model to evaluate the material constants, the results being presented in Table 3, while the graphical variation is presented in Figure 8.

Table 3. Parameters of the Ramberg-Osgood model for AZ91

σ_0^y [MPa]	n [-]	α [-]	R^2 [-]
13.64	4.127	0.000148	0.998

5. Simulations

The geometric model consisted of a cylinder of 5.5mm diameter and a length of 50 mm, and used a structured mesh consisting of 1600 prismatic elements of 1mm average size (Figure 9). Four types of elements were investigated: linear elements with reduced integration (one integration point per element), linear elements with full integration (8 integration points per element), quadratic elements with reduced integration (8 integration points per element) and quadratic elements with full integration (27 integration points per element).

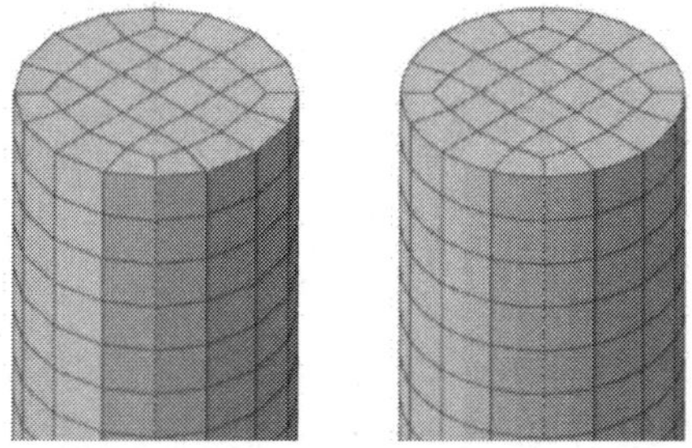

a) b)

Figure 9. Meshed model with linear (a) and quadratic (b) elements

For the tensile analyses, the bottom region used a Y-symmetry boundary condition while a displacement was applied to the top of the model. The reaction forces for the nodes of the bottom surface and the travel of the nodes of the top surface were recorded as field output. For the torsion analyses, the displacements and rotations of the nodes of both the top and bottom surfaces of the model were attributed to control points using a beam-type multi-point constraint. A rotation was applied to the top reference point while the bottom reference point was fixed. As field output, the top reference point rotation and the bottom reference point reaction moments were considered.

The simulation results for the multi-linear hardening model are presented in Figure 10, for the Johnson-Cook model in Figure 11 and for the Ramberg-Osgood model in Figure 12, for both tensile and torsional loadings.

The accuracy of the results (expressed through the coefficient of determination R^2) are presented in Table 4 for the tensile loading and in Table 5 for the torsional loading.

The comparison between the relative simulation times of each model are presented in Figure 13 for both tensile and torsional loadings.

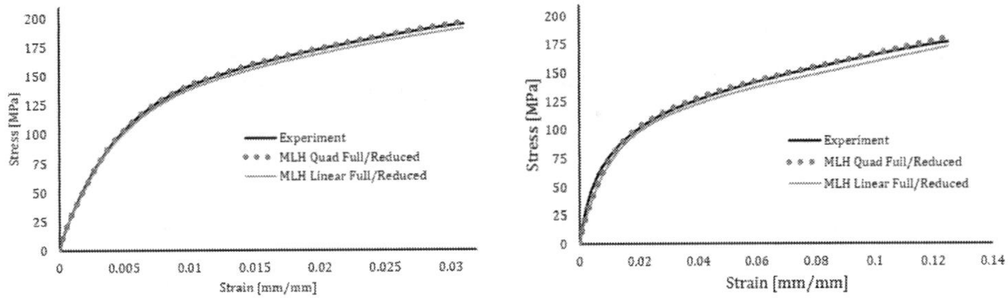

a) b)

Figure 10. Comparison between the experiment and numerical stress-strain responses for the multi-linear hardening model for tensile (a) and torsional (b) loadings

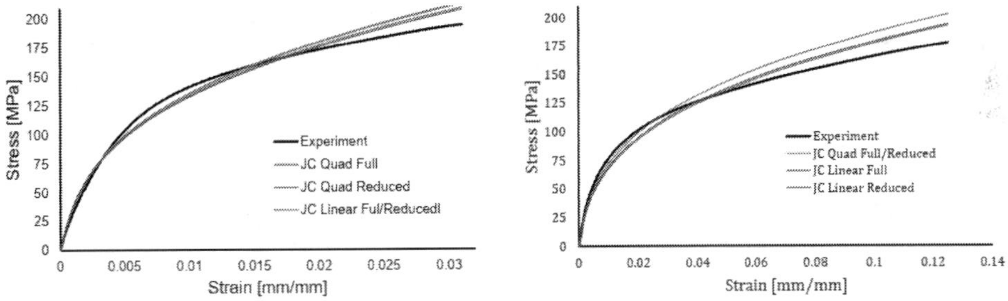

a) b)

Figure 11. Comparison between the experiment and numerical stress-strain responses for the Johnson-Cook model for tensile (a) and torsional (b) loadings

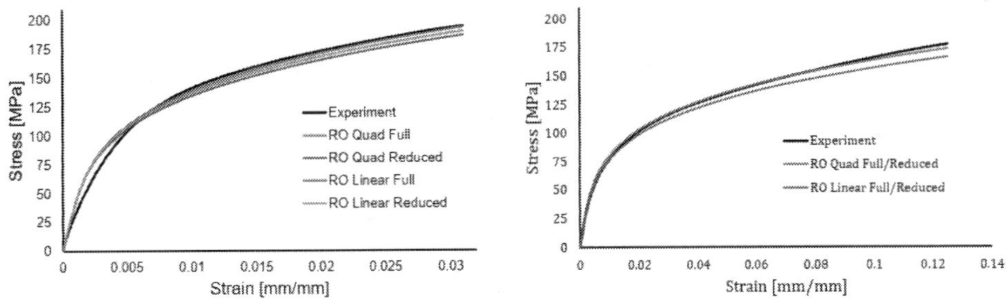

a) b)

Figure 12. Comparison between the experiment and numerical stress-strain responses for the Ramberg-Osgood model for tensile (a) and torsional (b) loadings

Table 4. Accuracy of the models for the tensile loadings

	Linear elements Reduced integration	Linear elements Full integration	Quadratic elements Reduced integration	Quadratic elements Full integration
Multi-linear hardening	0.9945	0.9945	0.9998	0.9998
Johnson-Cook	0.9744	0.9744	0.9771	0.9638
Ramberg-Osgood	0.9847	0.9847	0.9724	0.9902

Table 5. Accuracy of the models for the torsional loadings

	Linear elements Reduced integration	Linear elements Full integration	Quadratic elements Reduced integration	Quadratic elements Full integration
Multi-linear hardening	0.9691	0.9794	0.9964	0.9964
Johnson-Cook	0.9486	0.9397	0.8476	0.8474
Ramberg-Osgood	0.9586	0.9674	0.9969	0.9969

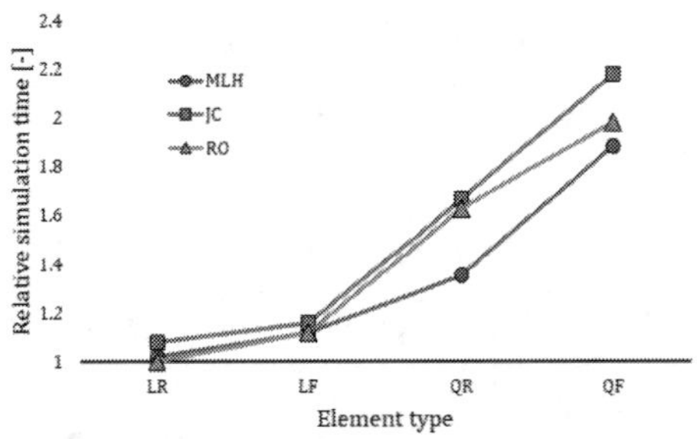

a)

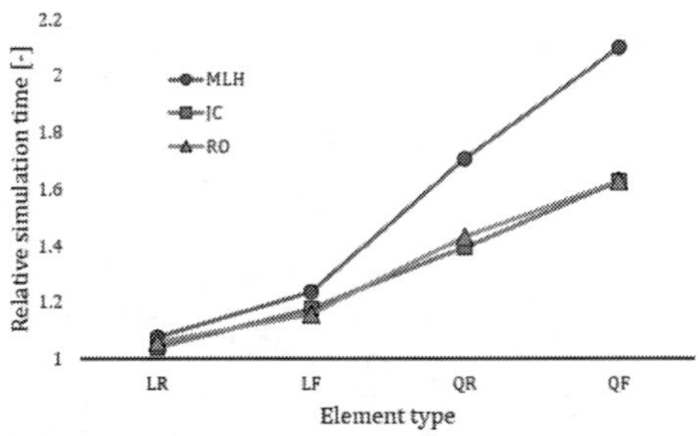

b)

Figure 13. Comparison between the relative simulation times of the models for tensile (a) and torsional (b) loadings

6. Discussion and conclusions:

Considering the simulation results, it can be concluded that the multi-linear hardening model produces the most accurate results for the tensile loading analyses, while the Ramberg-Osgood model is most suited for torsional loadings, by a very small margin. As anticipated due to the rather low coefficient of determination from the analytical model, the Johnson-Cook formulation produced the least accurate results for both tensile and torsional loadings. In addition, the multi-linear hardening model and the Ramberg-Osgood model showed to be the least sensitive to the element formulations. In terms of simulation times, the multi-linear hardening model produces the shortest simulations for tensile loadings while the Ramberg-Osgood model was optimal in torsion.

In conclusion, given the high degree of accuracy for both tension and torsion, relatively short simulation times and the storage of plastic deformations, the optimal material model for the simulation of the mechanical response of AZ91 magnesium alloy is the multi-linear hardening model.

7. References:

[1] H. E. Friedrich and B. L. Mordike, "Magnesium Technology. Metallurgy, Design Data, Applications," 2006.

[2] M. M. Avedesian and Hugh Baker, *ASM Specialty Handbook: Magnesium and Magnesium Alloys*. Almere,Netherlands: ASM International, 1999.

[3] J. Willekens, Proc. Intl. Conf. D.Magers, "Magnesium and their Applications".

[4] IL La Grange, "Product Design for Die Casting in Recyclable Aluminium, Magnesium, Zinc and ZAAlloys,Die Casting, Development Council," 1996.

[5] Data Birdge Market Research. Europe Magnesium Alloys Market - Industry Trends and Forecast to 2029. [Online]. https://www.databridgemarketresearch.com/reports/europe-magnesium-alloys- market

[6] T.Ruden D.L. Albright, "Magnesium Utilization in North American Automotive Industry," *Light Metal*,1994.

[7] C.S. Roberts, "Magnesium and its Alloys," *Wiley*, 1960.

[8] H.Baker M.M. Avedesian, "ASM Speciality Handbook,Magnesium and magnesium alloys,ASMInternation,Ohio,USA," 1999.

[9] European Standard, EN 1753:2019 - Magnesium and magnesium alloys - Magnesium alloy ingots andcastings, 2019.

[10] P.S. Frederick, "Die Casting Engineer," 1996.

[11] G.V. Raynor, "The Physical Metallurgy of Magnesium and its Alloys, London,NY,Paris,LA," *PergamonPress*, 1959.

[12] International Organization for Standardization, ISO 6892-1 Metallic materials - Tensile testing - Part 1:Method of test at room temperature, 2019.

[13] F. Dunne and N. Petrinic, *Introduction to Computational Plasticity*.: Oxford University Press, 2006.

[14] D.W.A. Rees, *Basic Engineering Plasticity*.: Elsevier, 2006.

[15] J. Bonet and R.D. Wood, *Nonlinear Continuum Mechanics for Finite Element Analysis*.: CambridgeUniversity Press, 1997.

[16] A. A. Shabana, *Computational continuum mechanics*.: Cambridge University Press, 2008.

[17] D.A. Şerban, L. Marşavina, L. Rusu, and R. Negru, "Numerical study of the behavior of magnesium alloyAM50 in tensile and torsional loadings," *Archive of Applied Mechanics*, vol. 89, no. 5, pp. 911-917, 2019.

[18] Abaqus, "User's Manual," vol. Analysis, 2017.

[19] G. Johnson and W. Cook, "A constitutive model and data for metals subjected to large strains, high strainrates and high temperatures," in *Proceedings of the Seventh International Symposium on Ballistics*, Hague, The Netherlands, 1983.

[20] W. Ramberg and W.R. Osgood, "Description of stress–strain curves by three parameters," in *NationalAdvisory Committee For Aeronautics.* Washington, D.C., 1943.

Influence of dealloying on wetting behavior and adhesion properties of Cu-Zr-Al-Ag glassy ribbons

Petru HIDIDIS[1, a], Mircea NICOLAESCU[2, b], Iosif HULKA[1, c], Viorel-Aurel ŞERBAN[1,3, d]

1 "Politehnica" University of Timisoara, P-ta Victoriei, No. 2, Timisoara

2 National Institute of Research and Development for Electrochemistry and Condensed Matter, Str. Paunescu Podeanu, Nr. 144, Timişoara, Romania

3 Technical Sciences Academy of Romania, Bd. Dacia nr. 26, sector 1, Bucuresti, Romania

a petru.hididis@student.upt.ro, b nicolaescu.mircea13@yahoo.com, c iosif.hulka@upt.ro, d viorel.serban@upt.ro

Keywords: Glassy alloy, amorphous structure, dealloying, adhesion

Abstract

This research aims to clarify the correlation between dealloying conditions, surface chemistry, and wettability behavior in Cu-based metallic glassy ribbons. Investigating the influence of dealloying on the wettability of Cu-based glassy ribbons is the focus of this study. Glassy ribbons composed of the Cu-Zr-Al(-Ag) alloy family were dealloyed in hydrofluoric acid (HF) solution, with molarities ranging from 0.01 M to 0.5 M for durations between 2 and 10 minutes, followed by a thorough cleaning in distilled water. The impact of dealloying parameters on water droplet adhesion was evaluated together with the droplets' ability to resist detachment after mechanical disturbances. Such insights are pivotal for the development of novel superhydrophobic or superhydrophilic materials, which could find applications in diverse fields such as microfluidics and oil-water separation technologies. The research demonstrates the substantial effects of dealloying on Cu-Zr-Al(-Ag) glassy ribbons, showing significant alterations in surface morphology, elemental composition, and wettability, all influenced by dealloying conditions

Introduction

Dealloying is a process that selectively dissolves specific components from a material, thereby altering its composition and resulting in a modified structure. By removing certain elements, dealloying can create porous networks with unique properties. Cu-based alloys often demonstrate favorable dealloying behavior, wherein specific alloying elements can be selectively dissolved under suitable conditions, resulting in a porous structure with modified composition and morphology [1]. This method is regarded as advantageous for porous structure fabrication due to several factors. Firstly, dealloying enables precise control over the composition and morphology of the resulting porous network by adjusting parameters such as alloy composition, dealloying time, and temperature. Secondly, the process typically occurs under mild conditions, minimizing the risk of structural damage or degradation. Additionally, dealloyed structures possess high mechanical stability and structural integrity.

In the last decade a major increase in research papers on the topic of oil and water separation was observed [2]. An effective and cost-efficient method for separating water from oil relies on superwetting [3] surfaces with extraordinary wetting properties, achieved using specific porous materials that selectively attract and retain one fluid phase while repelling the other. Researchers have explored strategies involving chemical modifications or the creation of porous structures to engineer materials with good wetting properties. By leveraging these advancements, novel materials capable of selectively filtering or absorbing oil-water mixtures can be developed [4], offering potential solutions for addressing environmental challenges. A widespread common interest was on developing such

IOP Conf. Series: Materials Science and Engineering **1319** (2024) 012015 doi:10.1088/1757-899X/1319/1/012015

materials based on copper [5], magnesium alloy [6], steel [7] and others. One method that allows creating structures used for separation of oil and water mixtures is dealloying or electrochemical etching which is a combination of photolitography and micro electrochemical machining [8] and also plasma etching that uses an element as a plasma source and flushes it on the desired surface to be etched [9]. With these methods, specific components can yield nanostructures that show good mechanical stability [10].

An exciting potential application of these findings is the development of mesh filters composed of dealloyed ribbons, which could be strategically placed to separate oil from water by selectively filtering the unwanted fluid phase. This application leverages the enhanced wettability properties and structural characteristics induced by the dealloying process, demonstrating the practical utility of the research in addressing real-world challenges in oil-water separation. The primary objective of this study was to investigate the effects of the dealloying process on Cu-Zr-Al(-Ag) based glassy ribbons with the objective of generating porous structures and assessing water adhesion upon immersion in HF. The selection of HF as the dealloying medium for Cu-based glassy ribbons in this research was influenced by prior studies that demonstrated its effectiveness in selectively dissolving alloying elements [1, 11].

Experimental procedure

A first stage in obtaining the glassy (amorphous structure) ribbons, which will later be subjected to the dealloying process, is the production of the "master" alloys. Master alloys, with the chemical compositions $Cu_{48}Zr_{47}Al_5$, $Cu_{45}Zr_{45}Al_5Ag_5$, $Cu_{40}Zr_{45}Al_5Ag_{10}$, ingots weighting 20g with a diameter of 140mm were synthesized using the vacuum arc melting technique (ArcMelter 500 equipment) of the mixture of pure elements Cu, Zr, Al respectively Cu, Zr, Al, Ag in a controlled atmosphere to prevent oxidation. The choice of these materials is based on their exceptional glass-forming ability (GFA), which is further enhanced by the addition of Ag. Knowing that Cu and Ag share similar electronic structure and are part of the same periodic group, thus these elements have the same number of valence electrons. In order to increase the mechanical strength and hardness Zr was added to the composition. By including Al the viscosity of the melt is controlled [12].

The Melt Spinning method was used to obtain the alloys in the form of continuous ribbons that are 20μm thick and 1.5mm wide (Figure 1). The master alloy ingots were remelted by induction in a quartz crucible (Ilmasil PN) placed in an inductor, part of the melt spinning technique equipment is visible in figure 2. The melt was ejected continuously on a rotating wheel of copper which ensures a very high cooling rate 10^5-10^6 °C/s. This prevents the formation of crystalline structures during solidification, thus preserving the amorphous structure of the melt .

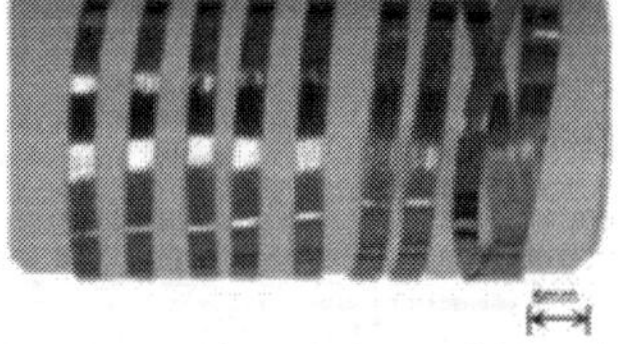

Fig 1: Examples of glassy ribbons [12]

Fig 2: Melt spinning technique [12]

After producing the glassy ribbons, the next step is subjecting them to the dealloying process. This was carried out by immersing the ribbons in HF solution with varying concentrations for different durations of time. It is worth noting that prolonged soaking times in the HF solution may pose a risk to the structural integrity of the glassy ribbons. Extended exposure to the acidic environment could lead to excessive dissolution of the alloy constituents, potentially resulting in undesired alterations to the ribbon's morphology or mechanical properties. Therefore, careful consideration was given to optimizing the dealloying parameters to strike a balance between achieving the desired surface modification and preserving the integrity of the ribbons. The dealloying process parameters are presented in table 1 below.

Table 1. Dealloying process

No.	Sample	HF concentration (M)							Soaking time (min)		
1	$Cu_{48}Zr_{47}Al_5$	0.5	0.4	0.3	0.2	0.1	0.05	0.01	2	5	10
2	$Cu_{45}Zr_{45}Al_5Ag_5$										
3	$Cu_{40}Zr_{45}Al_5Ag_{10}$										

A plastic container was used to hold the HF solution due to its chemical compatibility with HF. Unlike many metals, which can corrode or react with HF, plastic containers offer resistance to the corrosive effects of this highly acidic solution. Also, it ensures safety, prevents contamination of the solution, and facilitates efficient handling of the HF solution during the dealloying process.

Gas bubbles form on the ribbon's surface during soaking in HF due to the chemical reactions occurring between the HF solution and the alloy constituents of the ribbons. HF is known to react with certain metals, particularly metals like aluminum and zirconium present in the alloy composition, resulting in the release of hydrogen gas (H_2). This reaction can be represented by the following equation:

$$2Al + 6HF \rightarrow 2AlF_3 + 3H_2 \quad (1)$$

Results and discussion

After the completion of the dealloying process in HF, the glassy ribbons were carefully removed from the HF solution using tweezers, ensuring minimal agitation to prevent damage to the ribbons. Subsequently, the ribbons were gently rinsed in distilled water to remove any residual HF and other contaminants. Following the rinsing step, the ribbons were carefully dried to eliminate excess moisture. Once dried, the ribbons were placed on a support structure, with one end secured and the other end exposed. This positioning facilitated the subsequent testing and characterization of the ribbons' surface properties, particularly their interaction with water droplets.

Using a pipette, a water droplet was placed underneath the ribbon's exposed end. Once the water droplet was positioned, the timer was initiated to monitor the duration of the experiment for each sample. At predetermined intervals, typically every 5 minutes, nudges were applied to the support structure holding the ribbon. These nudges were intended to simulate mechanical disturbances and evaluate the adhesive properties of the water droplet to the ribbon's surface. Throughout the testing period, observations were made to determine whether the water droplet remained attached to the ribbon despite the applied nudges.

In this research paper, a selective presentation of results is employed to focus on specific experimental conditions that are deemed most relevant or representative of the study's objectives. The aim is to provide a comprehensive understanding of the effects of dealloying parameters on the wettability behavior of the glassy ribbons. This targeted approach allows for a detailed analysis of the trends and patterns observed under these conditions.

The samples are designated for easier reference and clarity throughout the study, by assigning shorthand names to each composition Al5 for $Cu_{48}Zr_{47}Al_5$, Ag5 for $Cu_{45}Zr_{45}Al_5Ag_5$ and Ag10 for $Cu_{40}Zr_{45}Al_5Ag_{10}$. Before dealloying, the wettability of the original glassy ribbons showed negligible effect, particularly evident as water droplets dropped immediately upon nudges, indicating poor adhesion and stability on the surface. In figure 3, the samples are shown placed with the water droplet underneath at 0.5, 0.4, 0.3, 0.2, 0.1 HF concentration after a soaking of 2 min. It can be observed that droplet bends the ribbons differently at different molar composition of HF. At higher concentrations of HF, the droplet induces a more pronounced bending of the ribbon, characterized by a higher angle of deflection. Conversely, at lower HF concentrations, despite the longer length of the ribbon, the extent of bending induced by the droplet is comparatively reduced. It also varies the placement of the droplet on the ribbon, e.g. it either goes to the exposed end of it or it remains attached to the middle of the ribbon or two thirds along the ribbons' length.

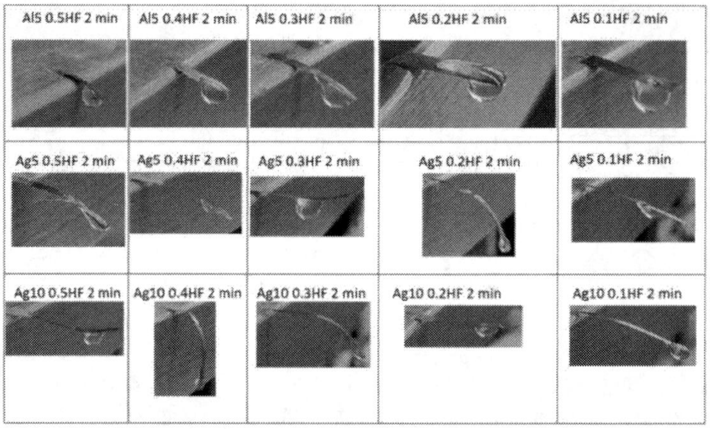

Fig. 3: Glassy ribbons after 2 min soaking in all HF concentrations

At higher concentrations of HF the ribbons tend to bend more due to the increased dissolution rate of the alloy constituents during the dealloying process. By immersing in a concentrated HF solution, the acidic environment accelerates the selective removal of specific alloying elements from the surface of the ribbons. This selective dissolution alters the surface composition and morphology of the ribbons, making them more susceptible to deformation when subjected to external forces, such as the weight of the water droplet. Higher concentration of HF results in a more aggressive etching action, leading to the creation of larger pores and surface irregularities on the ribbon's surface. The varying lengths of the ribbons can also affect how stable the water droplet remains when subjected to external disturbances. Longer ribbons, for example, might bend differently when nudged or exposed to vibrations, which can alter how well the droplet adheres to the surface over time. This variability in ribbon length adds complexity to the experimental conditions, potentially influencing the surface's morphology and composition due to uneven effects from the dealloying process along the length of the ribbon. These structural modifications contribute to a decrease in surface tension and an increase in surface area, which in turn enhances the adhesion of the water droplet to the ribbon. The increased adhesion force exerted by the droplet, combined with the altered surface morphology and composition induced by the concentrated HF solution, results in a greater bending of the ribbon.

There is a direct link in how the water droplet behaves on ribbons with shorter soaking time in 0.2 HF, for 2 and 5 minutes. There is a weaker adhesion to the ribbons' surface proven by the droplet detaching at mechanical disturbances. This weaker adhesion can be attributed to the limited surface modification achieved during the shorter exposure to HF, resulting in less roughness and modification of surface morphology. But in the case of soaking 10 minutes in 0.2 HF, pores form and the surface become rougher, thus the water droplet exhibits stronger adhesion to the ribbon's surface, making it more resistant to mechanical disturbances. This enhanced adhesion is likely due to the presence of surface features generated by the longer exposure to HF, which provide more sites for interaction between the water droplet and the ribbon's surface, thereby increasing the adhesive forces.

To evaluate how the ribbons surface morphology changed after the dealloying process, a scanning electron microscope (Quanta FEG 250 FEI) was used. It also allowed the observation of structural changes and to evaluate if pores did form. Upon magnification at 80,000 times, detailed examination of the Al5 sample reveals the emergence of pores scattered across its surface, as it can be seen in figure 4a. These pores exhibit a uniform distribution, extending throughout the entirety of the ribbon's surface. The presence of these small pores suggests the occurrence of localized dissolution or etching during the dealloying process, resulting in the formation of nanoscale features on the ribbon's surface.

IOP Conf. Series: Materials Science and Engineering **1319** (2024) 012015 doi:10.1088/1757-899X/1319/1/012015

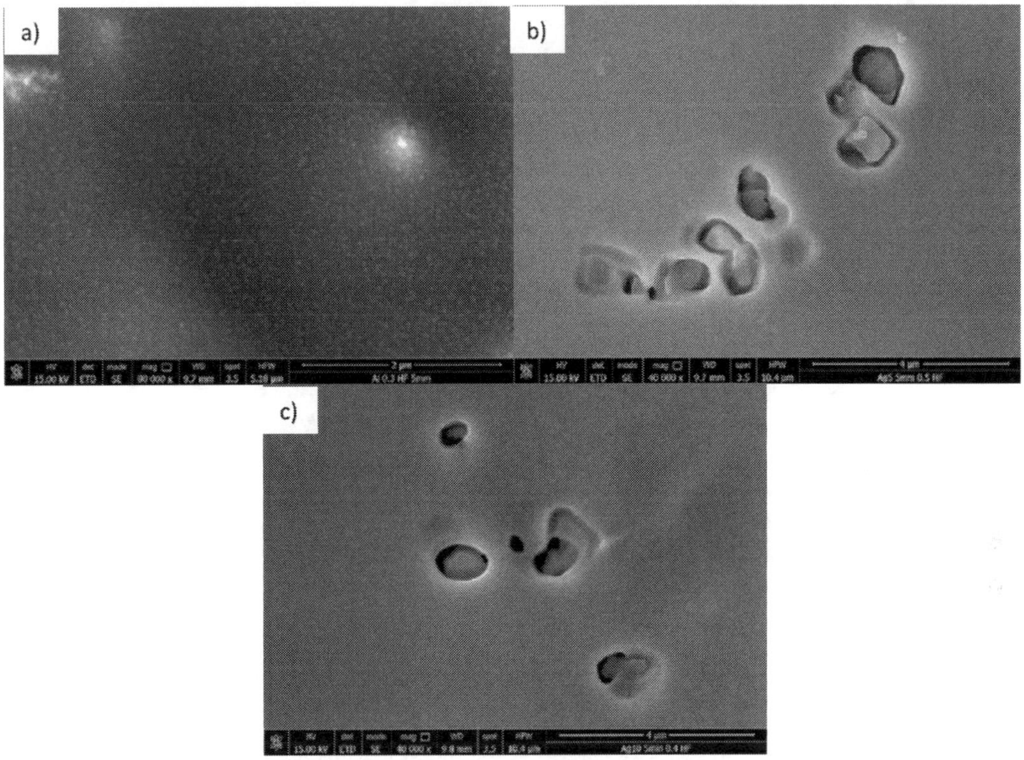

Fig. 4: Al5 glassy ribbons SEM analysis (a), Ag5 glassy ribbons SEM analysis (b), Ag10 glassy ribbons SEM analysis (c)

The Ag5 (figure 4b) and Ag10 (figure 4c) samples, exhibit the presence of pores distributed across their surfaces. However, what stands out is the observation that these pores are not uniformly dispersed; instead, they tend to cluster together in specific localized regions. This clustering phenomenon gives rise to distinct areas of increased porosity, where pores are densely packed near one another. This clustering phenomenon suggests a non-uniform distribution of pore formation, potentially influenced by variations in the alloy composition or the dealloying process parameters.

EDX analysis was integral to characterizing the elemental composition of the glassy ribbons before and after dealloying due to its ability to provide precise and quantitative information about the distribution of elements.

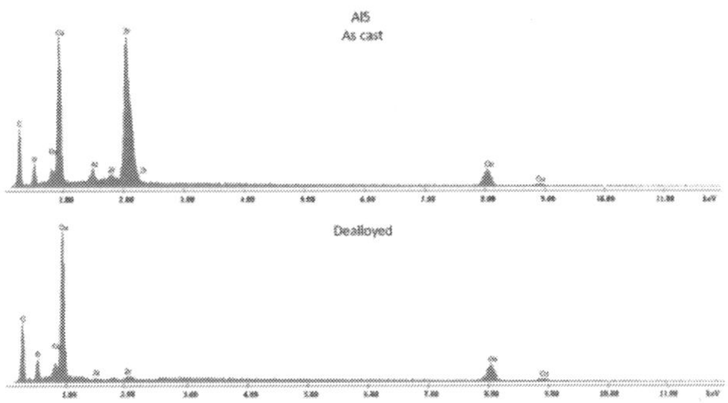

Fig. 5: EDX comparison of Al5 ribbon as cast and after dealloy process

In the EDX analysis of the Al5 sample, noticeable reductions in the peak intensities corresponding to Al and Zr were observed compared to the initial composition. As observed in figure 5 where the first figure show EDX for the as cast ribbons and bellow figure show EDX of the ribbon after dealloying process. This decrease in peak intensity suggests a preferential dissolution of Al and Zr from the surface of the glassy ribbons during the dealloying process. The selective removal of Al and Zr indicates the effectiveness of the dealloying treatment in altering the elemental composition of the ribbons. The observed reduction in Al and Zr peaks aligns with the formation of pores and surface roughness observed in SEM images, indicating a correlation between elemental composition changes and microstructural evolution induced by dealloying.

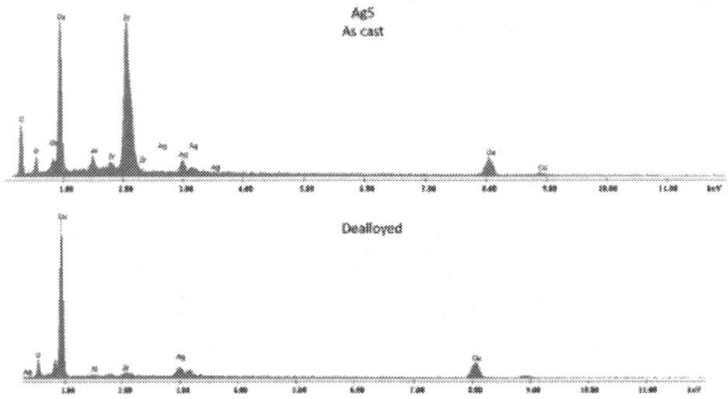

Fig. 6: EDX comparison of Ag5 ribbon as cast and after dealloy process

Regarding the Ag5 sample, in addition to the reduction in Al and Zr peaks, slight decreases in the peaks corresponding to silver were also noted. This reduction in Ag peaks suggests that, despite its lower concentration in the alloy composition compared to aluminum and zirconium, silver was also subject to dissolution to some extent during dealloying. The observed decrease in Ag peaks alongside Al and Zr peaks underscores the complex interplay between the alloying elements and the dealloying process, potentially influenced by factors such as chemical reactivity and dissolution kinetics.

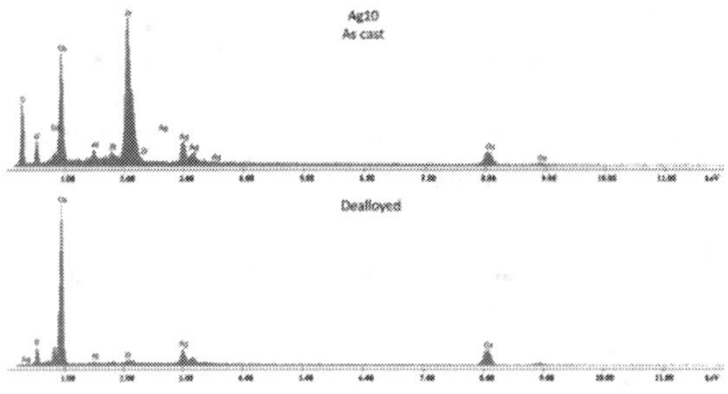

Fig. 7: EDX comparison of Ag10 ribbon as cast and after dealloy process

In the EDX analysis of the Ag10 sample, similar reductions in the peak intensities of aluminum and zirconium were observed, indicating their preferential dissolution during the dealloying process, consistent with observations from the Ag5 and Al5 samples. However, a notable difference was observed in the silver peaks, where the decrease in intensity was less pronounced compared to the Ag5 sample. This suggests that, despite the higher silver content in the Ag10 alloy composition, silver exhibited relatively lower susceptibility to dissolution during dealloying.

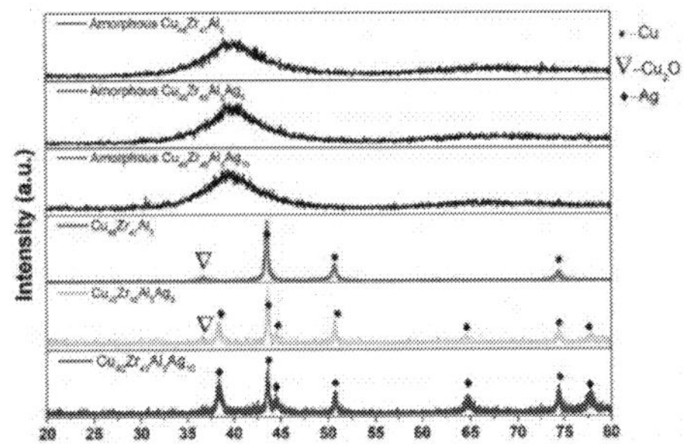

Fig. 8: XRD of as cast ribbons and dealloyed in 0.5 HF solution [11]

XRD analysis of the ribbons in their as-cast state reveals broad intensity peaks characteristic of an amorphous or glassy structure, as suggested in figure 8. However, upon dealloying in a 0.5 M HF solution, distinct diffraction peaks corresponding to metallic copper (111), (200) and (220) are present. The reaction between nanoporous copper and oxygen is prone to formation of crystalline phases of Cu_2O which is observed in Al5 and Ag5 samples. Cubic Ag peaks are observed in Ag5 and Ag10 samples. During the dealloying process, Al and Zr contribute to formation of hydrogen gas bubbles which is observable by direct visual inspection this also leads to an increase of the pH of the solution [11].

Conclusions

The research findings underscore the profound impact of the dealloying process on Cu-Zr-Al(-Ag) glassy ribbons, revealing significant alterations in surface morphology and elemental composition. SEM imaging and EDX analysis demonstrate selective dissolution of alloying elements, particularly Al and Zr, leading to the formation of pores and surface roughness. The degree of surface modification varies depending on the concentration and duration of the dealloying process, with higher concentrations and longer durations resulting in more pronounced changes. XRD analysis confirms the structural evolution of the ribbons, indicating a transformation from an amorphous to a partially crystalline structure with the emergence of metallic copper phases.

Wettability testing further reveals variations in water droplet adhesion behavior, with higher HF concentrations leading to increased adhesion and bending of the ribbons by water droplets. The presence of clustered pores on certain samples suggests a non-uniform distribution of pore formation, potentially influenced by alloy composition and dealloying parameters. Overall, these findings provide valuable insights into the intricate interplay between dealloying conditions, surface morphology, and wettability behavior, offering implications for the design and optimization of materials for diverse applications, including oil-water separation and surface engineering.

Furthermore, the insights gained from this study hold significant promise for surface engineering applications, where tailored surface morphologies and compositions are crucial. By optimizing dealloying conditions to achieve desired surface properties, these engineered surfaces could be utilized in various industries for anti-fouling, self-cleaning, and enhanced adhesion purposes, showcasing the broad potential of dealloyed Cu-Zr-Al(-Ag) glassy ribbons in advanced material design and application.

Refferences

1. Dan, Z., et al., *Fabrication of nanoporous copper by dealloying of amorphous Ti–Cu–Ag alloys.* Journal of Alloys and Compounds, 2014. **586**: p. S134-S138.
2. Gupta, R.K., et al., *Oil/water separation techniques: a review of recent progresses and future directions.* Journal of Materials Chemistry A, 2017. **5**(31): p. 16025-16058.
3. Zhou, Y., et al., *Fabrication of Superhydrophobic Porous Brass by Chemical Dealloying for Efficient Emulsion Separation.* Molecules, 2023. **28**(18): p. 6509.
4. Xue, Z., et al., *Special wettable materials for oil/water separation.* Journal of Materials Chemistry A, 2014. **2**(8): p. 2445-2460.
5. Yao, C.-W., et al., *Corrosion Resistance and Durability of Superhydrophobic Copper Surface in Corrosive NaCl Aqueous Solution.* Coatings, 2018. **8**(2): p. 70.
6. Qian, Z., et al., *Corrosion resistance and wetting properties of silica-based superhydrophobic coatings on AZ31B Mg alloy surfaces.* Applied Surface Science, 2018. **453**: p. 1-10.
7. Liu, H., et al., *Preparation of superhydrophobic coatings on zinc, silicon, and steel by a solution-immersion technique.* ACS Appl Mater Interfaces, 2009. **1**(9): p. 2086-91.
8. Chen, M., et al., *Development of superhydrophilic Al foil with micropore arrays via mask electrochemical machining and chemical immersion for efficient oil/water separation.* Journal of Dispersion Science and Technology, 2019. **41**: p. 1-11.
9. Chen, F., et al., *Atmospheric Pressure Plasma Functionalized Polymer Mesh: An Environmentally Friendly and Efficient Tool for Oil/Water Separation.* ACS Sustainable Chemistry & Engineering, 2016. **4**(12): p. 6828-6837.
10. Zhu, S., W. Deng, and Y. Su, *Recent advances in preparation of metallic superhydrophobic surface by chemical etching and its applications.* Chinese Journal of Chemical Engineering, 2023. **61**: p. 221-236.
11. Nicolaescu, M., et al., *Synthesis of nanoporous copper by dealloying CuZrAl and CuZrAlAg amorphous ribbons in acidic solution.* Materials Today: Proceedings, 2023. **72**: p. 565-571.
12. Hididis, P., et al., *Technological aspects regarding the production of Cu-Zr glassy ribbons. Difficulties and proposals.* Materials Today: Proceedings, 2021. **45**: p. 4371-4375.

IOP Conf. Series: Materials Science and Engineering **1319** (2024) 012016 doi:10.1088/1757-899X/1319/1/012016

Investigation of an Automotive Part Obtained by the WIG Process

Florin Chepetean [1], Anamaria Ioana Feier [1] and Banciu Felicia Veronica [1*]

[1] Materials and Manufacturing Engineering Department, Politehnica University Timisoara, Bd. Mihai Viteazu 1, 300222, Timişoara, Romania

*E-mail: felicia.banciu@upt.ro;

Abstract. The paper presents a case study from the automotive industry, namely a part made of Inconel 713C and 18/8 L stainless steel and further focuses on the study and analysis the applicability of the WIG process in the automotive industry. The WIG process chosen for the joining of the two components and all the steps from pre-assembly to the examination part will be presented. In a context where quality and performance requirements are critical, WIG welding is a key element in manufacturing processes.

1. Introduction

The automotive industry faces a pressing need for cleaner and more efficient engines. The rotor turbine is emerging as a promising solution, aiming to increase fuel combustion efficiency by capturing waste heat energy from exhaust gases. This captured energy is converted into mechanical energy, driving a compressor that increases air intake and optimizes fuel combustion. This powerful technology boasts a number of benefits, including a 15% increase in fuel combustion efficiency, reduced CO2 emissions, improved engine performance and lower fuel consumption.

The rotor turbine finds applications in various vehicle segments, including passenger cars, trucks, buses, agricultural machinery and even industrial equipment. The future looks bright for this technology, with projected growth in the automotive industry due to its significant contributions to both fuel efficiency and reduced emissions. While there are different types of rotor turbines, each with unique characteristics and performance levels, ongoing research is addressing challenges such as cost and system complexity, paving the way for wider adoption and a more sustainable future in the automotive industry.

Internal combustion engines, notorious for emitting CO2 and air pollutants, waste a significant amount of heat energy through exhaust. This low fuel combustion efficiency presents a pressing challenge that requires innovative solutions.

The rotor turbine, a technology poised to make a major impact. By intelligently capturing the heat energy wasted in the exhaust gases, it significantly improves engine energy efficiency and reduces CO2 emissions. This translates into several benefits: a 15% increase in fuel combustion efficiency, reduced CO2 emissions, improved engine performance and lower fuel consumption.

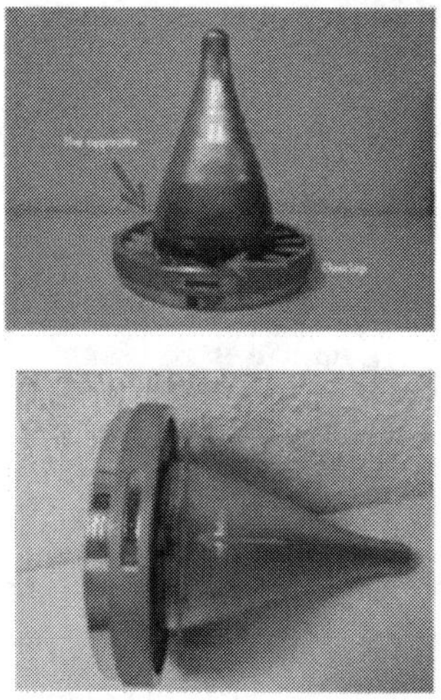

Figure 1. Automotive Part

The WIG welding process offers several significant advantages in the automotive industry. These benefits have contributed to the increased use of this process in automotive production and include:

- Mechanical strength: using the WIG process results in high quality welds this led to increased mechanical strength. These welds can withstand high stresses and vibrations, ensuring the reliability of automotive components.
- Corrosion resistance: thanks to the controlled atmosphere and the protection provided by the inert gas, the WIG process develops welds that are resistant to corrosion. This is a crucial feature for components exposed to extreme weather conditions or aggressive chemicals.
- Aesthetics: in the automotive industry, appearance matters. The WIG process can result in clean welds, without spatter or scratches, which contributes to the aesthetic appearance of automotive components. Also, this quality appearance helps the positive image of the brand and the product.
- Deformation control: Using the WIG process we can have better control of thermal deformation.
- This is important for automotive components with tight tolerances, where warpage can affect fitment and functionality.
- Reduced defects: Due to the high level of control and precision, using the WIG process minimizes defects such as pores or cracks, which leads to reduced repair costs and time lost.
- Varied applicability: The WIG welding process can be used for a wide range of materials including carbon steels, stainless steels, aluminum, copper and many others. This versatility makes the WIG welding process a suitable choice for various requirements in the automotive industry. These benefits highlight the significant contribution of the WIG welding process to the success and development of the automotive industry, highlighting the advantages it brings to the production and quality of automotive components.

2. Materials and welding parameters

The combination of Inconel 713C and 18/8 L stainless steel in the weld provides a robust solution for demanding applications. Inconel 713C is resistant to high temperatures (up to 982°C), corrosion, oxidation; it has durability in aggressive environments and excellent mechanical properties.

Stainless steel 18/8 L is corrosion resistant, has excellent formability, relatively low cost, and high biocompatibility.

Welding Inconel 713C with 18/8 L stainless steel provides a remarkable synergy of advanced materials with significant benefits in performance, versatility and cost.

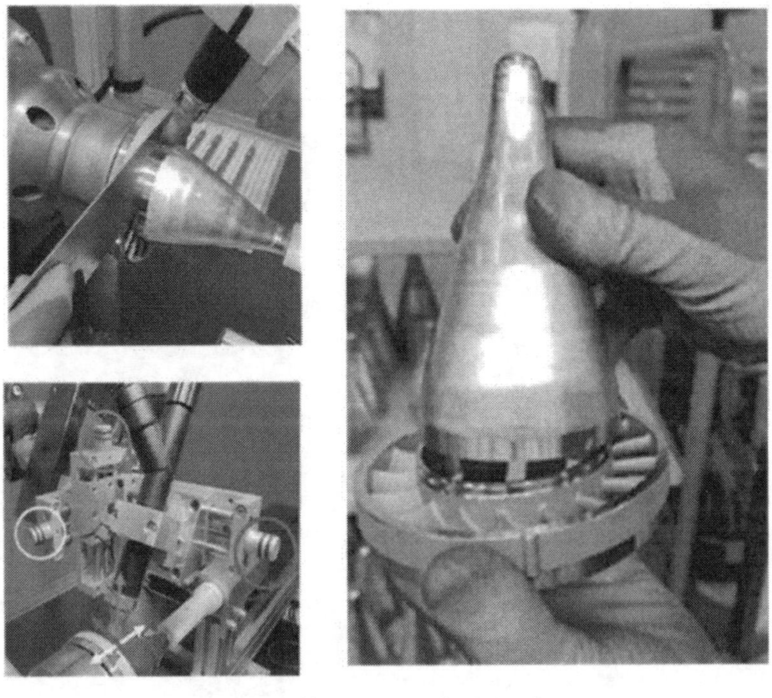

Figure 2. Welding steps in obtaining the part

Table 1. Welding parameters

Row	Welding process	AM Size	Current intensity A	Voltage V	Current type/ polarity	Welding speed (cm/min)	Thermal energy input (KJ/cm)
Tack	-	-	-	-	CC-	N. A	N. A
1	141	Ø1,2	30-35	10-11	CC-	0,65-0,75	2.40-3,55

Research: After cooling the parts, the welded joints are checked to see if they are defects and after they are tested for concentricity. The examination is carried out in the first stage visually, and if additional examinations are necessary, then the examination with penetrating liquids is also carried out. By random selection of parts from each batch, more detailed destructive examinations are carried out, i.e., metallographic examinations in the resulting welded joint area. In figures presented it can observe some defects resulting in welded joints, parts with defects as in figure presented are considered scrap. Most defects have the Wolfram electrode as the main factor.

3. Inspection

After cooling the parts, the welded joints are checked to see if they are defects and after they are tested for concentricity. The examination is carried out in the first stage visually, and if additional

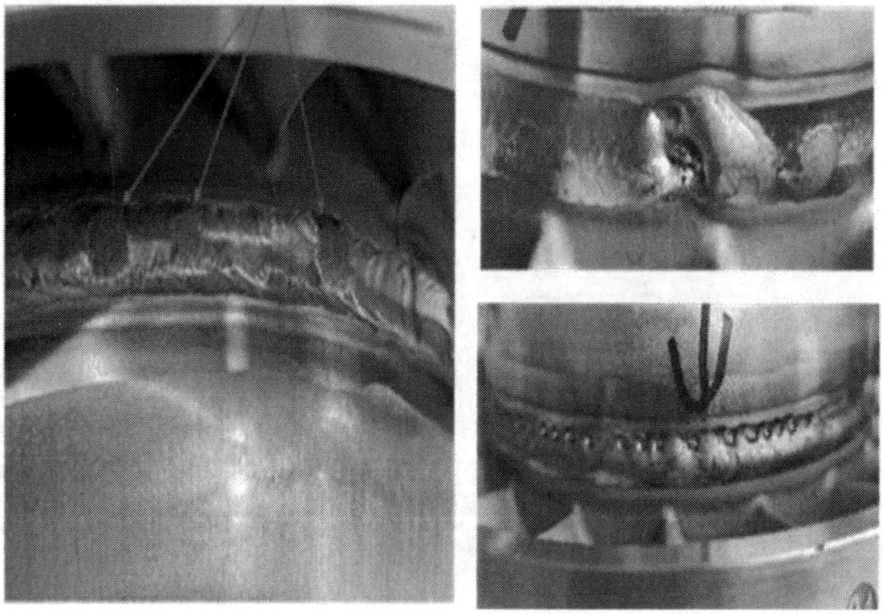

Figure 3. Different type of defects

examinations are necessary, then the examination with penetrating liquids is also carried out. By random selection of parts from each batch, more detailed destructive examinations are carried out, i. e. metallographic examinations in the resulting welded joint area.

In figure 3 are some defects resulting in welded joints, parts with defects as in figure 3 are considered scrap. Most defects have the Wolfram electrode as the main factor.

Metallographic examination: The samples are prepared, ground, chemically attacked. The pores are analyzed and compared with previous results.

Weld penetration: must be at least 0.3mm deeper than the edge of the part.

Cracks - No cracks are allowed in the contact area of the weld with the base material.

Cracks up to 0.3mm are allowed in the other areas.

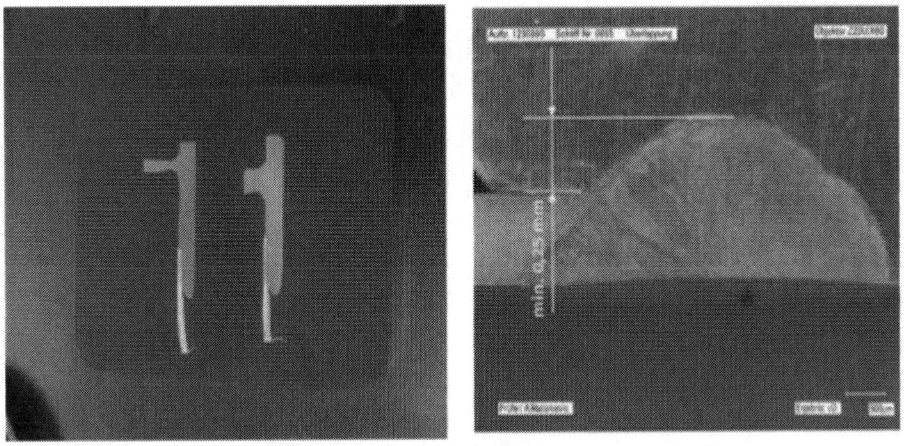

Figure 4. Lack of penetration.

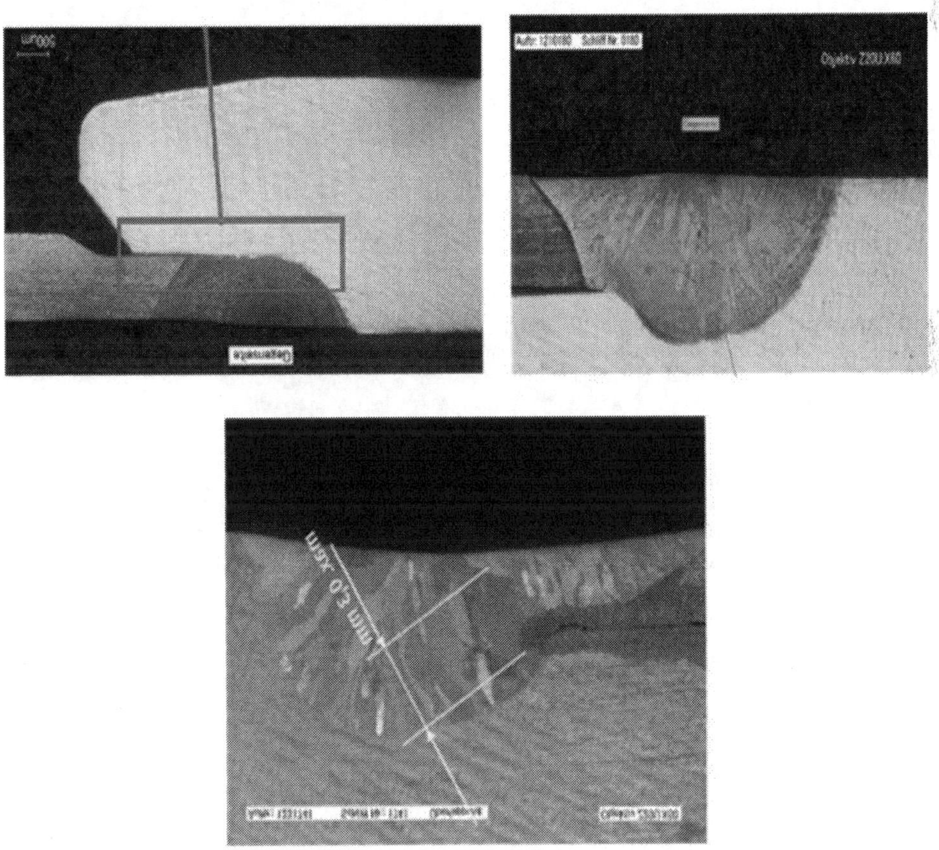

Figure 5. Weld depth: must be at least 0.3mm deeper than the edge of the part.

4. Conclusions

One of the key aspects we have highlighted is the superior quality of WIG welds. This technique stands out for the increased mechanical strength of the welds, corrosion resistance and the flawless aesthetic appearance of the car components. Meticulous control of the welding process also allows the reduction of deformations and defects, contributing to production efficiency.

In the contact area of the weld and the base material no joining errors are allowed.

In conclusion, WIG welding has become a key element in the production and manufacture of automotive components, providing significant benefits in terms of quality, reliability and aesthetics. With proper training and control, this welding technique remains a pillar in the automotive industry and an example of excellence in automotive manufacturing.

References

[1] Chepetean F, Studiu de caz în industria automotive Procedeu sudare Wolfram Inert Gas – proiect PSI, 2023
[2] G. M. Dumitru, C. Radu, B. Dumitru, „Recondiționarea şi repararea produselor", Editura Printech, Bucharest, 2010,111-115.
[3] P. Peças, E. Henriques, B. Pereira, M. Lino, M. Silva „Fostering the use of welding technology in the mould repair", 2006
[4] https://www.tracytec.com/products/welding/ttw-800-en
[5] Dr. Ing. NICOLAE JONI, Ing. TRAIAN CHEVEREȘAN, „Sudarea WIG", Editat de Societatea Comercială ROBCONTM SRLTimişoara
[6] Suport Curs de Inginer Sudor European/International, ASR, 2022
[7] Eneia Frici, Richard Molnar, Îmbunătățirea tehnologiei de sudare prin înlocuirea gazului de protecție, revista SUDURA nr 3/2020, ISSN 1453-0384, 2020
[8] János Dobránszky, Balázs Bebők, Balázs Varbai, Attila Szlancsik, Tibor Gerencsér, Árpád Németh, Modeling of welding distortions and flame straightening deformations, Budapest (Hungary) - 15th IMEKO TC10 Workshop on Technical Diagnostics Technical Diagnostics in Cyber-Physical Era, 2017
[9] https://www.twi-global.com/technical-knowledge/job-knowledge/welding-of-titanium-and- its-alloys-part-1-109
[10] https://www.arc-zone.com/Tungsten-Electrodes
[11] https://www.weldtrailingshields.com
[12] Cosmin Ion Belin, Ion Mitelea, Ilare Bordeasu, Ion-Dragos Utu, Corneliu Marius Crăciunescu, Ultrasonic cavitation erosion mechanism in TIG remelted surfaces of Nimonic 80 A alloy,Materials Today: Proceedings, 2021
[13] Mutașcu, D.; Karancsi, O.; Mitelea, I.; Crăciunescu, C.M.; Buzdugan, D.; Uțu, I.-D. Pulsed TIG Cladding of a Highly Carbon-, Chromium-, Molybdenum-, Niobium-, Tungsten- and Vanadium-Alloyed Flux-Cored Wire Electrode on Duplex Stainless Steel X2CrNiMoN 22-5-3. *Materials* **2023**, *16*, 4557. https://doi.org/10.3390/ma16134557
[14] Mitelea, I.; Bordeașu, I.; Cosma, D.; Uțu, I.-D.; Crăciunescu, C.M. Microstructure and Cavitation Damage Characteristics of GX40CrNiSi25-20 Cast Stainless Steel by TIG Surface Remelting. *Materials* **2023**, *16*, 1423. https://doi.org/10.3390/ma16041423
[15] *Samu, A., Feier, A., Optimization of WIG refurbishing procedures for automotive moulds, Sudura, nr. 3 (2023), year XXXIII, x-x*

Characterization of the properties of arc-welded aluminum matrix syntactic foam-filled tubes

Gábor Pados[1,2*], Alexandra Kemény[1,2] and Imre Norbert Orbulov[1,2]

[1] Department of Materials Science and Engineering, Faculty of Mechanical Engineering, Budapest University of Technology and Economics, Műegyetem rkp. 3., H-1111 Budapest, Hungary
[2] MTA-BME Lendület "Momentum" High-performance Composite Metal Foams Research Group, Műegyetem rkp. 3., H-1111 Budapest, Hungary

*E-mail: padosgabor@edu.bme.hu

Abstract. In this research, the authors aimed to manufacture and weld in-situ composite foam-filled tubes from expanded clay aggregate particles, AlSi12 aluminum matrix, and AlMgSi0.5 tubes using tungsten inert gas welding. Subsequently, the quality of the welded joints was examined after precipitation hardening heat treatment and without heat treatment, followed by the evaluation of the results. The welded joints were characterized by microstructural examinations, microhardness measurements, and hardness testing. It was found that the hardness of the tube can be increased with heat treatment after the welding by 23 % on average and that the heat input of the welding process causes silicon enrichment due to diffusion within the AlMgSi0.5 tube, which is accelerated by the heat treatment cycle.

1. Introduction

The proper selection of materials and bonding methods plays a critical role in safety and efficiency; therefore, continuous advancements are being made to create even more resistant and stronger structures while trying to minimize weight since reducing mass could offer advantages in many application areas. The latter-mentioned goal, reducing the density of the different structures, has been a prominent research area for nearly a century, and various materials have been foamed using different processes and technologies. The advantages of metal foams, such as low and adjustable density, high strength in specific directions, and distinctively good energy absorption capability, show promising future and potential for structures made from them; therefore, it is pivotal to find solutions for their quality manufacturing and bonding under industrial conditions, to be able to implement the promising potential of structures made from these materials [1,2].

Composite metal foams are a type of hybrid structure that offers numerous significant advantages, including high specific strength and low density, as well as exceptional mechanical energy absorption. Thermal and chemical stability are also important properties, which can play a distinguished role in many fields of engineering. [1–3].

These materials are usually characterized by their cellular structure. That could include closed [4] or open cells [5], depending on the cells' penetrability. Metal matrix syntactic foams (MMSFs) [2,6–8] are a subgroup of closed-cell foams, so there is no way for penetration among the cells since they are formed by hollow or porous filler material (ceramic or metal), where the size and shape have a crucial impact on the properties of the samples.

To an engineering issue, bare foam is usually not a decent solution; therefore, metal foams are usually placed or produced into hollow metallic sections or between dense sheets. These structures are more reliable in case of any type of load. The production of foam-filled tubes (FFTs) can be categorized as an ex-situ [9–11] or an in-situ [12–16] method, based on the mentioned fact that the foam has been placed into the hollow section after the production of the foam or the foam has been produced already in the closed section. The choice between the mentioned methods depends on what kind of application requirements, desired structure, and material characteristics are needed.

In a recent study, Rabiei et al. produced metal foams with a 316L matrix and stainless steel hollow spheres as filler material using a powder metallurgy route. The filler materials had a wall thickness of 0.1 mm and an outer diameter of ø2 mm or a wall thickness of 0.2 mm and an outer diameter of ø4 mm. Stir welding and induction welding with different parameters were used to join these composite metal foams. The first method was unsuccessful with any combination of parameters, while the latter proved to be feasible with long welding times and high-performance values [17].

Contrary to the mentioned welding processes, this current research aimed to implement and evaluate a type of fusion welding process and its feasibility, as there is no extensive history of using this process for cellular structures, especially on composite metal foams with ceramic fillers.

2. Experimental Section

2.1 Materials and methods

Aluminum matrix (AlSi12) MMSFs were infiltrated into AlMgSi0.5 alloy tubes with a wall thickness of 2 mm and an outer diameter of ø32 mm. The process was executed by low-pressure infiltration technique, similar to [18]. Lightweight expanded clay aggregate (LECA) spheres were used as filler material, which has been provided by a company named Liapor GmbH & Co. KG. The size of the filler material was ø2,5-3,0 mm, and their chemical composition has been given in weight percentage by the supplier as: SiO_2~60 %, Al_2O_3~17 %, Fe_2O_3~14 %, and ~9 % other oxides (CaO, MgO, Na_2O, K_2O). **Table 1** shows the standardized chemical composition of the applied aluminum alloys. The infiltration parameters have been detailed in a previous research paper by the authors [19].

Table 1. The standardized chemical composition of the used materials in wt. %(based on [20–22])

	Al	Si	Mg	Cu	Fe	Mn	Ti	Zn.	Other
AlSi12	Bal.	10.5 – 13.5	-	≤ 0.05	≤ 0.55	≤ 0.35	≤ 0.15	≤ 0.1	≤ 0.15
AlMgSi0.5	Bal.	0.3 – 0.5	0.35 – 0.6	≤ 0.1	0.1 – 0.3	≤ 0.1	≤ 0.1	≤ 0.15	≤ 0.15
AlSi5	Bal.	4.5 – 6.0	≤ 0.05	≤ 0.3	≤ 0.8	≤ 0.05	≤ 0.2	≤ 0.1	≤ 0.15

2.2 Welding of the FFTs

Before the welding, the FFTs underwent edge preparation to be able to access them with the welding torch. The edge-prepared pieces (**Figure 1**) were manually tungsten inert gas (TIG)-

welded by using a Syrius WIG 351 AC/DC PULSE equipment with alternating current. The welding was done in a flat position (PA) with prism support, with axial welding direction. First, a quasi-buttering layer was welded over the surface of the machined FFTs, and after that, filler layers were welded in the axial direction. The FFT was rotated within the prism after each welded bead. The circumstances are detailed in **Table 2**.

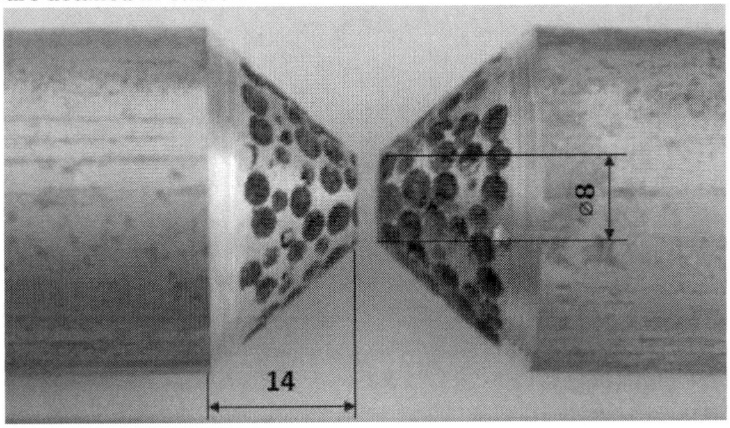

Figure 1. An image of the FFT which has been prepared and produced in one step.

Table 2. The circumstances of the TIG welding.

Parameter	Value
Shielding gas	99,996% pure argon, 10 l/min
Electrode	⌀1,6 mm grey (WC20 2% CeO2), rounded tip
Supply material	⌀2,4 mm 4043-AlSi5 alloy
Root gap	2 mm
Welding speed	~150-200 mm/min
Arc voltage and current	15-16 V and 80-90 A, AC (50 Hz)

2.3 Investigations

Two specimens were produced, but only one of them was treated with a precipitation hardening T6 heat treatment, used to increase the strength of certain aluminum alloys. The solution heat treatment step was carried out at 500°C for 2 hours, and the aging step at 200°C for 3 hours, according to previous results [23]. After that, the specimen was quenched in cold water and prepared for microstructural analysis. Both specimens were cut along the length axis of the FFTs using a precision cutting machine and then mounted in epoxy resin. The specimens were prepared by standard metallographic preparation, which involved grinding with continuous water rinsing up to P4000 emery SiC sandpaper. After the grinding the specimens were polished by diamond suspension with 3 μm and 1 μm grain sizes. Keller's reagent (95 ml H2O, 2.5 ml HNO3, 1.5 ml HCl, and 1 ml HF) was employed before microscopy examinations. Microstructural images were taken with an Olympus PMG-3 microscope. Hardness measurements were performed on the cross-sectional specimens, that had been prepared previously for metallography, using a Buehler IndentaMet 1101 micro-Vickers hardness testing machine with a 200 g load, while the indentation time was 11 seconds long.

3. Results and discussion

During the TIG welding, that was performed manually, some problems occurred that needed to be solved. The electrical conductivity of the filler particles was significantly worse than the used aluminum alloys; therefore, the welding arc was unstable, and the density of the clay aggregate particles was much lower than the density of the liquid melt pool; therefore, these particles began to move to the surface of the weld. For the arc movement and the moving of the LECA spheres the welding professional could manage and compensate effectively; therefore, a seemingly decent joint was produced. A sharp line between the matrix material of the metal foam and the wall of the tube can be clearly observed (**Figure 2a**) in the specimens, so the tube wall did not melt during manufacturing, and no cohesive bond formed between the two materials. In the case of the similarly manufactured, heat-treated, but not welded FFTs, dissolution was not observable; thereby, the assumption can be taken that the process of dissolution is induced by welding. Both of the evaluated specimens show silicon enrichment in the material of the tube (**Figure 2.**) but in the matter of the heat-treated specimen (**Figure 2b**), it is clearly visible that the matrix material has dissolved more efficiently into the material of the tube wall. This phenomenon may arise from the fact that deformations occur due to the heat load during welding, which could be the reason for the interruption of the connection between the two materials, but it is also a considerable possibility that the coefficient of thermal expansion in the metal foam is different from the material of the tube, so this can also be an explanation why the metal foam core separates from the wall of the tube.

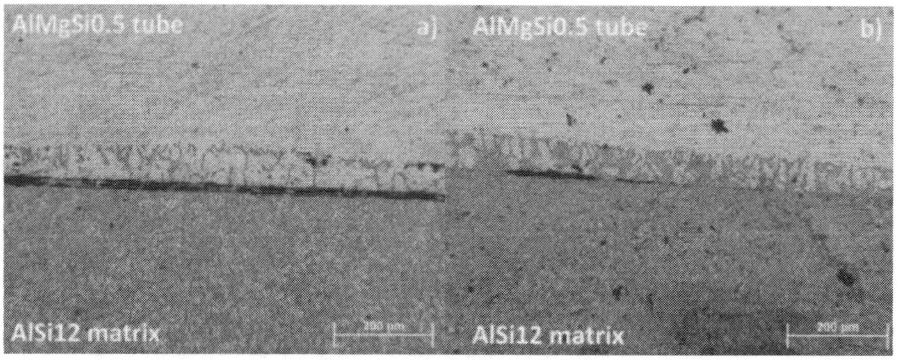

Figure 2. The border between the tube and the metal-foam core in the case of the specimen a) without heat treatment and b) with heat treatment.

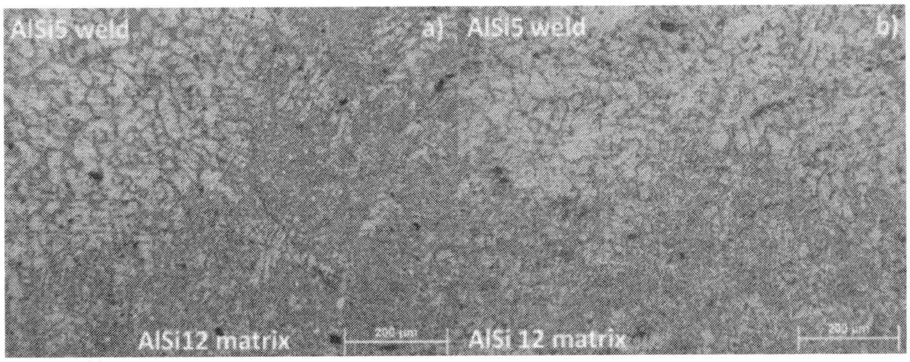

Figure 3. The observable silicon enrichments in the matrix and weld in case of a) the untreated and b) the heat-treated specimen.

Figure 3 shows the microstructure of the matrix and weld materials, which were both aluminum-silicon alloys; therefore, the silicon enrichments can be observed, which are presented in larger quantities in the matrix material than in the weld. No differences were observed between the differently treated specimens, based on the transition of the alloys, as the applied matrix and weld materials are not precipitation hardening alloys.

Microhardness measurements were conducted along the mid-line of the weld bead to assess the weld's homogeneity. The hardness values of the AlSi5 weld exhibited remarkably little variation for a weld, with an average of 49 ± 4 HV0.2. The average hardness of the AlMgSi0.5 tube was 57 ± 2 HV0.2, which could be further increased through precipitation-hardening heat treatment, as shown in **Figure 4**. After heat treatment, the hardness of the tube increased to 70 ± 2 HV0.2. The average hardness of the AlSi12 matrix was 66 ± 2 HV0.2, which is notably higher than the hardness achieved in the weld.

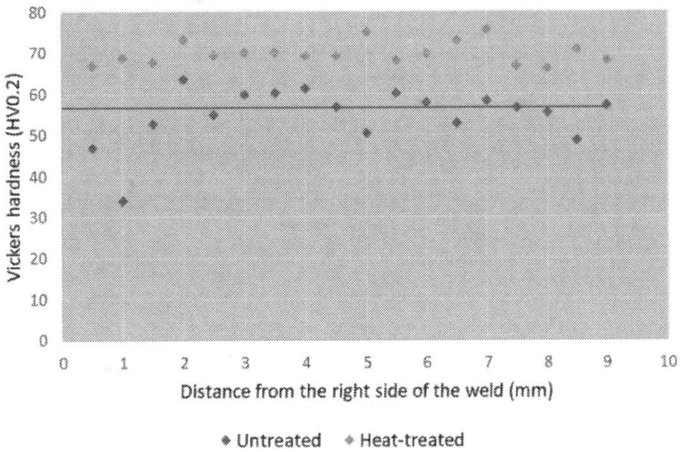

Figure 4. Hardness values in the midline of the tube wall, where the '0' distance corresponds to the right side of the weld.

4. Conclusion

In conclusion, TIG welding can be used to bond aluminum alloy FFT structures, even though the filler particles can cause involuntary arc movement. The filler particles could move in the melt pool since their density is lower than the melt's. In the microstructural images, silicon enrichments can be observed, which are presented in larger quantities in the matrix material than in the weld. The hardness of the weld is not sufficient, but the heat-treatment succeeded in increasing the hardness of the tube. The next step of this research is the mechanical testing of the specimens to consider the true potential of these structures.

Acknowledgement

This work was partially supported by the National Research, Development and Innovation Office (NKFIH), under grant agreement OTKA-FK_21 138505. The research reported in this paper is part of project no. TKP-6-6/PALY-2021, which has been implemented with the support provided by the Ministry of Culture and Innovation of Hungary from the National Research, Development and Innovation Fund, financed under the TKP2021-NVA funding scheme.

References

[1] Kemény A, Leveles B and Károly D 2021 Functional aluminium matrix syntactic foams filled with lightweight expanded clay aggregate particles *Mater. Today Proc.* **45** 4229–32
[2] Gupta N, Woldesenbet E and Mensah P 2004 Compression properties of syntactic foams: Effect of cenosphere radius ratio and specimen aspect ratio *Compos. Part A Appl. Sci. Manuf.* **35** 103–11
[3] Orbulov I N and Szlancsik A 2018 On the Mechanical Properties of Aluminum Matrix Syntactic Foams *Adv. Eng. Mater.* **20** 1–12
[4] Movahedi N and Linul E 2021 Radial crushing response of ex-situ foam-filled tubes at elevated temperatures *Compos. Struct.* **277** 114634
[5] Garai F 2020 Modern Applications of Aluminium Foams *Int. J. Eng. Manag. Sci.* **5** 14–21
[6] Bálint A and Szlancsik A 2015 Mechanical Properties of Iron Hollow Sphere Reinforced Metal Matrix Syntactic Foams *Mater. Sci. Forum* **812** 3–8
[7] Kubelka P, Kádár C and Jost N 2021 Effect of the interface on the compressive properties of magnesium syntactic foams *Mater. Lett.* **287** 129293
[8] Kádár C, Kubelka P and Szlancsik A 2023 On the compressive properties of aluminum and magnesium syntactic foams: Experiment and simulation *Mater. Today Commun.* **35** 106060
[9] Duarte I, Krstulović-Opara L and Vesenjak M 2015 Characterisation of aluminium alloy tubes filled with aluminium alloy integral-skin foam under axial compressive loads *Compos. Struct.* **121** 154–62
[10] Vesenjak M, Duarte I, Baumeister J, Göhler H, Krstulović-Opara L and Ren Z 2020 Bending performance evaluation of aluminium alloy tubes filled with different cellular metal cores *Compos. Struct.* **234** 111748
[11] Linul E, Movahedi N and Marsavina L 2017 The temperature effect on the axial quasi-static compressive behavior of ex-situ aluminum foam-filled tubes *Compos. Struct.* **180** 709–22
[12] Szovák B, Károly D, Kemény A, Kemény D M and Kádár C 2022 On the mechanical properties of functionally graded hybrid metal foams *Mater. Today Proc.* **78** 251–6
[13] Jenei P, Kádár C, Szabó Á, Hung S, Kuo C, Choe H and Gubicza J 2022 Mechanical behavior of freeze-cast Ti foams with varied porosity *Mater. Sci. Eng. A* **855** 143911
[14] Duarte I, Krstulović-Opara L and Vesenjak M 2018 Axial crush behaviour of the aluminium alloy in-situ foam filled tubes with very low wall thickness *Compos. Struct.* **192** 184–92
[15] Taherishargh M, Vesenjak M, Belova I V, Krstulović-Opara L, Murch G E and Fiedler T 2016 In situ manufacturing and mechanical properties of syntactic foam filled tubes *Mater. Des.* **99** 356–68
[16] Leveles B, Kemény A and Szijártó A 2021 Mechanical investigation of in-situ produced aluminium matrix syntactic foam-filled tubes *Mater. Today Proc.* **45** 4221–4
[17] Rabiei A, Cance J and Chacko Z 2024 A Study on Welding of Porous Metals and Metallic Foams *Adv. Eng. Mater.* **26** 2301430
[18] Pados G, Kemény A, Szovák B, Károly D and Orbulov I N 2023 Fabrication of In-situ Syntactic Aluminium Foam-Filled Steel Tubes *Acta Mater. Transylvanica* **6** 38–41
[19] Kemény A, Leveles B, Kincses D B and Károly D 2022 Manufacturing and Investigation of In Situ and Ex Situ Produced Aluminum Matrix Foam-Filled Tubes *Adv. Eng. Mater.* **24** 2100365
[20] EN 1706:2020+A1:2021 - Aluminium and aluminium alloys - Castings - Chemical composition and mechanical properties
[21] EN 573-3:2019+A1:2022 - Aluminium and aluminium alloys - Chemical composition and form of wrought products - Part 3: Chemical composition and form of products

[22] EN ISO 18273:2015 - Welding consumables - Wire electrodes, wires and rods for welding of aluminium and aluminium alloys - Classification
[23] Kemény A, Pados G, Károly D and Orbulov I N 2024 The Effects of Various Adhesives on Multiple Types of Quasi-Static Mechanical Properties of Rectangular and Square Aluminum Hollow Sections Filled with Composite Metal Foam Cores *Adv. Eng. Mater.* **26** 2302066

IOP Conf. Series: Materials Science and Engineering **1319** (2024) 012018 doi:10.1088/1757-899X/1319/1/012018

Effect of different parameters on mode I fracture toughness of resin samples manufactured by DLP

Marian Vasile Baban[1], Alexandru Viorel Coşa[2] and Emanoil Linul[1*]

[1]Department of Mechanics and Strength of Materials, Politehnica University Timisoara, 1 Mihai Viteazu Avenue, 300 222, Timisoara, Romania
[2]Department of Mechatronics, Politehnica University Timisoara, 1 Mihai Viteazu Avenue, 300 222, Timisoara, Romania

*E-mail: emanoil.linul@upt.ro

Abstract. Additive manufacturing is spreading rapidly in almost all industries, from household to advanced engineering. Components produced by Digital Light Processing (DLP) are not comprehensively characterized, but exceed the capabilities of many AM processes. Its advantages include the ability to produce highly complex designs, superior precision, fast printing and lower operating costs. The present paper investigates the mode I fracture toughness (K_{IC}) of UV-sensitive resin specimens obtained through DLP process. Single edge notch bending (SENB) specimen were 3D printed and tested according to the ASTM D5045 standard. The influence of printing orientation-PO (0°, 45° and 90°), resin color-RC (white, black and transparent) and corrosive environment-CE (air, water and saline solution) was studied. It was observed that all the studied parameters have some effect on the fracture properties. The highest K_{IC} values were obtained for 45°-PO, white-RC (1.88 MPa·$m^{0.5}$) and saline solution-CE (2.24 MPa·$m^{0.5}$). However, the greatest influence is highlighted by RC (~92%), while CE (~32%) shows a minimal effect. The fracture surface of the investigated samples is influenced by the printing parameters.

1. Introduction

Additive manufacturing (AM), also known as 3D printing, is a method for constructing physical objects by sequentially adding material in layers following a digital blueprint [1-3]. This approach diverges from conventional manufacturing techniques, which typically involve subtracting material from a solid block to shape the final product [4, 5].

AM provides distinct advantages over these traditional methods by significantly reducing the complexity of the supply chain, which can often be lengthy and costly, particularly in high-volume production settings [6-8]. Within the realm of AM, various specialized techniques each follow their own set of standards; Digital Light Processing (DLP) is one such technique [9, 10]. While parts created using DLP may not be comprehensively characterized, they generally exceed the performance of those produced by many Fused Filament Fabrication (FFF) 3D printers [11-13]. DLP excels in producing parts with complex geometries and fine details, which is especially beneficial for fields like 3D printed jewelry and prototyping. Key benefits of DLP include its capacity for crafting exceptionally detailed designs, achieving greater precision than FDM or Selective Laser Sintering (SLS), and offering faster production times compared to Stereolithography (SLA) [14]. Additionally, DLP's use of a shallower resin vat contributes to reduced operational costs and less material waste [15, 16].

There is a series of emphasis on fracture mechanics (mode I and mode II fracture toughness – K_{IC}, K_{IIC}), however, researchers refer to samples obtained through FFF or SLS processes. Stoia et al. [17] studied the effect of PO on the fracture behavior of Polyamide (PA) and Alumide sintered samples. They found that the 0-degree orientation gave the K_{IC} values. The authors also noted that K_{IC} of SLS printed PA samples has almost double values compared to mode I critical energy release rate [18]. They also reported that the Maximum Energy Release Rate and Equivalent Stress Intensity Factor criteria most accurately predict the experimental data [19]. Marsavina et al. [20] and Valean et al. [21] noticed that the notch insertion mode, growing direction, PO, printer type, layer thickness, specimen thickness and filament color significantly affect K_{IC} and K_{IIC} values of FFF-printed PLA samples. Thiede et al. [22] report that DLP typically shows better performance than Multi Jet Fusion (MJF) and SLS in terms of print resolution (detailed prints), increased surface quality and less anisotropy. Brighenti et al. [23] use DLP additive manufacturing technology and conclude that process parameters have an effect on the mechanical characteristics of 3D printed specimens. Fracture toughness is studied in detail by Rahmatabadi et al. [24] on PLA-TPU materials, highlighting particularities of FFF-printing technology.

In the present paper, the focus is solely on DLP process, which is utilized for the fabrication of the Ultraviolet (UV) specimens. The investigation aims to evaluate the fracture behavior of 3D printed specimens, which are crucial for applications requiring precise dimensional stability and fine-detail resolution under varying environmental conditions (e.g. aerospace, automotive, and biomedical devices). This study employs standardized testing methods to measure the K_{IC} of DLP-printed specimens fabricated under different printing conditions and exposed under different environmental conditions.

2. Materials and methods

2.1 Materials and specimens

For the present specimens, standard UV sensitive resin was used, in three different colors, namely White, Black and Clear. The main properties of the resin highlighted by the seller (ANYCUBIC Online Store 900246095 Guangdong, China) are presented in Table 1

Table 1. Main properties of the used UV resin.

Property	Value	Unit
Capacity	500	g/1Kg
Shelf life	1	year
UV wavelength	365-410	nm
Flexural strength	50-65	MPa
Notched impact Strength	25	J/m
Viscosity(25 °C)	150-200	mPa·s
Liquid density	1.05-1.25	g/cm^3
Hardness (Shore D)	82	D
Solid density	1.05-1.25	g/cm^3
Flexural modulus	1200-1600	MPa
Elongation	8-12	%
Tensile strength	36-45	MPa
Heart distortion temperature	65-70 (0.45 MPa)	°C

The geometry of the samples was according to the ASTM D5045 standard [25]. The geometric parameters of the Single edge notch bending (SENB) specimen are presented in Figure 1.

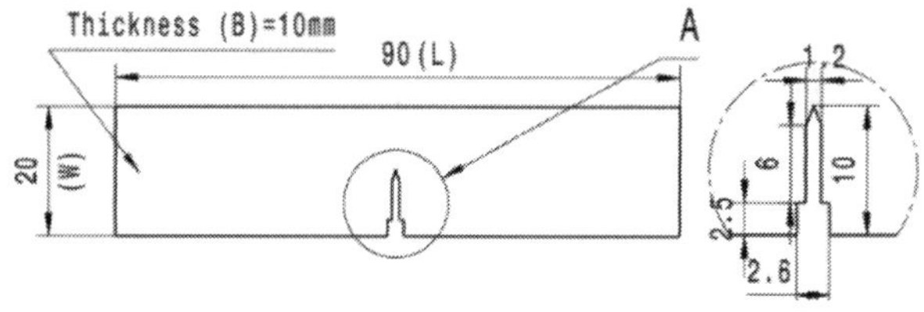

Figure 1. Specimens geometry, dimensions in mm

2.2 Manufacturing process

Printing of the specimens was realized with a Anycubic photon mono DLP printer (Figure 2a) and washing together with curing where done on a Anycubic wash and cure 2.0 machine. The Anycubic Photon Mono is a resin-based 3D printer that falls within the family of SLA machines, using a monochrome LCD as its light source to cure photopolymer resin. The Anycubic Wash and Cure 2.0 is a combined post-processing station designed for cleaning and curing 3D printed parts, especially those created using resin-based 3D printers such as the Anycubic Photon series (Figure 2b). This device streamlines the post-processing steps, making it an essential tool for anyone using resin printers to achieve optimal final results in their prints.

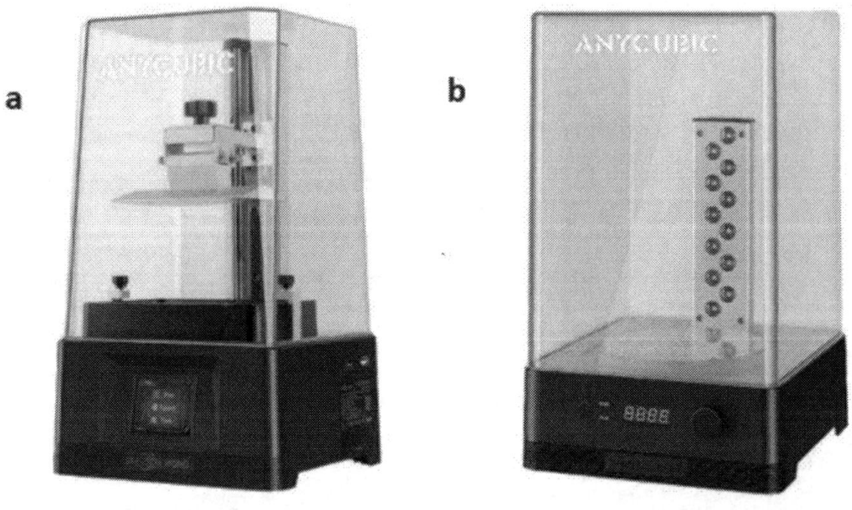

Figure 2. Anycubic photon mono (a) and Anycubic wash and cure 2.0 (b)

Besides the color of the resin (see samples from Figure 3), the orientation of the specimens (Figure 4) and the environment (air, water and saline water) were two other factors investigated.

IOP Conf. Series: Materials Science and Engineering **1319** (2024) 012018 doi:10.1088/1757-899X/1319/1/012018

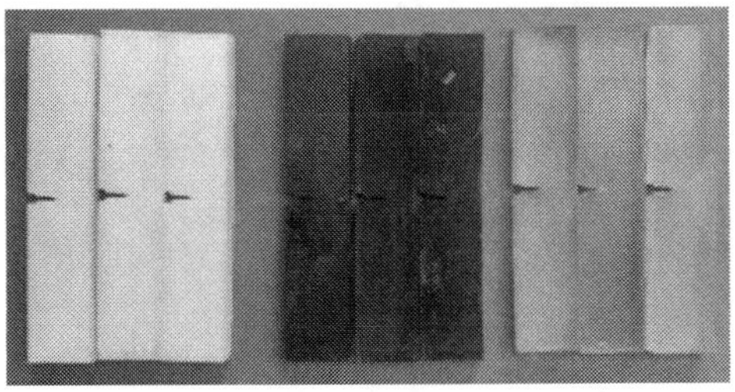

Figure 3. Specimens printed with different resin color

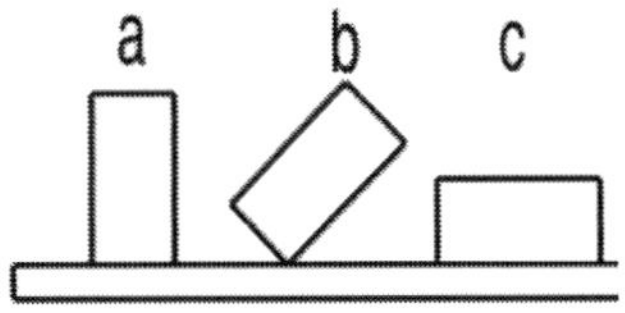

Figure 4. Specimens orientation on built platform

In the manufacturing process of the samples, a layer height of 0.05 mm was used, 6 bottom solid layers, 4 s normal layer exposure time and 40 s bottom layer exposure time. Subsequent, for each specimen, the post-process consisted of a 7-minute wash in Isopropyl alcohol and a 5-minute curing at room temperature (25 °C) using 405 nm UV light.

Regarding the corrosive environment, a group of specimens was submerged in a saline solution with a concentration of 6 grams of salt per liter of distilled water. This medium simulates a harsh environment, similar to seawater or bodily fluids, which can accelerate corrosion and material degradation. Also, another group of samples was exposed to plain distilled water to evaluate the baseline effects of moisture absorption and any potential hydrolytic degradation. For both groups, the corrosive medium was stagnant for 72 hours. This duration was chosen to provide a significant amount of time for the mediums to interact with the specimens. This period allows for observable changes in mechanical properties. For each variation parameter, 5 specimens were tested.

2.3 Experimental setup

Experimental tests were performed on Zwick Roell 5 kN universal testing machine. Figure 5 shows images of the sample in the machine's testing device. The tests were performed at room temperature, with a test speed of 2 mm/min. The distance between the lower rollers (supports) was 80 mm, and the diameter of one roller was 100 mm.

Figure 5. Experimental setup (a) and tested specimen (b)

3. Results and discussions

3.1 Effect of print orientation

Figure 6 shows the force-displacement curves and the K_{IC} variation for the three studied print orientations. All SENB samples were printed with white resin.

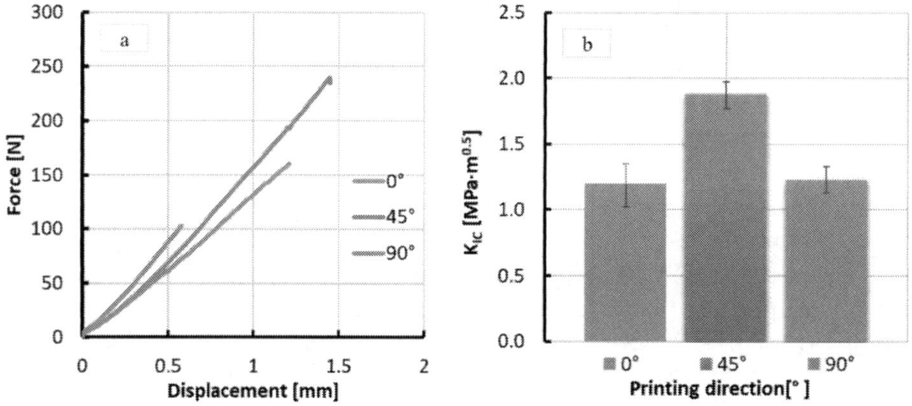

Figure 6. Effect of print orientation: force-displacement curves (a) and variation of mode I fracture toughness values (b).

The results of this test suggest that specimens printed at a 45° orientation exhibit the highest forces (Figure 6a), therefore and mode I fracture toughness values (Figure 6b). This orientation is followed by 90 and 0° orientations. Printing at a 45° orientation likely results in a more favorable alignment of the printed layers, enhancing interlayer adhesion. This alignment helps distribute stresses more evenly throughout the structure, leading to improved fracture toughness and overall maximum force. Specimens printed at a 90° orientation, where the layers are stacked vertically, can create points of weakness along the layer lines, potentially leading to incipient fractures under loading. Printing at a 0° orientation results in layers that are oriented horizontally. The 0°orientation (layers are oriented horizontally) is most susceptible to delamination and interlayer failure, especially under tensile or bending loads. The horizontal layering can create significant weak points where interlayer bonding is not as robust, leading to lower fracture toughness and reduced overall mechanical performance.

3.2 Effect of resin color

The effect of resin color on fracture properties of DLP printed samples is shown in Figure 9. The influence of resin color is highlighted for all three orientations (0°, 45° and 90°).

Test results show that the white resin enhances the highest forces at every printing orientation followed by black and transparent at 0° and 90° (Figures 7a and 7c), respectively by transparent and black at 45° (Figures 7b). Regarding K_{IC} values, white resin has the highest values on every printing orientation (Figure 7d-f). Black and transparent resin have almost the same K_{IC} values at 0° (Figure 7d) and 90° (Figure 7f), the black samples presenting slightly higher values than the transparent ones. Total internal reflection index influences directly the material's structural integrity and mechanical performance increasing or decreasing the exposure intensity in amount of time. Consequently, resin color plays a significant role in determining the final properties of the printed parts.

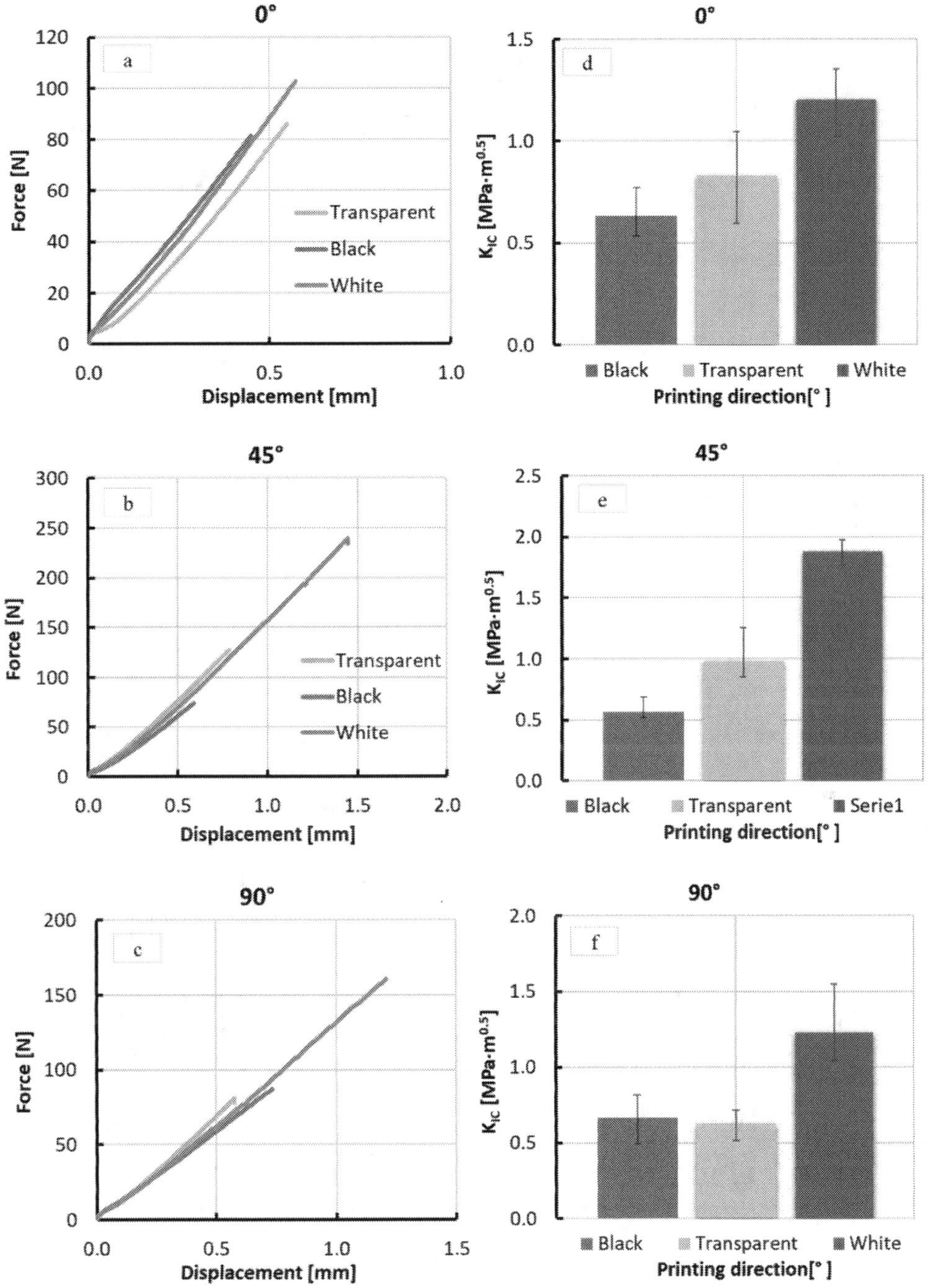

Figure 7. Effect of resin color: force-displacement curves (a-c) and variation of K_{IC} values (d-f).

3.3 Effect of environment

Figure 8 shows the influence of the environment (air, water and saline solution) on the load-displacement and K_{IC} values of the white resin 3D printed samples.

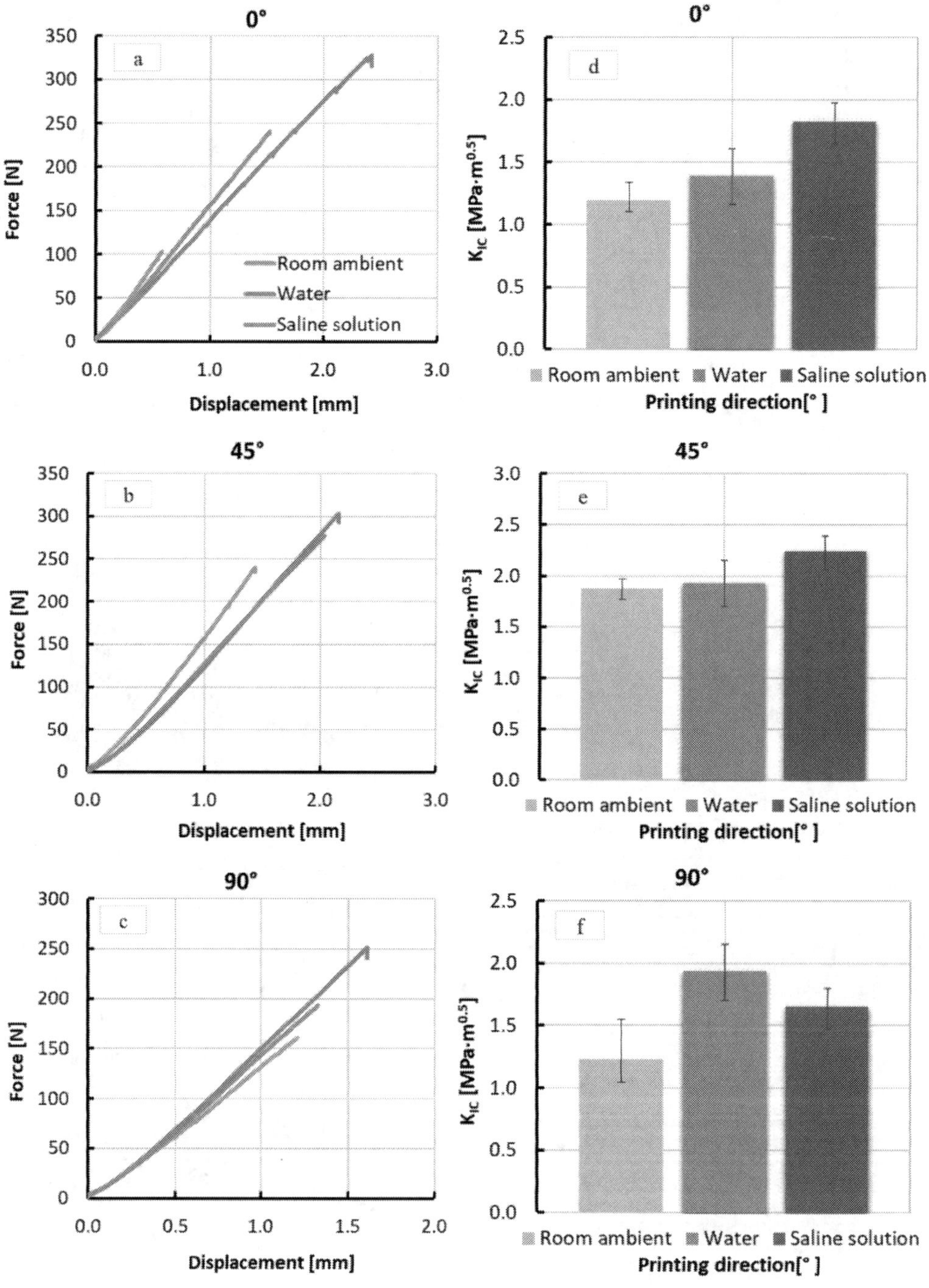

Figure 8. Effect of corrosive environment: force-displacement curves (a-c) and K_{IC} values (d-f).

Environmental exposure of the specimens is highlighting unexpected results. Corrosive medium has a positive influence on the K_{IC} values of the 3D printed specimens for all POs. These are followed by water immersed and room ambient exposed specimens. Highest K_{IC} value is shown by saline solution immersed specimen, printed at 45° (Figure 7e).

Figure 9 shows the fracture surfaces of the tested specimens for all three corrosive environments.

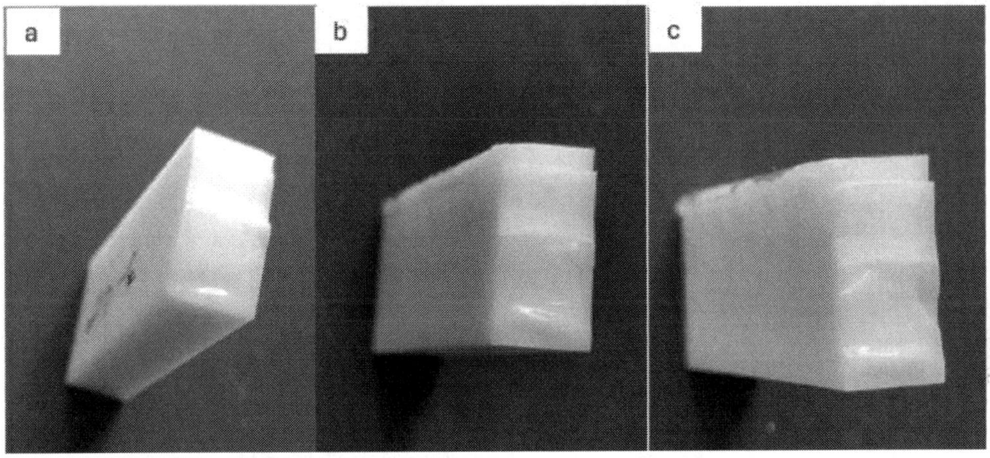

Figure 9. Fractured specimens for air ambient (a), water immerged (b) and saline solution (c)

As can be seen, differences in medium exposure are also evident on the fracture surfaces of the specimens. The more aggressive the environment, the more uneven the fracture surface.

4. Conclusions

The present paper investigates the impact of various parameters on mode I fracture toughness (K_{IC}) of Digital Light Processing (DLP) printed Ultraviolet-sensitive resin specimens. The investigated parameters are printing orientation (0°, 45° and 90°), resin color (white, black and transparent) and environment (air, water and saline solution).

- The orientation of the samples at 45 gives the highest K_{IC} value (1.88 MPa·m$^{0.5}$), 52.8% higher than the next one (90°). For the orientations of 0 and 90°, almost identical values are obtained (about 3% difference).
- Among the investigated colors, regardless of the printing orientation, the white-colored samples show the highest K_{IC} values: 1.20 MPa·m$^{0.5}$ (0°), 1.88 MPa·m$^{0.5}$ (45°) and 1.23 MPa·m$^{0.5}$ (90°). The white samples have K_{IC} values with 90.5% (0°), 91.8% (45°) and 83.6% (90°) higher than the next color.
- Notable effects on K_{IC} were also identified by the corrosive environment applied to the samples. The highest value of K_{IC} was found in the case of samples subjected to saline solution (1.83 MPa·m$^{0.5}$ for 0°, 2.24 MPa·m$^{0.5}$ for 45° and 1.65 MPa·m$^{0.5}$ for 90°), followed by those in water and finally in air.
- Depending on the studied parameter, the fracture surface of the samples exhibits a different characteristic.

IOP Conf. Series: Materials Science and Engineering **1319** (2024) 012018 doi:10.1088/1757-899X/1319/1/012018

References

[1] D.A. Şerban, A.V. Coşa, G. Belgiu, R. Negru, 2022, Polymers, 14(18), 3822.
[2] C. Vălean, D.I. Stoia, C. Opriş, et al., 2024, Procedia Structural Integrity, 56, 97-104.
[3] D.G. Zisopol, M. Minescu, D.V. Iacob, 2024, Engineering, Technology & Applied Science Research, 14(2), 13487-13492.
[4] D.I. Stoia, E. Linul, 2024, Theoretical and Applied Fracture Mechanics, 132, 104478.
[5] D.A. Şerban, C. Marşavina, A.V. Coşa, et al., 2021, Mathematics, 9(13), 1495.
[6] N. Vidakis, M. Petousis, E. Karapidakis et al., 2023, Advances in Industrial and Manufacturing Engineering, 6, 100119.
[7] J.D. Kechagias, N. Vidakis, M. Petousis, N. Mountakis, 2023, Materials and Manufacturing Processes, 38(8), 941–953.
[8] C. Vălean, I.N. Orbulov, A. Kemény, et al., 2024, Engineering Failure Analysis, 161, 108335.
[9] D.G. Zisopol, M. Minescu, D.V. Iacob, 2024, Technology and Applied Science Research, 14(2), 13592-13597.
[10] N. Vidakis, M. Petousis, N. Mountakis, et al., 2023, Applied Sciences, 13, 6471.
[11] C. Vălean, E. Linul, G. Palomba, G. Epasto, 2024, Journal of Materials Research and Technology, 30, 1470-1481.
[12] C. Vălean, M. Baban, D.K. Rajak, E. Linul, 2024, Construction and Building Materials, 414, 135015.
[13] D.G. Zisopol, I. Nae, A.I. Portoacă, I.N. Ramadan, 2022, Technology & Applied Science Research, 12(4), 8978-8981.
[14] M. Petousis, N. Vidakis, N. Mountakis, et al., 2023, Polymers, 15, 1232.
[15] J. Schneider, S. Kumar, 2023, Journal of Materials Research and Technology, 26, 7182-7198.
[16] M. Xin, Z. Liu, B. Wang, et al., 2023, Journal of Materials Research and Technology, 25, 5833-5844.
[17] Stoia, D.I.; Marsavina, L.; Linul, E., 2020, Polymers, 12, 640.
[18] D.I. Stoia, L. Marsavina, E. Linul, 2021, Theoretical and Applied Fracture Mechanics, 114, 102968.
[19] D.I. Stoia, E. Linul, L. Marsavina, 2022, Theoretical and Applied Fracture Mechanics, 121, 103527.
[20] L. Marşavina, C. Vălean, M. Mărghitaş, et al., 2022, Engineering Fracture Mechanics, 274, 108766.
[21] C. Vălean, L. Marşavina, M. Mărghitaş, et al., 2020, Procedia Structural Integrity 28, 1134-1139.
[22] S. Thiede, M. Wiese, C. Herrmann, 2021, Procedia CIRP, 104, 653-658.
[23] R. Brighenti, L. Marsavina, M.P. Marghitas, et al., 2023, Materials Today: Proceedings, 78, Part 2, 331-336.
[24] D. Rahmatabadi, I. Ghasemi, M. Baniassadi, K. Abrinia, 2022, Journal of Materials Research and Technology, 21, 3970-3981.
[25] ASTM D5045-99, Standard test methods for plane-strain fracture toughness and strain energy release rate of plastic materials, 1999.

Compressive behavior of closed-cell metal foams under cryogenic conditions

Răzvan Bercuci[1], Liviu Marşavina[1,2], Jaroslav Kováčik[3] and Emanoil Linul[1*]

[1]Department of Mechanics and Strength of Materials, Politehnica University Timisoara, 1 Mihai Viteazu Avenue, 300 222, Timisoara, Romania
[2]CCTFA Romanian Academy - Timisoara Branch, 24 Mihai Viteazu Avenue, 300223, Timisoara, Romania
[3]Institute of Materials and Machine Mechanics, Slovak Academy of Sciences, Dúbravská cesta 9, 845 13, Bratislava, Slovak Republic

*E-mail: emanoil.linul@upt.ro

Abstract. Cellular materials in general and foams (polymeric, metallic and ceramic) in particular have seen a major development in recent years. Most of the data are reported at room temperature (RT), while the area of extreme temperatures is little studied. This work presents the quasi-static compression behavior of metallic foams (MFs). The MFs have closed cells and are manufactured from aluminum alloys (AlSi10) through powder metallurgy route. The mechanical tests are performed at cryogenic temperature (CT), and the results are compared with those at RT. It was found that as the temperature drops (CT), the samples become more brittle and withstand higher loads. The collapse mechanisms differ depending on the used test condition. The highest energy absorption performances are highlighted at CT. Also, the characteristic strains associated with the strength properties are influenced by the test temperature.

1. Introduction

Because the aluminum foams (AFs) have a wide range of applications and outstanding physical and mechanical properties, they were studied in the past decades intensively by researchers [1-3]. Their low weight and high compressive strength alongside other properties like low thermal conductivity, high energy absorption performance, they all together have made industries (like automotive, aerospace, maritime and so many others) start using them more and more in their specifics fields [4-6]. These materials can be classified as a special class of advanced composites and due to their cellular structure, they obtain extraordinary advantages compared to other fully dense materials [7-9]. If properties of the metallic foams are to be compared with other porous materials (ceramic or polymeric), they present clearly superior mechanical properties [10-12]. Therefore, because of the mentioned exceptionally properties and broad applications in different industries, the Al matrix is popular and is still being studied today [13, 14].

The main characterization of metallic foams is done at room temperature (RT) and under quasi-static mechanical loads [15-17]. Metallic foams dissipate the energy in a controllable form and therefore makes them exceptionally good in compressive applications [18, 19]. In the literature, there are various studies conducted on the compression behavior of metallic foams at RT [20-22]. But, there are also a series of studies that report their behavior at low (LT) and high (HT) temperatures [23, 24]. Compression performances of composite AF tubes was studied by Zhang et al. [25]. The authors found that the deformation process is more controllable than for thin-walled tubes. Moreover, they also found that when the diameter ratio is increasing the energy absorption is also increasing. Wang et al. [26] studied the effect of carbon nanotubes (CNTs) content on compressive properties and deformation behaviors of Al matrix composite foams (CFs). They concluded that the pore size can be reduced, and compressive properties of CFs can be enhanced. With the incorporation of CNTs in the Si phase the pore size significantly reduced. The yield stress, plateau stress and energy absorption have been improved by 23.24%, 98.41% and 84.70%, respectively. The quasi-static compressive behavior and energy absorption properties of Al matrix syntactic foams were studied by Sonti et al. [27] under RT and HT conditions. This research shows that particle size is affected by different test temperature (TT). The yield stress value at a lower density has a 1% drop. At a density of 0.86 g/cm^3, the energy absorption efficiency is at 75.77%. Movahedi et al. [28] studied the quasi-static compressive characteristics of the ex-situ Al-alloy foam-filled tubes under HT conditions. Under this research, it has been found out that AF behaves like a ductile metal at 300 °C under compression, and the energy absorption can be improved by 23% at 300 °C by using closed-cell AF as a filler.

As can be seen, a lot of research has been done at RT and HT, while LTs are a bit neglected. In this work, the compressive behavior of metallic foam is investigated at CT and a comparison is made with the RT results. The effect of negative temperature on strength properties and energy absorption performance is highlighted.

2. Materials and methods

2.1 Materials and samples

The AlSi10 foams were prepared through the powder metallurgy route. Initially, large plates of aluminum foam were obtained (200 mm × 200 mm × 30 mm). Later, these plates were cut with a Water Jet Cutting (WJC) machine to obtain cubic specimens (30 mm × 30 mm × 30 mm). Figure 1 shows an overview of the cut foam plate.

Figure 1. Aluminum foam plate cut by the WJC process.

The cell-walls of the studied metallic foam being made of Al alloys; they deform very easily even at fairly low loads. Therefore, the WJC process was chosen over others (milling, saw) in order to avoid the destruction of the foam cells. In this way, the cellular structure of the samples was preserved in its initial shape. The manufactured foams showed closed cells, and the tested samples had a density of around $0.48^{\pm0.05}$ g/cm^3 (relative density of about 0.18).

IOP Conf. Series: Materials Science and Engineering **1319** (2024) 012019 doi:10.1088/1757-899X/1319/1/012019

2.2 Experimental setup

The cubic specimens were subjected to an experimental program. It consisted of quasi-static compression tests with a test speed of 10 mm/min. The 100 kN LBG machine was used as a test device. Figure 2 shows images of the initial sample (undeformed) and the one during the mechanical test (deformed). The tests were carried out according to the ISO standard [29].

Figure 2. The specimen before (left) and during (right) the compression test.

Liquid nitrogen (LN) and a special enclosure were used for the CT tests. The LN used had a temperature of -196 °C. Each sample was immersed in LN and kept for 10 minutes to equalize the temperature throughout its mass. In order not to lose their predefined temperature (-196 °C), the cubic metal foam samples were tested in LN.

3. Results and discussions

The characteristic compression stress-strain curves are shown in Figure 3 for the two studied TTs (RT and CT). It seems that lowering the TT leads to a hardening of the material from which the foam is made. It can be easily seen that the curve obtained at CT is clearly higher than the one at RT. In addition, due to the brittleness of the metal matrix, the resulting CT curve shows some oscillations compared to the RT one. Both TTs give a typical behavior to cellular materials, easily finding linear-elastic, plateau and densification areas [30, 31].

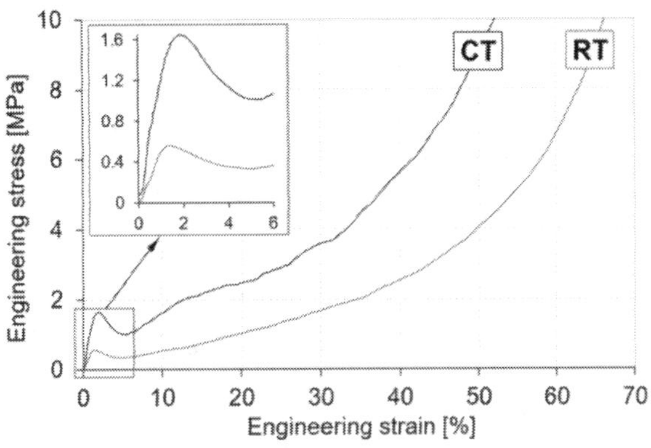

Figure 3. Compressive stress-strain curves of AlSi10 foams.

IOP Conf. Series: Materials Science and Engineering **1319** (2024) 012019 doi:10.1088/1757-899X/1319/1/012019

At low strains, below 2%, the foams behave elastically, however, the CT-tested foam exhibits a wider area of the linear slope. The linearly elastic zone ends with the foam's yield point, which differs significantly depending on the TT. This point is associated with the compressive strength of the foams. After reaching the first local maximum, the cell walls of the samples yield, obtaining the first stress drop in the stress-strain curve. This drop is called the stress amplitude and is given by the difference between the compressive strength and the first minimum stress after the yield point. This property is already found on the plateau region that extends up to strains of approximately 40%. It can be seen that the plateau area is not a constant one, but shows an increase (hardening) along its entire extent. The shape of this area is dictated by the way the foam cells deform or fail. The main failure mechanisms at the level of cells are their yield, fracture, buckling or bending. The most important strength property called plateau stress can be identified from this area. In addition, the plateau is important for structural applications where high energy absorption is required. For this reason, it is recommended that the length of the plateau be as large as possible. The last region, the densification, does not show an obvious transition in the stress-strain curve. The densification stress, in the case of metallic foams, is 1.3 times the plateau stress. This region stands out for densification strain, a specific property of cellular materials. Through it, the length of the plateau can be determined as the difference between densification strain and strain at strength. The last part of the stress-strain curve, at high strains (>60% in this case), is associated with complete compaction of the specimen.

The same characteristic regions are also found in the energy absorption-strain curves from Figure 4. From this representation it can be noted that the linearly elastic zone absorbs a very small amount of energy. The largest amount of energy is absorbed in the plateau area, as previously mentioned.

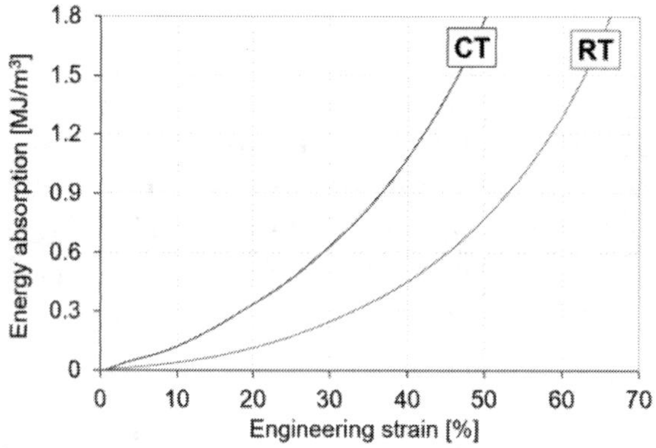

Figure 4. Compressive energy absorption-strain curves of AlSi10 foams.

Figure 5 shows the variation of the main compression properties of the investigated foam for the two TTs (CT and RT). The properties discussed in the following are strength (plateau stress, compressive strength and densification stress), strain (densification strain and strain at strength) and energy absorption.

Regardless of the type of property (strength, strain or energy), the samples tested at CT show higher values than those at RT. Thus, for plateau stress, compressive strength and densification stress at CT, the 1.64, 4.05 and 5.27 MPa values are noted (see Figure 5a). The biggest percentage difference between CT and RT for strength properties is found for compressive strength (192.9%),

IOP Conf. Series: Materials Science and Engineering **1319** (2024) 012019 doi:10.1088/1757-899X/1319/1/012019

while plateau stress and densification stress show almost identical differences (126.3% for plateau and 126.2% for densification).

The linear elastic zone also gives a 112.1% higher compressive modulus for CT than for RT (Figure 5b). Easy to identify aspect in the stress-strain curves in Figure 3. In this sense, the two obtained values of the modulus are 99.39 MPa for CT and 46.86 MPa for RT.

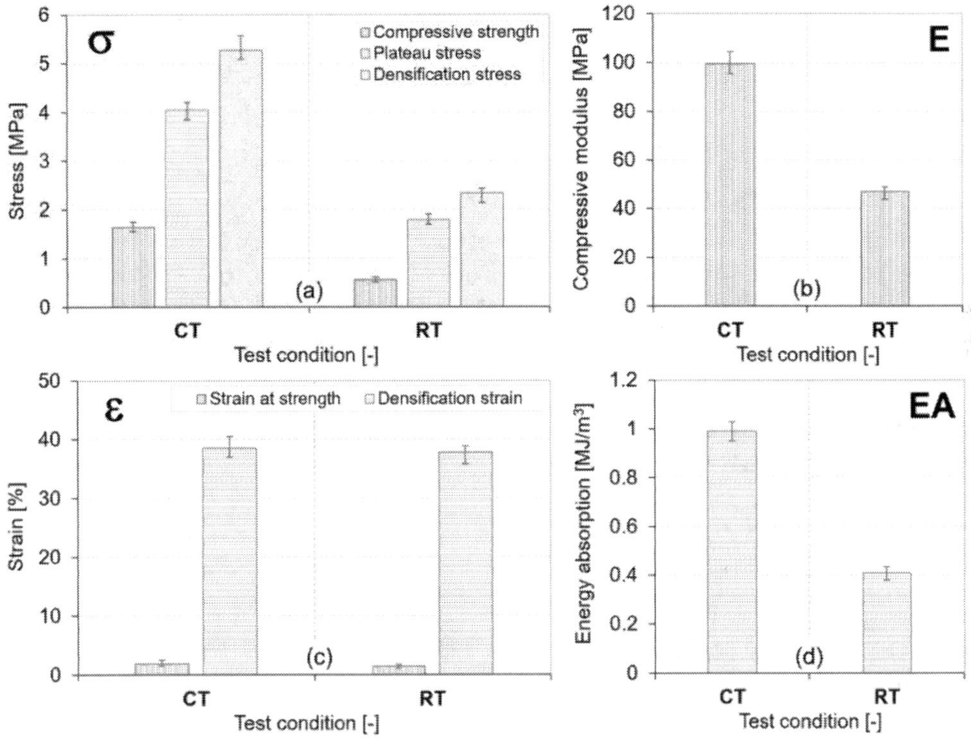

Figure 5. Main compressive properties of AlSi10 foams.

Among the strain properties, only strain at strength shows an obvious increase at CT compared to RT (35.5%), while densification strain is only 1.8% higher at CT (see Figure 5c). However, of the two properties, densification strain is much more important than strain at strength, especially for energy absorption performances.

The second most influenced compression property, by the TT, is energy absorption (Figure 5d). The maximum value is at CT (0.99 MJ/m^3), which is 141.5% higher than that at RT (0.41 MJ/m^3).

4. Conclusions

The paper presents a comparison of the compression properties of the closed-cell AlSi10 metallic foams obtained at RT and CT. The following were noted:

- The TT influences the compression behavior of metallic foams, the specimens tested at CT showing superior characteristics to those at RT.
- Among all the properties, compressive strength highlights the biggest difference (192.9%) between CT-RT. At CT, a strength value of 1.64 MPa was obtained.

- The other two strength properties (plateau stress and densification stress) exhibit equal CT-RT differences (~126%).
- The specimens tested at CT show 141.5% better energy absorption performance than those from RT. It was found that the samples subjected to CT absorb an energy of 0.99 MJ/m^3.
- The specimens tested at CT are more than twice as stiff (112.1%) compared to RT. The specimens exposed to CT have a compressive modulus of 99.39 MPa.
- Densification strain shows almost identical values between CT and RT (1.8%), and strain at strength was noted to be superior for CT (35.5%).

References

[1] JE Maróti, IN Orbulov, 2023, Materials Science and Engineering: A 869, 144817.
[2] G Epasto, F Distefano, L Gu, et al., 2020, Materials & Design 196, 109120.
[3] E Linul, L Marsavina, J Kováčik, et al., 2017, Proceedings of the Romanian Academy - Series A 18 (4), 361-369.
[4] J Kováčik, J Jerz, N Mináriková, et al., 2016, Frattura ed Integrita Strutturale 10 (36), 55-62.
[5] A Fehér, JE Maróti, DM Takács, et al., 2024, International Journal of Heat and Mass Transfer 226, 125446.
[6] A Mauko, M Sarıkaya, M Güden, et al., 2023, Journal of Materials Research and Technology 25, 1208-1221.
[7] D Pietras, E Linul, T Sadowski, A Rusinek, 2020, Composite Structures 249, 112548.
[8] C Kádár, R Gorejová, P Kubelka, et al., 2024, Advanced Engineering Materials, 2301496.
[9] D Heitor, I Duarte, J Dias-de-Oliveira, 2021, Metals 11 (6), 925.
[10] G Thalmaier, NA Sechel, A Csapai, et al., 2022, Materials 15 (15), 5446.
[11] A Kemény, N Movahedi, T Fiedler, et al., 2022, Materials Science and Engineering: A 852, 143706.
[12] E Linul, L Marsavina, 2011, Iranian Polymer Journal (English) 20 (9135), 735-746.
[13] SC Pinto, PAAP Marques, R Vicente, et al., 2020, Metals 10 (6), 768.
[14] A Szlancsik, IN Orbulov, 2021, Materials Science and Engineering: A 827, 142081.
[15] J Aquino, I Duarte, J Dias-de-Oliveira, 2018, Science and Technology of Materials 30 (1), 43-49.
[16] L Marsavina, DM Constantinescu, E Linul, et al., 2014, Procedia Materials Science 3, 1342-1352.
[17] G Thalmaier, NA Sechel, I Vida-Simiti, 2020, Jom 72, 3377-3382.
[18] A Kemény, B Leveles, T Bubonyi, et al., 2021, Composites Part A,140, 106152.
[19] I Vida-Simiti, N Jumate, E Bruj, et al., 2011, Materials Science Forum 672, 141-144.
[20] G Epasto, F Distefano, H Mozafari, et al., 2021, Applied Sciences 11 (3), 1148
[21] N Movahedi, IN Orbulov, A Kemény, et al., 2020, Materials Science and Engineering: A 798, 140255.
[22] I Vida-Simiti, N Jumate, E Bruj, et al., 2010, Optoelectronics and Advanced Materials-Rapid Communications 4, 1174-1177.
[23] N Movahedi, E Linul, 2021, Composite Structures 277, 114634.
[24] E Linul, O Khezrzadeh, 2021, Composite Structures 271, 114156.
[25] Z Zhang, N Liu, Z Zhang, et al., 2023, International Journal of Mechanical Sciences, 242, 108039.
[26] S Wang, K Yang, M Xie, et al., 2024, Materials Science and Engineering: A, 898, 146391.
[27] KSM Sonti, S Vincent, SKR Narala, 2023, Materials Today Communications, 35, 105580.
[28] N Movahedi, E Linul, 2017, Materials Letters, 206, 182-184.
[29] ISO13314, 2011, Mechanical testing of metals - ductility testing - compression test for porous and cellular metals.
[30] E Linul, D Pietras, T Sadowski, et al., 2021, Composites Part A: Applied Science and Manufacturing 149, 106516.
[31] H Mozafari, F Distefano, G Epasto, et al., 2022, Journal of Marine Science and Engineering 10 (10), 1385.

IOP Conf. Series: Materials Science and Engineering **1319** (2024) 012020 doi:10.1088/1757-899X/1319/1/012020

Influence of curing time and temperature on the mechanical properties of resins manufactured through stereolithography

Alexandru-Viorel COŞA[1*], Marian-Vasile BABAN[2], Dan-Andrei ŞERBAN[1]

[1] Department of Mechatronics, University Politehnica Timişoara, Timisoara, Romania
[2] Department of Mechanics and Strength of Materials, University Politehnica Timişoara, Timisoara, Romania

*E-mail: alexandru.cosa@studet.upt.ro

Abstract. Additive manufacturing technologies are widely used and provide unique advantages over other manufacturing techniques. However, to achieve the optimal properties of an additive manufactured product, knowledge about the specific tool and parameters is required. A suitable approach to understand the parameters and the influence on the final properties is to investigate the variations of a mechanical property based on the fluctuation of the manufacturing parameters. This study focuses on determining the influence of the curing temperature and time under UV light (405nm wavelength) on the tensile properties of specimens obtained through stereolithography (SLA). This aspect was attained by manufacturing batches of specimens from a urethane dimethacrylate (UDMA) based resin (commercial name Formlabs Tough 2000®) and then curing them for different periods of time (from uncured up to 180 minutes). The samples were tested and, based on the resulting stiffness and strength, the optimal curing time of 90 minutes was established. The value of this parameter was used to investigate the influence of the temperature at which the specimens are cured, with a variation from 25 °C to 80 °C, resulting with an optimal curing temperature of 60 °C. The study also concluded that after a certain temperature, the stiffness is declining, even though the strength of the specimens reaches a plateau-like region.

1. Introduction

SLA (stereolithography) is one of the most common 3D printing processes, that uses light-reactive thermosetting materials (resins). When exposed to certain wavelengths of light, the light-reactive resins produce short molecular chain bonds, polymerizing monomers and oligomers into solidified rigid or flexible geometries. SLA is increasing its popularity among individual users as the equipment and the resins are becoming more affordable. The ability to produce high-accuracy prototypes and low volume production parts that present isotropic properties and are watertight makes it more desirable in the ever-evolving market. For now, the costs of high-quality resins are still rather high, (around 10-times higher per kilogram than the fused deposition modelling filaments for example), but better behaviour and a clean surface finish, among other benefits, are taking the highlight and direct it to SLA technology. Comparing all additive manufacturing technologies, SLA has the fastest manufacturing speed, highest resolution and accuracy, sharpest details, and smoothest surface finishes of all [1].

In consequence, its applications vary from mechanical parts (housings, cases, functional parts) to biomedical applications [2, 3, 4] and aero-spatial applications [5] among others. Many of the SLA printers have the light source positioned below the resin tank, using what is called inverted stereolithography technology, as the build platform is in the superior region of the printer, and the resin tank is on the lower side. The printing process is done by lowering the build platform into the resin tank until it is covered by resin and curing of the layers one-by-one. When one layer is fully printed, the build platform elevates and allows the liquid resin to flow underneath, exposing the new layer. In recent years, different processes that are using resins as base material for additive manufactured parts were developed, which are differentiated by the type of light source they use. These include laser-powered stereolithography (SLA), digital light processing (DLP), or masked stereolithography (MSLA, also often used interchangeably with LCD 3D printing) [6, 7]. After the manufacturing process is finished, the specimens are removed from the build platform and washed with specific solutions (i.e. isopropyl alcohol) and a curing treatment is applied, which consists of subjecting the specimens to ultraviolet light, combined with temperature in some cases.

The post-printing curing process plays a pivotal role in determining the mechanical properties and overall performance of the printed parts. Understanding the influence of curing conditions, specifically varying temperatures and curing times on the final properties of SLA-printed parts is crucial for optimizing their mechanical strength, surface finish, and dimensional accuracy. While extensive research has been conducted on curing parameters in SLA, some gaps remain to be further investigated regarding the effects of curing under Variable Ultraviolet (VUV) light [8, 9, 10, 11]. This study aims to perform a systematic analysis of the impact of different curing temperatures and curing times under VUV light on the mechanical properties of SLA-printed parts. By investigating the intricate relationships between curing conditions and part performance, this research seeks to provide valuable insights for enhancing the quality and functionality of SLA-printed components across diverse applications, ranging from aerospace to biomedical engineering. Although, the manufacturer recommendation of UV curing settings for the investigated resin are 60minutes at 70°C [12], there are some enhancements that may be obtained after discovering the behaviour of the material at different curing times and temperatures.

2. Materials and Methods

For this study, a Formlabs Form 3+ printer was used (Figure 1), which features a resin tank that holds the liquid photopolymer resin, which can be provided in a variety of types, optimized for different applications. The build platform is situated above the resin tank and it moves vertically during the printing process, gradually lifting the printed object out of the resin tank as each layer is cured.

IOP Conf. Series: Materials Science and Engineering **1319** (2024) 012020 doi:10.1088/1757-899X/1319/1/012020

Figure 1. SLA Formlabs 3+ printer (formlabs.com)

The investigated specimens were manufactured from a photosensitive resin consisting of a mixture of Urethane Dimethacrylate (45-65%), Methacrylate Monomer (15-25%), Isobornyl methacrylate (10-20% and phenyl bis(2, 4, 6-trimethylbenzoyl)-phosphine oxide (<0.6%), with the commercial name Tough 2000, having the properties detailed in Table 1 [13, 14]. These mechanical properties recommend this compound for a relatively large range of applications, such as functional prototypes, jigs and fixtures, and end-use parts. A literature survey showed that the actual mechanical properties of the material exhibit variations from study to study (in the rage of 10-20%) [8, 15, 16, 17, 18], so a further analyses are suitable to better understand the causes.

Table 1. Material Properties of Tough 2000 resin

Tensile Properties	Green	Post-Cured
Ultimate Tensile Strength	29 MPa	46 MPa
Tensile Modulus	1.2 GPa	2.2 GPa
Elongation at Break	74%	48%
Flexural Properties		
Flexural Strength	17 MPa	65 MPa
Flexural Modulus	0.45 GPa	1.9 GPa
Temperature Properties		
Heat Deflection Temp. @ 1.8 MPa	42 °C	53 °C
Heat Deflection Temp. @ 0.45 MPa	48 °C	63 °C
Thermal Expansion (0-150°C)	107 μm/m/°C	91 μm/m/°C

*Data was obtained from parts printed using Form 2, 100 μm, Tough 2000 settings and post-cured with a Form Cure for 120 minutes at 80 °C.

IOP Conf. Series: Materials Science and Engineering **1319** (2024) 012020 doi:10.1088/1757-899X/1319/1/012020

3. Experimental procedures

To assess the tensile properties of Formlabs Tough 2000 resin, specimens were designed according to the ISO 527-2 standard for plastics [19]. The dimensions of the specimens are shown in Figure 2.

For each set of curing conditions, five specimens were tested to ensure statistical reliability of the results. A total of 75 specimens were tested using various curing conditions, 50 of which being used to determin the optimal curing time while 25 were used to determine the optimal temperature.

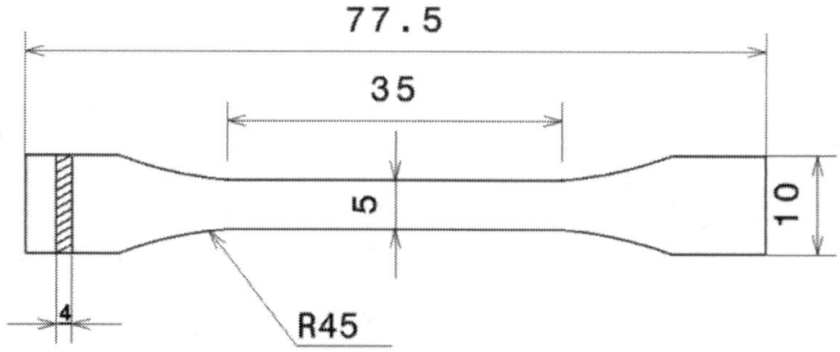

Figure 2. Tensile specimen dimensions in mm

Tensile tests were performed using a Zwick Roell Z005 testing machine equipped with a 5 kN load cell at room temperature (23°C). The crosshead speed was set at 5 mm/min and the strains were measured with an incremental extensometer with an initial gauge length of 30 mm.

Proper arrangement of specimens in the UV curing machine is critical to ensure uniform exposure and optimal curing of SLA printed parts. According to manufacturer instructions and literature studies, specimens must be evenly spaced and positioned in such a manner as to avoid the occurrence of shadows and to ensure consistent UV light penetration [20, 21, 22, 23]. This involves placing the specimens at an appropriate distance from the UV light source and arranging them on a rotating platform, to ensure all surfaces receive equal exposure. Proper arrangement helps achieve uniform mechanical properties across all specimens, preventing variations in strength and stiffness that can arise from uneven curing. A piece of sealant tape was used to fix the specimens vertically inside de UV curing machine (Figure 3).

(a) (b)

Figure 3. Example of arrangement of specimens on the turntable of the UV curing machine

IOP Conf. Series: Materials Science and Engineering **1319** (2024) 012020 doi:10.1088/1757-899X/1319/1/012020

4. Results and discussions

For the curing time influence investigations, the specimens were maintained at ambient temperature (23°C) for periods of time ranging from 10 minutes up to 180 minutes (with the inclusion of a "green" specimen, that was not cured at all). The stress-strain variations depicted in Figure 4 show a visible influence of the curing time on the properties of the specimens. The overall shape of the stress-strain curves of the tested specimens are similar (all exhibiting strain hardening after yielding up to a certain point, beyond which strain softening is observed) with variation in the elastic modulus and mechanical strength.

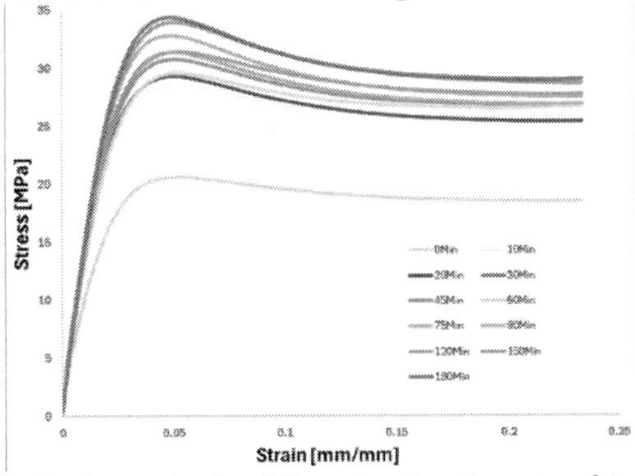

Figure 4. Stress-Strain results for different curing time at ambient temperature

The resulting mechanical properties and their standard deviations are presented in Table 2.

Table 2. Young's modulus and Strength values of the specimens for different curing times

Curing Time	Average Young's Modulus (MPa)	Standard deviation (%)	Average Strength (MPa)	Standard deviation (%)
0 minutes	1137.97	2.84	20.28	5.12
10 minutes	1531.1	5.8	28.51	2.74
20 minutes	1450.25	7.48	28.82	1.29
30 minutes	1607.3	6.55	30.34	1.97
45 minutes	1561.64	4.83	30.61	1.47
60 minutes	1634.19	5.49	29.75	1.25
75 minutes	1545.63	2.41	30.9	1.98
90 minutes	1703.04	4.77	32.22	1.87
120 minutes	1703.23	0.94	30.82	2.67
150 minutes	1528.02	2.95	33.51	2.22
180 minutes	1613.21	5.45	33.51	0.85

Considering the close nature of the results ranging from 90 minutes to 180 minutes, the optimal curing time was chosen using a decision matrix, considering the normalized values for the average stiffness and strength (current value divided by the maximum recorded value). As the authors considered that, for future applications, the stiffness of the material is slightly more important than its strength, the property weight for stiffness w_E was chosen 0.55 and the property weight for the strength w_σ was chosen 0.45. The decision matrix is presented in Table 3.

Table 3. Young's modulus and Strength values of the specimens for different curing times

Curing Time	Normalized stiffness (-)	Weighted normalized stiffness (-)	Normalized strength (-)	Weighted normalized strength (-)	Total score (-)
0 minutes	0.6681	0.3675	0.5921	0.2665	0.6339
10 minutes	0.8989	0.4944	0.8326	0.3747	0.8691
20 minutes	0.8515	0.4683	0.8415	0.3787	0.8470
30 minutes	0.9437	0.519	0.886	0.3987	0.9177
45 minutes	0.9169	0.5043	0.8941	0.4023	0.9066
60 minutes	0.9595	0.5277	0.869	0.391	0.9188
75 minutes	0.9075	0.4991	0.9026	0.4062	0.9053
90 minutes	0.9999	0.5499	0.941	0.4234	0.9734
120 minutes	1	0.5500	0.9001	0.4050	0.9550
150 minutes	0.8971	0.4934	0.9787	0.4404	0.9338
180 minutes	0.9472	0.5209	1	0.4500	0.9709

Following the decision matrix analysis, it can be observed that the optimal curing time is 90 minutes, the total score being narrowly ahead of the one determined by the 180 minutes curing time.

The second part of the study was performed by curing specimens at four different temperatures: 40 °C, 60°C, 70°C and 80 °C (the upper limit of the curing machine). Representative stress-strain curves for each parameter (along with the curve corresponding to the ambient temperature test for 90 minutes curing time from the previous investigations) are presented in Figure 5.

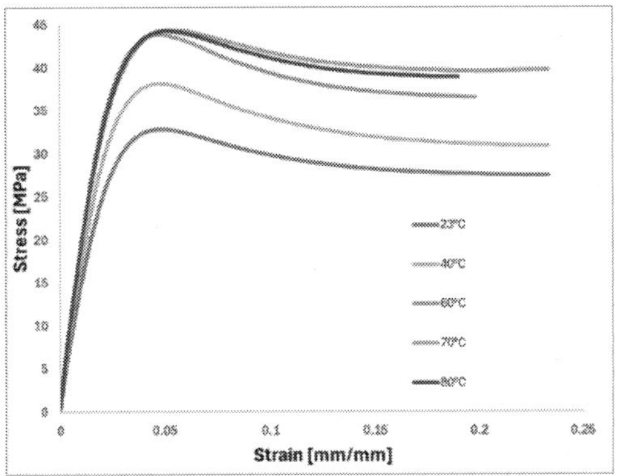

Figure 5. Stress-Strain results for different curing temperature

The results show a significant increase in the mechanical properties of the specimens with the temperature that is used for curing, the average Young's modulus and tensile strength values along with their standard deviations being presented in Table 4.

Table 4. Young's modulus and Strength values of the specimens for different curing temperature

Curing Time	Average Young's Modulus (MPa)	Standard deviation (%)	Average Strength (MPa)	Standard deviation (%)
23°C	1705.37	4.96	32.46	2.17
40°C	1904.3	2.69	37.86	1.78
60°C	2198.92	2.61	44.09	1.81
70°C	2166.47	7.02	44.18	0.96
80°C	2000.45	13	44.15	0.24

Figure 6 a) illustrates the relationship between curing time and the mechanical properties of the material (stiffness and strength). As curing time increases, there is an initial rise in both stiffness and strength, reaching local maximum before beginning to decrease. This trend suggests that insufficient curing time does not allow the material to fully crosslink, resulting in suboptimal mechanical properties. Conversely, excessive curing time can lead to over-curing, causing brittleness and a reduction in both stiffness and strength [15]. Similarly, Figure 6 b) depicts the influence of the temperature on the stiffness and strength of the material for a curing time of 90 minutes. Both properties improve with an increase in temperature, up to a certain point. Beyond this optimal temperature, the material properties begin to decrease. This indicates that, while higher temperatures can accelerate the curing process and improve crosslinking efficiency, there is a threshold beyond which thermal degradation or over-curing may occur, compromising the material's integrity. Compared to a green part, a 49.6% increase in stiffness was observed by curing the specimens at ambient temperature for 90 minutes, with a total of 93.23% increase for the curing temperature of 60°. In addition, the strength had a 58.9%

increase by curing the specimens at ambient temperature for 90 minutes, with a total of 117.4% increase for the added temperature, mirroring the results found in literature for other materials [24].

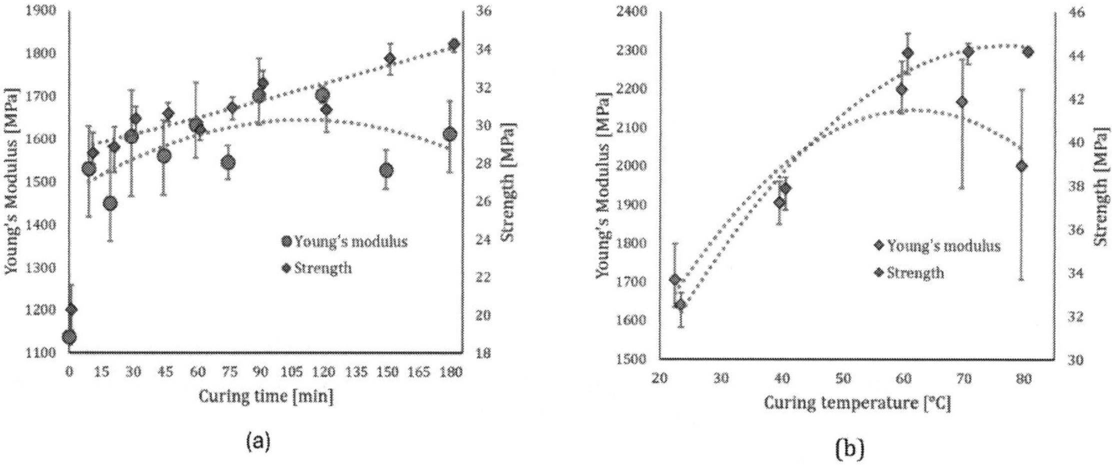

Figure 6. Stiffness and strength variation with curing time (a); Stiffness and strength variation with temperature for 90min curing time (b)

5. Conclusions

This study investigated the impact of UV curing parameters on the mechanical properties of Formlabs Tough 2000 resin, specifically focusing on tensile strength and stiffness. Through systematic experimentation, the ways in which the variation in curing time and temperature influenced the material properties of SLA prototyped specimens was assessed.

The experiments revealed a clear trend where both stiffness and strength increased with curing time up to a certain point. Beyond this optimal duration, mechanical properties began to deteriorate, likely due to over-curing effects that caused brittleness in the material. Similarly, curing temperature showed a significant impact on material properties. Both stiffness and strength improved with increasing temperature until an optimal level was reached. Exceeding this optimal temperature resulted in a decline in mechanical performance, indicating potential thermal degradation or over-curing. The results highlight the critical importance of precisely controlling the UV curing parameters to achieve the best mechanical performance of SLA printed parts.

Concluding, this study underscores the necessity of meticulous optimization of UV curing parameters to fully exploit the mechanical potential of SLA specific photosenisitive resins. By achieving the right balance in curing time and temperature the performance and applicability of SLA printed components can be significantley improved, paving the way for more robust and reliable 3D printed solutions in various industries.

Acknowledgements

This work was partially funded by the Research Grant PN-III-P2-2.1-PED-2021-1134, contract number 586PED/2022.

IOP Conf. Series: Materials Science and Engineering **1319** (2024) 012020 doi:10.1088/1757-899X/1319/1/012020

References

[1] K. Szykiedans and W. Credo, "Mechanical properties of FDM and SLA low-cost 3-D prints," in *The 20th International Conference: Machine Modeling and Simulations, MMS 2015*, Terchova, 2016.

[2] Q. Chen, B. Zou, Q. Lai and K. Zhu, "SLA-3d printing and compressive strength of PEGDA/nHAP biomaterials," *Ceramics International,* vol. 48, pp. 30917-30926, 2022.

[3] M. Jusufbegovic, A. Pandzic, A. Sehic, R. Jasic, F. Julardzija, S. Vegar-Zubovic and A. Beganovic, "Computed tomography tissue equivalence of 3D printing materials," *Radiography,* vol. 28, pp. 788-792, 202.

[4] J. Wang, A. Goyanes, S. Gaisford and A. Basit, "Stereolithographic (SLA) 3D printing of oral modified-release dosage forms," *International Journal of Pharmaceutics,* vol. 503, pp. 207-212, 2016.

[5] Z. Miller, B. Stidham, T. Fairbanks and C. Maldonado, "The Use of Stereolithography (SLA) Additive Manufacturing in Space-Based Instrumentation," in *2023 IEEE Aerospace Conference*, Big Sky, 2023.

[6] H. Bikas, P. Stavropoulos and G. Chryssolouris, "Additive manufacturing methods and modelling approaches: a critical review," *Int J Adv Manuf Technol,* vol. 83, pp. 389-405, 2016.

[7] Z. Zguris, "How Mechanical Properties of Stereolithography 3D Prints are Affected by UV Curing," formlabs.com.

[8] S. Bieler and K. Weinberg, "Behavior of additively manufactured lattice structures unter compressive loading," in *Applied Mathematics & Mechanics*, Aachen , 2023.

[9] R. Chand, V. Sharma and T. Rajeev, "Investigating mechanical properties of 3D printed parts manufactured in different orientations on Multijet printer," *Research Square,* 2021.

[10] F. Cosmi and A. Dal Maso, "A mechanical characterization of SLA 3D-printed specimens for low-budget applications," *Materials Today: proceedings,* vol. 32, no. 2, pp. 194-201, 2020.

[11] G. Cwikla, C. Grabowik, K. Kalinowski, I. Paprocka and P. Ociepka, "The influence of printing parameters on selected mechanical properties of FDM/FFF 3D-printed parts," in *IOP Conf. Series: Materials Science and Engineering*, 012033, 2017.

[12] Formlabs, "Formlabs Tough 2000 resin details," Formlabs, 2024. [Online]. Available: https://formlabs.com/eu/store/materials/tough-2000-resin/. [Accessed 02 June 2024].

[13] Formlabs, "Tough 2000 Datasheet," Formlabs, 2022.

[14] Formlabs, "Safety Data Sheet of Tough 2000 resin," Formlabs, EU, 2022.

[15] C. Riccio, M. Civera, O. Grimaldo Ruiz, P. Pedulla, M. Rodriguez Reinoso, G. Tommasi, M. Vollaro, V. Burgio and C. Surace, "Effects of Curing on Photosensitive Resins in SLA Additive Manufacturing," *Appl. Mech.*, vol. 2, pp. 942-955, 2021.

[16] J. Peloquin, Y. Han and K. Gall, "Printability and mechanical behavior as a function of base material, structure, and a wide range of porosities for polymer lattice structures fabricated by vat-based 3D printing," *Additive Manufacturing,* vol. 78, no. 103892, 2023.

[17] S. Wise, L. Bobbio, A. Russo, T. Simpson and A. Besse, "Strength Comparison of Topology Optimized Lattice From Printed SLA Resin, Electroplated Resin and PBF Aluminum Alloy," in *33rd Annual International Solid Freeform Fabrication Symposium – An Additive Manufacturing Conference*, Austin, 2022.

[18] R. Samples, R. Mukoyama, J. Shaffer, J. Mikucki and L. Giddings, "OpenASAP: An affordable 3D printed atmospheric solids analysis probe (ASAP) mass spectrometry system for direct analysis of solid and liquid samples," *HardwareX,* vol. 16, no. e00490, 2023.

[19] *ISO 527-1, Plastics-Determination of Tensile Properties-part 1: General Principles. International Organization for Standardization; ISO: Geneva, Switzerland, 23,* 2012.

[20] M. Staffova, F. Ondreas, J. Svatik, M. Zboncak, J. Jancar and P. Lepcio, "3D printing and post-curing optimization of photopolymerized structures: Basic concepts and effective tools for improved thermomechanical properties," *Polymer Testing,* vol. 108, no. 107499, 2022.

[21] J. Martin-Montal, J. Pernas-Sancez and D. Varas, "Experimental Characterization Framework for SLA Additive Manufacturing Materials," *MDPI-Polymers,* vol. 13, no. 7, p. 1147, 2021.

[22] Y. Yang, L. Li and J. Zhao, "Mechanical property modeling of photosensitive liquid resin in stereolithography additive manufacturing: Bridging degree of cure with tensile strength and hardness," *Materials and Design,* vol. 162, pp. 418-428, 2019.

[23] K. Ryan, M. Down and C. Banks, "Future of additive manufacturing: Overview of 4D and 3D printed smart and advanced materials and their applications," *Chemical Engineering Journal,* vol. 403, p. 126162, 2021.

[24] R. Brighenti, L. Marsavina, M. Marghitas, M. Cosma and M. Montanari, "Mechanical characterization of additively manufactured photopolymerized polymers," *Mechanics of Advanced Materials and Structures.*, vol. 30, no. 9, pp. 1853-1864, 2022.

Obtaining antimicrobial polymer composites based on copper oxide nanoparticles for sports medical devices.

V Dzitac[1,a], G I Rusu[1,b], V Gherman[1,c], E Valean [1,d], M Vodă[1,e], V-A Şerban [1, 2, f]

[1] Politehnica University of Timisoara, P-ta Victoriei, No. 2, Timisoara, Romania
[2] Technical Sciences Academy of Romania, Timisoara branch, B-dul. Vasile Parvan nr.28

[a]vasile.dzitac@student.upt.ro, [b]gerlinde.rusu@upt.ro, [c]vasile.gherman@upt.ro
[d]estera.valean@student.upt.ro, [e]mircea.voda@upt.ro, [f]viorel.serban@upt.ro

Keywords: Antimicrobial, Copper oxide, polymers, composite
Abstract. The beneficial effects of using antimicrobial materials in eliminating or inhibiting microbial proliferation on surfaces are well known, thus contributing to reducing the risk of diseases resulting from accidents involving pathogenic microorganisms.

The addition of certain metal particles in various proportions, (for example, silver, copper, or their oxides), in polymer matrices represents an effective method for producing new high-quality antimicrobial materials. In this study, the aim was to obtain a polymer/copper oxide composite with superior biocidal behaviour by using proportions ranging from 3-10% of copper oxides and polymers with different characteristics. The composites were prepared by melt mixing at various temperatures, durations, and homogenization speeds. Subsequently, the necessary samples for investigations were obtained through thermoforming. To evaluate the composite, tests were performed to determine mechanical characteristics, to verify the uniformity of particles in the composite material, and to test antimicrobial efficacy. The results demonstrated that the PP/CuO composite exhibited improved properties compared to pure polymers. These obtained characteristics recommend its potential use in medical devices for prevention or protection (examples include finger, ankle, or tendon prostheses), which can be produced through additive manufacturing or thermoforming.

1. Introduction

The field of antimicrobial materials represents a field of interest at the moment, continuously expanding and strongly debated, lately more and more work has been reported about materials of this type (Figure 1.) [1].

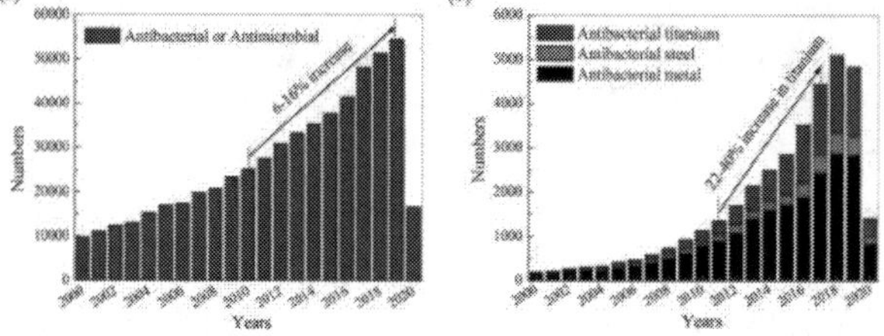

Figure 1. Antibacterial research publications searched by Web of Science. a) keywords: antibacterial or antimicrobial on topic, b) keywords: antibacterial metal, antibacterial steel and antibacterial titanium on topic [2].

Materials used in the past for medical devices were stainless steel, aluminum, titanium with non-toxic, non-corrosive and durable properties [3]

One of the new ways to further expand the range of applications of antimicrobial metals or their compounds is by adding them to a polymer to obtain a composite material. Particles of silver oxide or copper oxide can be either embedded on the surface of a polymer or dispersed throughout it. The integration of copper oxide nanoparticles into a polypropylene matrix represents a significant advancement in the development of antimicrobial and bacterial sport medical devices. This innovative composite material harnesses the inherent antimicrobial properties of copper to effectively combat bacterial growth and infection. By embedding copper oxide nanoparticle within the polypropylene matrix, these medical devices gain enhanced resistance to microbial colonization, thereby reducing the risk of healthcare-associated infections (HAIs) and promoting better patient outcomes [4], [5], [7]. Today, medical devices and instruments are obtained through special manufacturing processes through mechano-thermal casting processing and recently through additive manufacturing (Figure 2. 3D printing)[8], (Figure 3.)

Research on polymer (polypropylene) composites based on copper oxide nanoparticles revealed that increasing the percentage of copper oxide particles in the polypropylene matrix can inhibit the spread of Agar Plate Count (a universal medium for heterotrophs) as well as leads to a significant increase in mechanical strength properties.

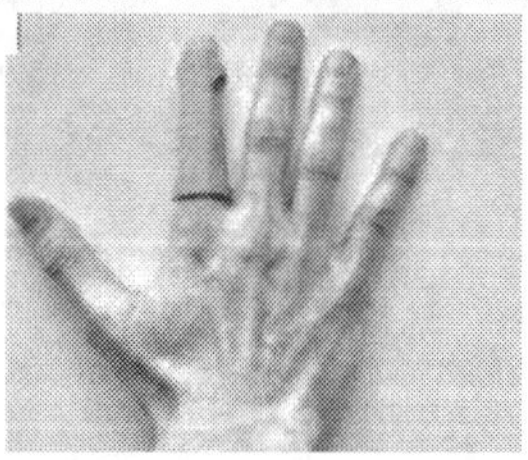

Figure 2 Finger orthosis [9]

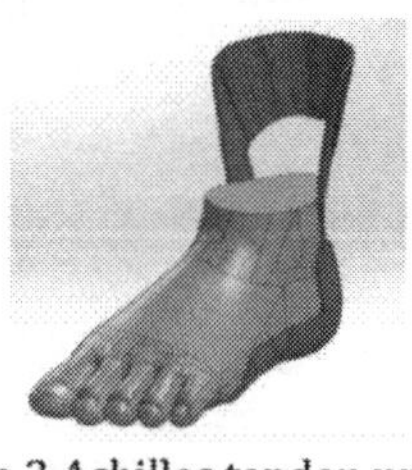

Figure 3 Achilles tendon prosthesis

2. Obtaining of the CuO based polymer composite

2.1 Materials and method of obtaining

2.1.1 Materials

Polypropylene (PP), discovered in 1954, quickly became highly popular due to having high rigidity and lowest density (0.90-0.91 g/cm3) among commodity plastics. This polymer, which is catalytically prepared from propylene, exhibits outstanding chemical resistance [10].

PP possesses several useful properties like high heat distortion temperature, transparency, dimensional stability which widen its application. As a matrix material, PP is widely used because it has some excellent characters for composite fabrication. PP is also very suitable for filling, reinforcing and blending. PP with natural fibrous polymers is one of the most promising routes to create natural–synthetic polymer composites [11]. Polypropylene was purchased from Sabic (white granules).

Copper oxide was used in the form of nanoparticles because, it is well known that nanoparticles have a larger surface area that interacts with microbes. The nanoparticles are more likely to penetrate the cell walls of bacteria and fungi, causing disruptions from within [12].

Copper(II) oxide (CuO) under the ROTI®nanoMETIC brand, used by us, is a nanomaterial used for various advanced application. CuO nanoparticles have a molecular weight of 79.545 g/mol and appear as black or brownish-black powder. As a nanomaterial, they typically have a particle size in the 1-100 nm range, significantly increasing surface area compared to bulk materials. They are characterized by high purity, usually above 99%, ensuring minimal impurities that could affect performance in applications. Their higher specific surface area enhances their reactivity and catalyctic proprietes

2.1.2 Preparation of polypropylene and copper oxide composites

The method used is Ex-Situ. The composite materials were obtained by the incorporation of copper oxide nanoparticles into the polypropylene using melt compounding via twin-screw extrusion. For this method we used the Rheomix R3000p/R3010p mixing machine (Figure 4).

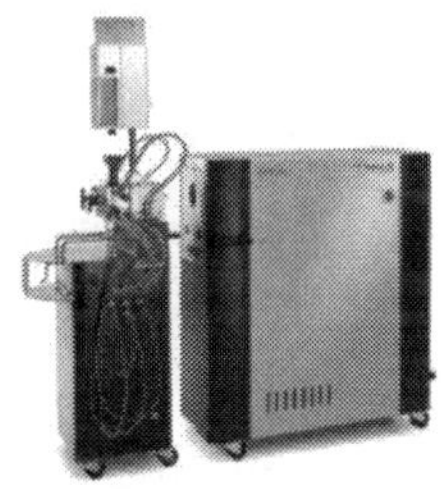

Figure 4 Rheomix R3000p/R3010p mixing machine

Figure 5 Wickert heated hydraulic press

2.2 Experiments for development of the composite

For these experiments (see table 1,) 200 grams of PP were used, and different amount of CuO (wt 3%, 5%, 7%, 10%) were introduced at rotational speed 20 rpm, a temperature of 220° Celsius for 10 m. For this paper, not all results are presented, only four samples are shown, each with 200 grams of PP, the rotation speed (n) of 20 rpm, the temperature of 220° Celsius for 10 minutes, but with different CuO percentages of 3%, 5%, 7%, and 10%.

Table 1. Preparation of composite material

Nr. crt	PP [g]	CuO [%]	n [rpm]	Temp [°C]	Time [min]
1	200	3	20	220	10
2	200	5	20	220	10
3	200	7	20	220	10
4	200	10	20	220	10

Following the mixing process, the composite material was removed from the Rheomix machinery and subsequently allowed to cool. Subsequent to the cooling phase, a quantity of 20 grams of the composite material was selected and introduced into a mould

This material was then compacted using a press (Figure 5.) heated to 220 degrees Celsius, resulting in a finalized sample measuring 80x80x5mm.

3. Characterization of the obtained composite

3.1 Determination of CuO residual mass

Thermogravimetric analysis were performed by using TG 209 F1 Libra thermogravimetric analyzer (NETZSCH-Gerätebau GmbH, Selb, Germany). The measurements were carried out in nitrogen atmosphere, in the temperature range 20–650°C, heating rate of 10 °C/min. The data were processed with the Netzsch Proteus—Thermal Analysis program version 6.1.0. (NETZSCH-Geraetebau GmbH, Selb, Germany).

From the TG thermograms (Figure 6.) can be seen that the decomposition of the polymer takes place between 400 and 500°C. At 650°C the polymer matrix is completely decomposed, and the residual mass is represented by the filer (CuO).

It can be seen that the values are similar to those used to obtain the composites.

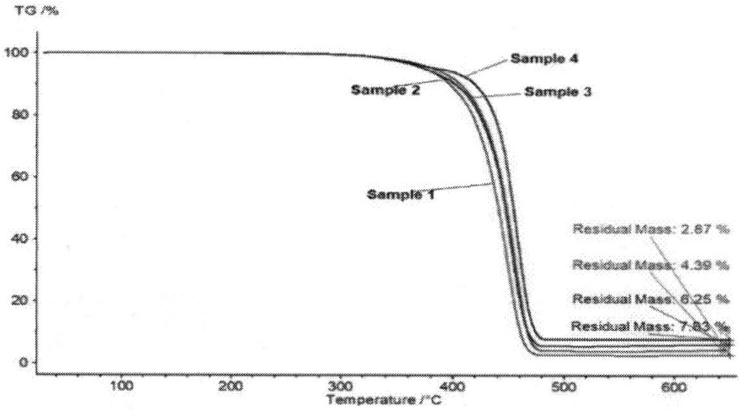

Figure 6 TG thermograms for the decomposition of the composites samples

3.2 Determination of mechanical characteristics

3.2.1 Strength and elongation variation

The graph (Figure 7.) displays a clear differentiation in the mechanical behavior of polypropylene (PP) composites with varying concentrations of copper oxide (CuO).

The tensile stress tests were performed using a Mecmesin MultiTest 5i at a loading speed of 5 mm/min. Strength and elongation variation was performed on sample Figure 8. As the CuO content increases (Table 2.) from 3% to 10%, there is a notable oscillation in both the force at break and the elongation capacity of the samples. The sample with 10% (Figure 9.) CuO demonstrates the highest force resistance but not the maximum elongation, indicating improved toughness, it is important to consider that the primary goal of integrating CuO in a PP matrix is to have better antimicrobial properties. The effectiveness of CuO as an antimicrobial agent is well-documented, but its impact on the mechanical properties, as evidenced by the results, also highlights the dual benefits of this composite formulation.

The optimal CuO concentration should therefore be determined not only based on mechanical performance but also considering the efficacy in inhibiting microbial growth.

Table 2. Elongation and force data for PP-CuO composites at different concentrations

Material	Maximum elongation [mm]	Maximum elongation strength [N]
PP - CuO 3%	3.1825	227.5786
PP - CuO 5%	2.185	206.688
PP - CuO 7%	2.12	205.9401
PP - CuO 10%	2.48	245.4866

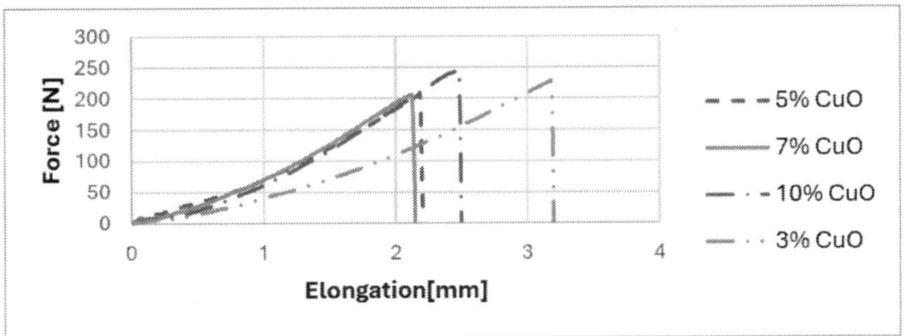

Figure 7. Representation of elongation as a function of force for PP-CuO composites at different concentrations

Figures 8 and 9 depict the sample containing 10% CuO, on which we conducted the determination of mechanical characteristics.

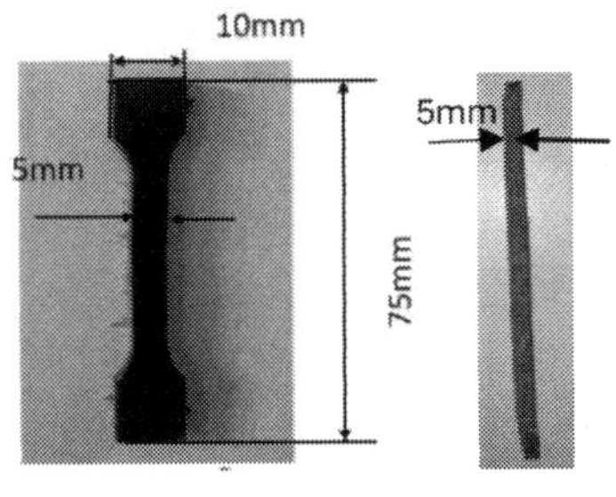

Figure 8. Samples test

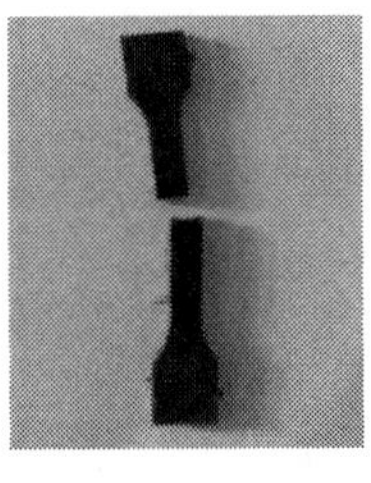

Figure 9. Tensile break sample with 10% CuO

3.2.2 Yield strength and yield point

The yield strength of materials with varying CuO concentrations, depicted in Figure 10 and listed in Table 3, shows a relationship with CuO content. The slope of the initial linear portion of the stress-strain curve represents the elastic modulus, which describes the material's stiffness. The yield strengths are as follows: **PP-CuO 3%:** 306.67 N/mm^2, **PP-CuO 5%:** 400 N/mm^2, **PP-CuO 7%:** 228.57 N/mm^2, **PP-CuO 10%:** 400 N/mm^2. The data indicates that higher CuO concentrations generally increase yield strength, though the 7% CuO sample disrupts this trend

IOP Conf. Series: Materials Science and Engineering **1319** (2024) 012021 doi:10.1088/1757-899X/1319/1/012021

possibly due to particle agglomeration or uneven dispersion affecting mechanical properties. The ultimate tensile strength and ductility are shown at the peak and subsequent drop of each curve. Higher CuO concentrations may increase strength but affect ductility based on the composite's formulation and processing conditions. The elongation at break varies, reflecting differences in toughness due to CuO distribution.

The yield point (0.2% strain offset), marking the onset of plastic deformation, deviates from the linear trend and occurs at stress values around 40 to 50 MPa. Specifically, for the green line, the yield point is noticeable around 50 MPa, indicating significant plastic deformation.

In summary, the yield points for the materials in the graph occur around stress values of 40 to 50 MPa, marking the end of the linear stress-strain relationship and the beginning of plastic deformation.,

Table 3. Yield strength and yield point for PP-CuO composites at different concentrations

Material	Elastic Modulus [GPa]	Yield Strength [MPa]
PP - CuO 3%	2	40
PP - CuO 5%	3	45
PP - CuO 7%	3	45
PP - CuO 10%	3.5	50
PP	1.4 - 1.7	25 -35

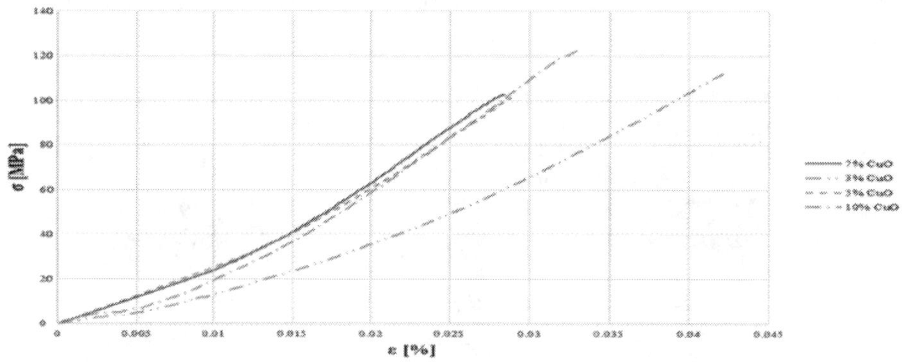

Figure 10 Representation of yield strength as a function of yield point for PP-CuO composites at different concentrations

In conclusion, increasing the CuO concentration in the PP matrix notably enhances the material's mechanical properties while also imparting valuable antimicrobial characteristics compared to pure PP material.

3.3 Determination of antimicrobial character

Bacterial growth on the tested material was monitored, with the material impregnated with different concentrations of CuO (3, 5, 7, 10%). The unimpregnated material (PP) and a Petri dish with only the culture medium were used as controls (Figure 13).

IOP Conf. Series: Materials Science and Engineering **1319** (2024) 012021 doi:10.1088/1757-899X/1319/1/012021

Sterilization of Test Materials: The pieces of material designated for testing were sterilized to ensure that no pre-existing microorganisms could affect the results.
This was achieved by placing the materials in an autoclave and heating them to 120°C for a duration of 20 minutes. The high temperature effectively killed any potential contaminants, ensuring that the subsequent bacterial growth could be attributed solely to the introduced inoculum.
Preparation of Bacterial Inoculum: a bacterial inoculum was prepared using the Gram-positive bacteria Staphylococcus aureus.
The bacteria were cultured and then suspended in sterile distilled water to create a uniform bacterial suspension. This suspension was carefully prepared under sterile conditions to avoid any external contamination and ensure that the inoculum was pure.
Inoculation of Test Materials: under sterile conditions within a microbiological hood,13250 CFU/200µl of the prepared bacterial inoculum was carefully pipetted onto each piece of material. This precise volume ensured a consistent starting concentration of bacteria across all test samples, allowing for accurate comparisons of bacterial growth on materials impregnated with different concentrations of CuO.
Application of Culture Medium: following the inoculation, 3ml of liquefied Plate Count Agar (PCA) medium, which serves as a universal medium for heterotrophic bacteria, was poured over the bacterial inoculum on each material.
The PCA medium, in its liquid state, evenly covered the inoculum and solidified upon cooling, providing a nutrient-rich environment that supports bacterial growth.
Incubation and Observation: the prepared samples were then incubated at 36°C for 48 hours Figure 11. This temperature is optimal for the growth of Staphylococcus aureus, ensuring robust bacterial proliferation.

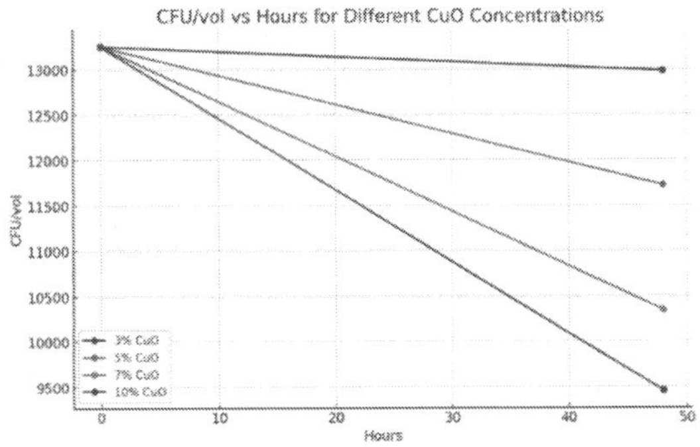

Figure 11 CFU/vol vs hours for different CuO concetration

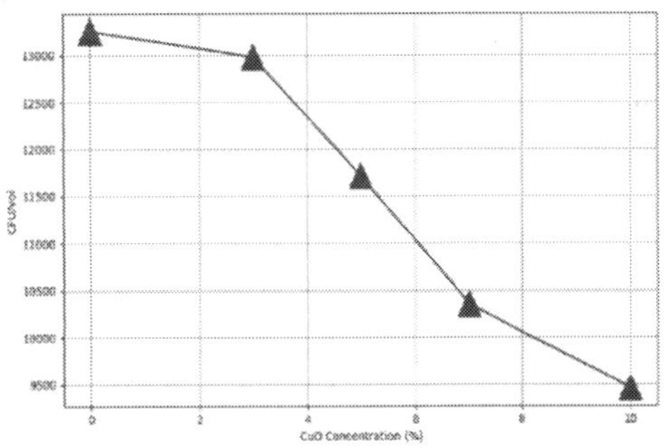

Figure 12 CFU/vol vs CuO concetration

After the incubation period, bacterial growth was assessed. It was observed that the growth was generally good, but there was a noticeable decreasing trend in bacterial proliferation directly proportional to the increasing concentrations of CuO in the materials, for the sample with 3% CuO the bacterial inoculum was 12980 CFU/vol, the sample with 5% CuO the bacterial inoculum was 11720 CFU/vol, sample with 7% CuO the bacterial inoculum was 10350 CFU/vol and the sample with 10% CuO the bacterial inoculum was 9460 CFU/vol.
Higher concentrations of CuO resulted in reduced bacterial growth, demonstrating the antimicrobial effectiveness of CuO, Figure 12.

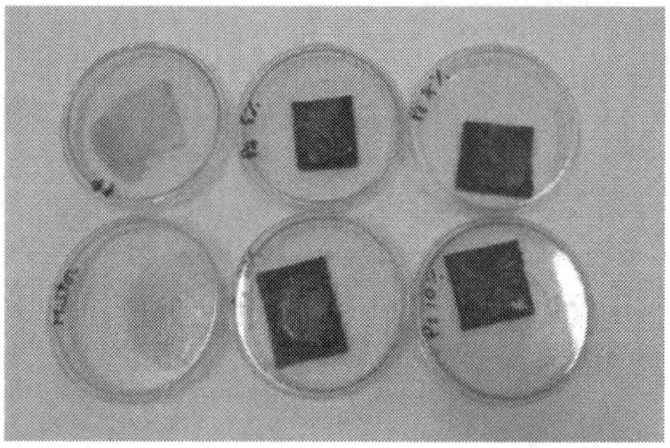

Figure 13 Antimicrobial Test Samples After Incubation

4. Conclusions

During the experiments, antimicrobial polymer composites based on copper oxide nanoparticles samples were obtained with different concentrations of 3%, 5%, 7%, 10% CuO, which were later subjected to investigations.
The thermal stability of the composites was confirmed through thermogravimetric analysis, which indicated that the decomposition of the PP matrix occurred between 400°C and 500°C, with complete decomposition at 650°C. The residual mass, primarily composed of CuO, validated the incorporation and stability of the nanoparticles within the composite.
Mechanical testing revealed that increasing CuO concentration generally enhanced the tensile strength and yield strength of the composites. Specifically, the elastic modulus of the composites significantly increased with higher CuO content. The composite with 3% CuO showed an approximately 29.03% increase in the elastic modulus compared to pure PP. The most substantial enhancement was observed in the composite with 10% CuO, which exhibited a 125.81% increase in the elastic modulus. These results indicate that incorporating CuO into PP not only improves the material's toughness but also substantially increases its stiffness.
The antimicrobial properties of PP-CuO composites were evaluated against Staphylococcus aureus. The results showed a significant reduction in bacterial growth with increasing CuO concentrations. If the 3% CuO composite showed a 2.04% decrease in bacterial growth, the highest reduction of 28.61% was obtained with the 10% CuO composite.
The effectiveness of CuO as an antimicrobial agent is thus certified..
Future work may focus on optimizing the dispersion of CuO nanoparticles and further exploring the balance between mechanical performance and antimicrobial effectiveness to tailor composites for specific applications.
In conclusion, it can be said that the experiments carried out highlight the potential of PP-CuO composites for applications making these composites suitable for applications demanding both strength and antimicrobial properties., such as sports medical devices.

References

[1] Laura Nicoleta CATANĂ, “CERCETĂRI ÎN DOMENIUL MATERIALELOR POLIMERICE ANTIMICROBIENE,” ŞCOALA DOCTORALĂ INTERDISCIPLINARĂ, Braşov, 2018.

[2] E. Zhang, X. Zhao, J. Hu, R. Wang, S. Fu, and G. Qin, “Antibacterial metals and alloys for potential biomedical implants,” *Bioact Mater*, vol. 6, no. 8, pp. 2569–2612, Aug. 2021, doi: 10.1016/j.bioactmat.2021.01.030.

[3] V. Geanta, I. Voiculescu, R. Stefanoiu, and E. R. Rusu, “Stainless Steels with Biocompatible Properties for Medical Devices,” *Key Eng Mater*, vol. 583, pp. 9–15, Sep. 2013, doi: 10.4028/www.scientific.net/KEM.583.9.

[4] H. Palza, R. Quijada, and K. Delgado, “Antimicrobial polymer composites with copper micro- and nanoparticles: Effect of particle size and polymer matrix,” *J Bioact Compat Polym*, vol. 30, no. 4, pp. 366–380, Jul. 2015, doi: 10.1177/0883911515578870.

[5] N. Jardón-Maximino *et al.*, “Antimicrobial Property of Polypropylene Composites and Functionalized Copper Nanoparticles,” *Polymers (Basel)*, vol. 13, no. 11, p. 1694, May 2021, doi: 10.3390/polym13111694.

[6] “Additive manufacturing in medical devices - Today’s Medical Developments.” Accessed: Jan. 18,2022.[Online].Available: https://www.todaysmedicaldevelopments.com/article/additive-manufaturing-medical-devices-dental/

[7] H. Palza, “Antimicrobial Polymers with Metal Nanoparticles,” *Int J Mol Sci*, vol. 16, no. 1, pp. 2099–2116, Jan. 2015, doi: 10.3390/ijms16012099.

[8] C. González-Henríquez, M. Sarabia-Vallejos, and J. Rodríguez Hernandez, “Antimicrobial Polymers for Additive Manufacturing,” *Int J Mol Sci*, vol. 20, no. 5, p. 1210, Mar. 2019, doi: 10.3390/ijms20051210.

[9] Jorge Zuniga, “Antimicrobial Materials for 3D Printing Medical Devices,” https://3dheals.com/antimicrobial-materials-for-3d-printing-medical-devices/.

[10] Hisham A. Maddah, “Polypropylene as a Promising Plastic: A Review,” *American Journal of Polymer Science 2016, 6(1): 1-11* , vol. 6, no. 1, pp. 1–11, 2016.

[11] Q. T. Shubhra, A. Alam, and M. Quaiyyum, “Mechanical properties of polypropylene composites,” *Journal of Thermoplastic Composite Materials*, vol. 26, no. 3, pp. 362–391, Apr. 2013, doi: 10.1177/0892705711428659.

[12] F. N. S. Raja, T. Worthington, and R. A. Martin, “The antimicrobial efficacy of copper, cobalt, zinc and silver nanoparticles: alone and in combination,” *Biomedical Materials*, vol. 18, no. 4, p. 045003, Jul. 2023, doi: 10.1088/1748-605X/acd03f.

[13] Carl Roth GmbH + Co. KG, “Specification,” https://www.carlroth.com/com/en/.

IOP Conf. Series: Materials Science and Engineering **1319** (2024) 012022 doi:10.1088/1757-899X/1319/1/012022

Impact of Aging on Tensile Properties of EPDM, TPV, and eco-TPV: A Comparative Study Across Diverse Environments

Daniel Foltuț[1*], Georgiana-Iulia Șoșoi[2], Estera Vălean[3], Jan Pospisil[4], Dragoș-Toader Pascal[5] and Viorel-Aurel Șerban[1,6]

[1] Department of Materials and Manufacturing Engineering, Politehnica University of Timișoara, 1 Mihai Viteazul Blv., Timisoara 300222, Romania
[2] Department of Applied Chemistry, Organic and Natural Compounds Engineering, Politehnica University of Timișoara, Carol Telbisz 6, Timisoara 300001, Romania
[3] Department of Mechanics and Strength of Materials, Politehnica University of Timișoara, 1 Mihai Viteazul Blv., Timisoara 300222, Romania
[4] Vitesco Technologies, Na Rovince 879, 720 00 Ostrava 20, Czechia
[5] Vitesco Technologies Engineering Romania, Calea Buziașului 11A, Timișoara 300699, Romania
[6] Technical Sciences Academy of Romania, B-dul Dacia 26, Bucuresti 030167, Romania

*E-mail: daniel.foltut@student.upt.ro

Abstract. This study examines the effects of aging on the tensile characteristics of Ethylene Propylene Diene Monomer (EPDM), Thermoplastic Vulcanizates (TPV), and Eco-friendly Thermoplastic Vulcanizates (Eco TPV) in different conditions. The samples underwent thermal aging at a temperature of 90°C and were immersed in sulfuric acid solutions with varying concentrations (1M, 0.1M, and 0.001M) for a duration of 1000 hours at the same temperature. The findings revealed that EPDM demonstrated the maximum tensile strength and a moderate level of elongation and breakage. Furthermore, it retained substantial mechanical integrity after undergoing aging, thereby establishing it as the most resilient material among those evaluated. The thermoplastic vulcanizate (TPV) exhibited intermediate tensile strength and elongation, but experienced significant deterioration when exposed to very acidic environments while Eco TPV had lower tensile strength but good chemical stability. The stress-strain behavior and compression set tests emphasized EPDM's exceptional durability and ability to recover elastically. The pH measurements of the immersion solutions indicated an elevation caused by the release of alkaline chemicals from the rubber components, especially in the 1M solution.

1. Introduction

Extensive study has been conducted on the durability and mechanical qualities of elastomers due to the increasing need for durable and high-performance materials in complex engineering applications. Ethyl propylene diene monomers (EPDM) and thermoplastic elastomer vulcanizate (TPV) are widely used elastomers because of their exceptional mechanical qualities, resilience to environmental factors, and ease of processing. These materials are suitable options for sealings and gaskets in Proton Exchange Membrane Fuel Cells (PEMFC). [1–3]

Thermal and chemical aging poses significant challenges for the longevity and reliability of these rubbers. Thermal aging involves prolonged exposure to high temperatures, resulting in changes in the structure and properties of materials such as rigidity, tensile strength, and elasticity. Chemical aging is caused by exposure to aggressive environments, such as acid solutions, which can degrade materials at the molecular level and cause surface degradation, hardness changes and mechanical failure. Understanding the impact of these aging processes on the mechanical properties of elastomers is vital to predicting their maintenance life and ensuring their performance in demanding applications. [4–8]

Previous research has explored the mechanical properties and aging behavior of EPDM, TPV, and Eco TPV. Research has shown that thermal aging can substantially modify the tensile strength and elasticity of certain materials. In the work conducted by Lin et al., the researchers extensively examined the dynamic mechanical properties of EPDM in simulated and accelerated PEMFC conditions. EPDM saw substantial alterations in its mechanical properties following the aging process. More precisely, the storage modulus (E') of EPDM reduced by around 30% following exposure to an accelerated aging environment. This environment involved subjecting the material to a humidified air stream at a temperature of 120°C, combined with a pressure of 1.5 atmospheres. This decrease in the storage modulus indicates a significant reduction in both stiffness and elasticity. In addition, the loss modulus (E'') and tan delta (tan δ) values of EPDM also exhibited significant deterioration, indicating a noticeable decrease in the material's capacity to dissipate energy and sustain mechanical integrity during dynamic loading. [9]

Liu et al. conducted a study to examine the deterioration characteristics of EPDM rubber in an acidic environment. The study specifically compared the degradation of peroxide-cured EPDM with sulfuric-cured EPDM. The study showed that peroxide-cured EPDM undergoes substantial deterioration as a result of molecular main chain breakage, resulting in the creation of double bonds and carbonyl (C=O) structures. Nevertheless, the inclusion of a coagent called triallyl isocyanurate (TAIC) in the peroxide curing system causes a postponement in the deterioration of the EPDM molecular structure. The reason for this is that the TAIC crosslinking spots undergo hydrolysis selectively, which improves the stability of the rubber. Over time, without the presence of TAIC, the peroxide-cured EPDM undergoes additional crosslinking reactions during the final phases of aging, resulting in the development of an ether-crosslinked structure. [10]

Finally, Naskar et al. examined the impact of various stabilizers on the ageing of dicumyl peroxide (DCP)-cured thermoplastic vulcanizates made from polypropylene (PP) and EPDM. The research highlighted that stabilizers significantly influence the ageing performance of TPVs. Without stabilizers, the TPVs showed poor ageing resistance, with notable degradation in tensile strength and elongation at break [11,12].

Our work seeks to expand on previous research by looking more into the aging characteristics of EPDM, TPV, and Eco TPV in relation to their use in dynamic sealing for PEMFC system valves. It is still necessary to comprehend the combined and long-term effects of such aging processes under conditions that closely resemble real-world operational environments, even though earlier research—such as that conducted by Lin et al. and Naskar et al.—has given important insights into the effects of thermal and chemical aging on the mechanical properties of such materials.

To close this gap, we investigate the effects of different concentrations of sulfuric acid solutions and long-term heat aging at 90°C for 1000 hours. To evaluate the chemical aging changes, the study also includes pH measurements of the aging solutions before and after exposure. Furthermore, as-delivered samples undergo compression set tests to assess their initial mechanical robustness and appropriateness for dynamic sealing applications in PEMFC systems.

2. Experimental setup

2.1. Materials

Developed to give thermoplastics the processability and resilience of thermoplastics while preserving their rubber-like properties, advanced elastomeric materials characterized as Thermoplastic Vulcanizate (TPV) and Eco Thermoplastic Vulcanizate (Eco TPV) is produced by mixing polypropylene (PP) and cross-linked EPDM rubber, therefore providing good mechanical and thermal stability. With at least 15% post-consumer recycled materials, the Eco TPV variant keeps same performance standards and promotes sustainability. Eco TPV's carbon impact is significantly lower compared to that of regular EPDM since recycled materials are used. The peroxidic cured ethylene propylene diene monomer rubber has strong resistance against oxidation, weathering, heat, and ozone.

Some of the material properties are presented in Table 1.

Table 1. Material Properties[a].

Material	EPDM	TPV	ECO-TPV
Shore A Hardness	77	87	83
Tensile Stress at 100% elongation	7.5 MPa	4.61 MPa	4.5 MPa
Stress at break	19.2 MPa	10.4MPa	8.7 MPa
Elongation at break	213%	526%	560%
Density	1.12 g/cm^3	0.92 g/cm^3	0.96 g/cm^3
Work Temperature	-40 to 150°C	-40 to 135 °C	-40 to 135 °C

[a] Datasheet values.

2.2. Chemical and Heat Aging Setup

The aging process was designed to replicate the ambient conditions found in fuel-cell systems, with an emphasis on studying the materials robustness and resistance to deterioration.

During the chemical aging experiments, the materials were submerged for a thousand hours at 90°C in solutions containing different concentrations of sulfuric acid (H_2SO_4). The pH values of the sulfuric acid solutions varied, with pH values ≈ 0.1 for 1M H_2SO_4, 0.84 for 0.1M H_2SO_4, and 2.24 for 0.001M H_2SO_4 being the most acidic, moderately acidic, and mildly acidic, respectively. This method sought to evaluate the long-term effects of varying acidity levels on the mechanical properties of the materials. Previous research has indicated that acidic environments can have a substantial impact on the mechanical integrity and longevity of elastomeric materials. This can result in modifications to their properties, including elongation at break and tensile strength. [9,13].

The materials were subjected to thermal aging in addition to chemical aging, which involved 1000 hours at 90°C in a thermal chamber. This part of the research sought to assess the materials' thermal resilience during extended exposure to high temperatures. [14,15].

2.3. Tensile Testing Setup

Tensile testing was performed on EPDM, TPV, and Eco TPV materials to assess their mechanical characteristics, including tensile strength, elongation at break, and tensile stress at 100% elongation. The specimens were produced according to the ISO 37 (2011) standard, utilizing dumbbell-shaped samples (Type 2). The specimens were obtained from sheet material using a precise die cutter and tested at a temperature of 23°C and a relative humidity of 50% for a minimum of 24 hours before conducting the tests.

The testing was conducted with an Instron universal testing equipment that was fitted with mechanical grips to minimize slippage and provide a consistent distribution of stress. Throughout the test, a video extensometer was used to have an accurate measurement of the stretch. The velocity of the crosshead was adjusted to 500 mm per minute.

2.4. Compression Set Measurements

The compression set testing of the studied materials was performed to assess their ability to maintain their elastic properties after prolonged compression at an elevated temperature. The tests followed the ISO 815-1A standard, specifically set at 125°C for 72 hours.

Specimens for the compression set tests were prepared in cylindrical shapes with standard dimensions as specified by ISO 815-1A, type B (6.3mm thickness and a diameter of 13mm).

The compression set was calculated as the percentage of the original deflection that remained as permanent deformation after the recovery period. The formula used for this calculation is:

$$Compression\ Set\ (\%) = \frac{h_0 - h_1}{h_0 - h_s} \times 100 \qquad (1)$$

where:

h_0 is the initial thickness of the test piece, in mm
h_1 is the thickness of the test piece after recovery, in mm
h_s is the height of the spacer, in mm

3. Results

3.1. Tensile testing

The tensile testing results are summarized in **Table 2**. For EPDM, the tensile strength ranges from 16.7 MPa to 18.84 MPa across different aging treatments. The highest tensile strength is observed in the as-delivered state (AD) at 18.84 MPa, while the lowest is in the heat-aged condition (HE) at 16.7 MPa. Elongation at break percentages vary significantly, with the as-delivered state exhibiting the highest elongation of 187% and the lowest observed in the HE condition at 163%. The modulus at 100% elongation remains relatively consistent, with values The results of tensile testing for different elastomeric materials, including EPDM, TPV, and ECO TPV, under various aging conditions are presented in Table 2. Across different aging treatments, the tensile strength of EPDM ranges from 16.7 MPa to 18.84 MPa. The as-delivered state (AD) exhibits the highest tensile strength at 18.84 MPa, whereas the heat-aged condition (HE) shows the lowest strength at 16.7 MPa. The percentages of elongation at break show significant variation, with the highest observed in the as-delivered state at 187% and the lowest in the HE condition at 163%. The modulus at 100% elongation remains consistently stable, with values ranging from 8.47 MPa to 9.2 MPa.

Table 2. Tensile testing results

MATERIAL			EPDM		
Sample State	AD[a]	HE[b]	1M[c]	0.1M[d]	0.001M[e]
Tensile Strength (MPa)	18.84	16.7	17.8	18.1	17.95
Elongation at break (%)	187	163	180	182	183
Modulus at 100% elongation (MPa)	9.05	9.2	8.57	8.86	8.47
MATERIAL			TPV		
Tensile Strength (MPa)	10.449	10	10.58	10.3	9.34
Elongation at break (%)	500	472	511	501	487
Modulus at 100% elongation (MPa)	4.937	5.109	4.917	4.651	4.481
MATERIAL			ECO TPV		
Tensile Strength (MPa)	7.85	7.32	7.35	7	7.148
Elongation at break (%)	519	467	479	481	473
Modulus at 100% elongation (MPa)	3.77	3.87	3.76	3.43	3.503

[a]Samples as delivered; [b]Heat aged at 90°C for 1000 hours samples; [c]samples aged in 1M H_2SO_4 at 90°C for 1000 h; [d]samples aged in 0.1M H_2SO_4 at 90°C for 1000 h; [e]samples aged in 0.001M H_2SO_4 at 90°C for 1000 h

Regarding TPV, the tensile strengths vary from 9.34 MPa to 10.58 MPa. The AD sample exhibits the highest strength. The elongation at break ranges from 470% to 500%, with the highest value observed in the AD state and the lowest value observed in the 0.001M H_2SO_4 aged condition. The modulus at 100% elongation for TPV exhibits a minor fluctuation, ranging from 4.481 MPa to 5.109 MPa.

In the case of ECO TPV, the tensile strengths range from 7.14 MPa to 7.85 MPa, with the highest strength observed in the as-delivered state. The range of elongation at break percentages is from 467% to 519%, with the as-delivered sample displaying the highest elongation. The modulus at 100% elongation for ECO TPV varies between 3.43 MPa and 3.87 MPa.

Figure 1 shows clear differences in mechanical properties among the samples of TPV, ECO TPV, and EPDM materials. The stress-strain curves highlight the contrasting characteristics, as EPDM's curve reaches a sharp peak, demonstrating its strong resistance to stress, whereas the curves for TPV and ECO TPV display their exceptional ability to stretch.

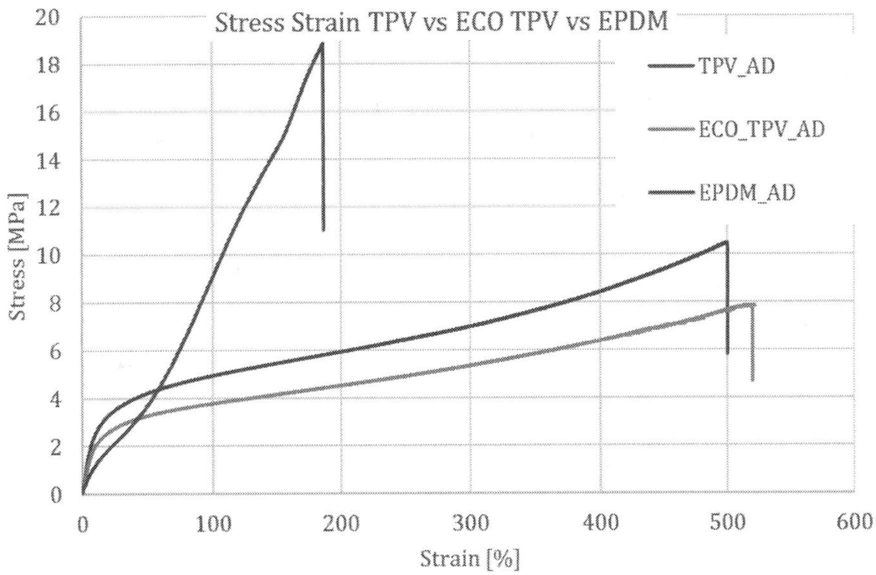

Figure 1 Stress-strain curves for as-delivered (AD) samples of TPV, ECO TPV, and EPDM

Figure 2 demonstrates the impact of aging on the tensile properties of TPV materials. This figure illustrates the stress-strain behavior of TPV samples under different aging conditions. The samples were subjected to aging in sulfuric acid solutions of varying molar concentrations (1M, 0.1M, and 0.001M), as well as heat aging (HE) and air drying (AD).

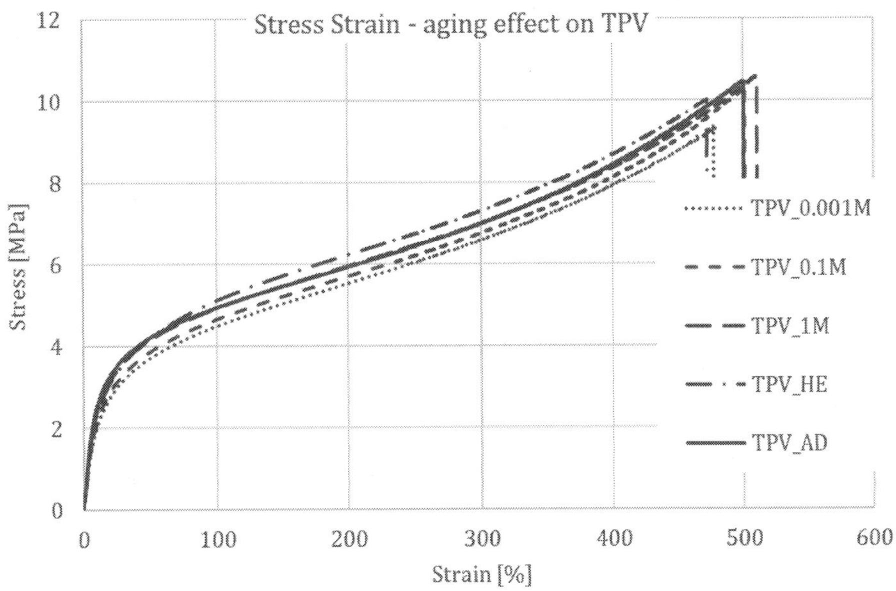

Figure 2 Stress-Strain behavior of TPV samples under different aging conditions, including as delivered (TPV_AD), heat aged at 90°C for 1000h(TPV_HE), and aged in 1M (TPV_1M), 0.1M (TPV_0.1M), and 0.001M (TPV_0.001M) H_2SO_4 for 1000h

The stress-strain curves indicate that the aging process has an impact on the mechanical properties of TPV to different extents. There are slight fluctuations in the tensile strength of the aged TPV samples compared to the as-delivered state. As the severity of the aging conditions increases, the maximum tensile strength slightly decreases. Regardless of the modifications, the elongation at break remains consistently stable across all samples, with values hovering around 500%. It appears that the aging process has a lesser impact on the flexibility and ductility of TPV, as opposed to its tensile strength.

When looking at the different samples, the AD sample has the highest tensile strength among the aged samples. This suggests that the aging conditions have caused a slight decrease in the material's ability to bear stress.

Figure 3 shows a significant decrease in the tensile strength of ECO TPV after chemical aging, particularly in environments with higher concentrations of sulfuric acid. However, the elongation at break remains relatively consistent.

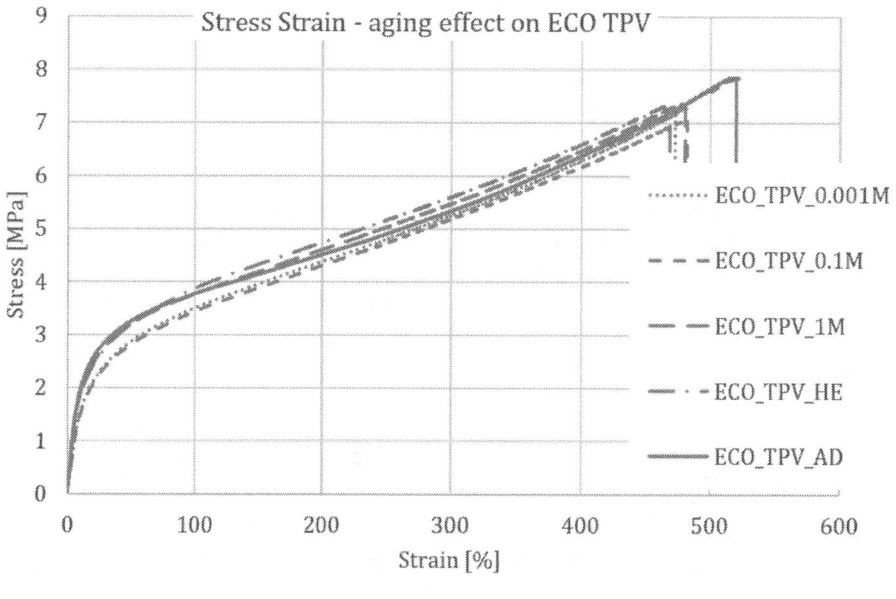

Figure 3 Stress-Strain behavior of ECO TPV samples under different aging conditions

Figure 4 shows a demonstration of the change in behavior of EPDM through tensile testing. The findings suggest that EPDM exhibits a remarkable ability to maintain its tensile strength and elongation even when subjected to strains of up to 100%, with only slight differences observed across various aging treatments. The as-delivered samples exhibit the highest tensile strength, but there is a slight decrease in tensile strength due to chemical aging, particularly in higher concentrations of sulfuric acid.

The tensile testing results of EPDM, TPV, and ECO TPV materials demonstrate unique mechanical properties and aging responses. The table and figures clearly show that among the three materials, EPDM exhibits the highest tensile strength and elongation at break, with values reaching up to 18.84 MPa and 187%, respectively. The stress-strain curves provide additional evidence of the exceptional performance of EPDM, demonstrating its significant strength even when subjected to extended periods of high temperatures and varying concentrations of sulfuric acid. TPV demonstrates moderate tensile strength and elongation, with a noticeable decrease in mechanical properties over time, especially in sulfuric acid environments, suggesting a greater vulnerability to chemical degradation. ECO TPV exhibits the lowest tensile strength and elongation, but it maintains consistent performance under various aging conditions, indicating superior chemical resistance compared to TPV.

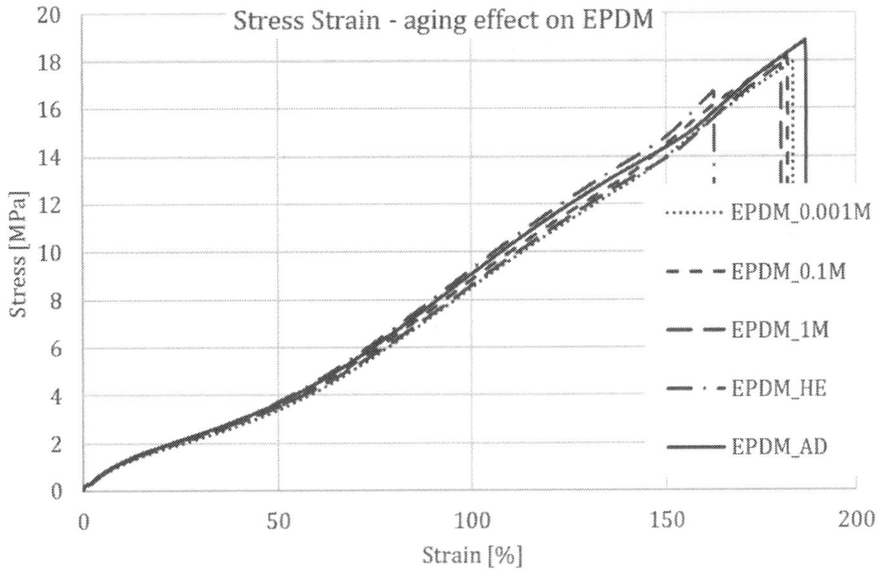

Figure 4 Stress-Strain behavior of EPDM samples under different aging conditions

3.2. Compression-Set results

Figure 5 indicates that EPDM exhibits the lowest compression set, suggesting its excellent capability to regain its initial shape even after prolonged deformation. The characteristic observed is consistent with the tensile test outcomes, where EPDM exhibited exceptional tensile strength and elongation at break, indicating remarkable elasticity and long-lasting performance.

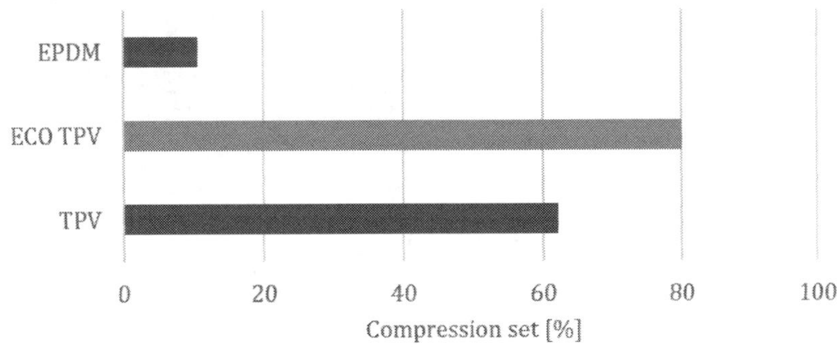

Figure 5 Compression set values of EPDM, ECO TPV, and TPV

On the other hand, ECO TPV shows the highest compression set, indicating a lower ability to bounce back after being deformed. This aligns with its comparatively lower tensile strength and elongation values. TPV exhibits a well-balanced performance in terms of compression set, showcasing its intermediate tensile strength and elongation characteristics, which are comparable to those of EPDM and ECO TPV.

3.3. pH measurements on the solutions

Table 3 presents the pH levels of the three different acidic solutions (0.001M, 0.1M, and 1M) measured initially and after 1000 hours of ECO TPV, TPV, and EPDM immersion at 90°C. Initial pH values were 2.24 for the 0.001M solution, 0.84 for the 0.1M solution, and 0.1 for the 1M solution. Post-immersion, immersion of ECO TPV showed slight increases in pH across all solutions, suggesting moderate chemical stability. Specifically, the 0.001M solution increased to 2.42, the 0.1M solution to 1.1, and the 1M solution to 0.37. TPV immersion led to a marginal decrease in the 0.001M solution's pH to 2.2, with slight increases in the 0.1M solution to 1.12 and the 1M solution to 0.36. EPDM immersion resulted in the highest increases in pH: 2.46 for the 0.001M solution, 1.13 for the 0.1M solution, and stabilization at 0.36 for the 1M solution.

Table 3 pH values after samples immersion[a]

Solution	Initial	after ECO TPV immersion (% increase)	after TPV immersion (% increase)	after EPDM immersion (% increase)
0.001M	2.24	2.42 (8%)	2.2 (-1.8%)	2.46 (9.9%)
0.1M	0.84	1.1 (31%)	1.12 (33.3%)	1.13 (34.5%)
1M	0.1	0.37 (270%)	0.36 (260%)	0.36 (260%)

[a] Samples aged in solutions at 90°C for 1000 h

The notable increase in pH for all materials in the 1M solution can be attributed to the dilution effect of the acidic medium due to leaching or extraction of additives and degradation by-products from the rubber materials [16]. These by-products likely include basic or neutral compounds that neutralize some of the acidity, thereby raising the pH.

4. Conlusions

The study comparing the effects of aging on the tensile properties of EPDM, TPV, and Eco TPV materials offers valuable insights into how these materials perform when exposed to high temperatures and acidic conditions for extended periods of time. The tensile strength of EPDM is superior to that of the other materials studied, although its elongation at break is lower. Despite this, EPDM maintains significant mechanical integrity even after aging, making it an excellent choice for demanding applications like PEMFC. The TPV displayed a decent level of tensile strength and a high elongation, but it showed significant degradation when exposed to sulfuric acid environments, suggesting that it is more prone to chemical aging. On the other hand, Eco TPV exhibited consistent performance under different aging conditions, indicating chemical resistance, despite its lower tensile strength compared to EPDM and TPV.

EPDM stress-strain results showed good performance, particularly in its ability to maintain tensile strength even after aging. On the other hand, TPV and Eco TPV exhibited different responses under various aging conditions. The results of the compression set tests highlighted the exceptional elastic recovery of EPDM, as it exhibited the lowest compression set values. On the other hand, Eco TPV demonstrated the highest compression set value. The pH measurements of the immersion solutions indicated that the pH levels were raised by all materials, particularly in the 1M solution, as a result of the release of basic compounds during degradation. In terms of thermal and chemical aging, EPDM demonstrated the highest level of resilience among the materials tested. TPV and Eco TPV also showed good performance, each with their own unique advantages depending on the specific application needs.

Acknowledgments

Special thanks to Vitesco Technologies Group AG for providing the necessary resources and support for this study.

Declaration

In preparing this paper, we utilized AI tools including ChatGPT (based on OpenAI's GPT-4 architecture) and QuillBot for paraphrasing, results interpretation, and drafting. We have reviewed, edited, and verified the content to ensure its accuracy and integrity.

References

[1] J.E. Klein, G.M. Divoux, H.K. Singh, S.W. Case, D.A. Dillard, J.G. Dillard, W. Kim, R.B. Moore, J.B. Parsons, Assessing durability of elastomeric seals for fuel cell applications, Conference Proceedings of the Society for Experimental Mechanics Series 5 (2011) 175–181. https://doi.org/10.1007/978-1-4419-9798-2_21.
[2] A. Boubakri, N. Haddar, K. Elleuch, Y. Bienvenu, Impact of aging conditions on mechanical properties of thermoplastic polyurethane, Mater Des 31 (n.d.) 4194–4201. https://doi.org/10.1016/j.matdes.2010.04.023.
[3] J. Garche, L. Jörissen, Applications of fuel cell technology: Status and perspectives, Electrochemical Society Interface 24 (2015) 39–43. https://doi.org/10.1149/2.F02152IF/XML.
[4] N. Ning, S. Li, H. Wu, H. Tian, P. Yao, G.H. HU, M. Tian, L. Zhang, Preparation, microstructure, and microstructure-properties relationship of thermoplastic vulcanizates (TPVs): A review, Prog Polym Sci 79 (2018) 61–97. https://doi.org/10.1016/J.PROGPOLYMSCI.2017.11.003.
[5] J. Liu, L.W. Zhang, B. Wang, Z.Q. Xu, W. Wang, Effect of Thermal Aging on Mechanical Properties of EPDM Rubber, Applied Mechanics and Materials 727–728 (2015) 318–321. https://doi.org/10.4028/WWW.SCIENTIFIC.NET/AMM.727-728.318.
[6] R. Wang, Y. He, Z. Wang, H. Kang, J. Ma, H. Jin, Effect of Thermal Aging on Breakdown Strength of Epdm Rubber, IOP Conf Ser Mater Sci Eng 768 (2020) 062091. https://doi.org/10.1088/1757-899X/768/6/062091.
[7] W. Shuang-Hong, Z. Chang-Sheng, H. Chao-Fan, The aging property and storage life prediction of EPDM, J Phys Conf Ser 2478 (2023) 032046. https://doi.org/10.1088/1742-6596/2478/3/032046.
[8] J. Song, Z. Wang, H. Ji, R. Wang, Y. Zhou, Y. He, H. Jin, Mechanical Behavior of Aged EPDM Insulation of High-Voltage Cable Joints in Thermal-Oxidative Environment, IOP Conf Ser Mater Sci Eng 768 (2020) 062088. https://doi.org/10.1088/1757-899X/768/6/062088.
[9] C.W. Lin, C.H. Chien, J. Tan, Y.J. Chao, J.W. Van Zee, Dynamic mechanical characteristics of five elastomeric gasket materials aged in a simulated and an accelerated PEM fuel cell environment, Int J Hydrogen Energy 36 (2011) 6756–6767. https://doi.org/10.1016/J.IJHYDENE.2011.02.112.
[10] Q. Liu, J. Li, Y. Jiang, C. Cong, L. Xu, Y. Zhang, X. Meng, Q. Zhou, Effect of crosslinked structure on the chemical degradation of EPDM rubber in an acidic environment, Polym Degrad Stab 185 (2021) 109475. https://doi.org/10.1016/J.POLYMDEGRADSTAB.2020.109475.
[11] K. Naskar, D. Kokot, J.W.M. Noordermeer, Influence of various stabilizers on ageing of dicumyl peroxide-cured polypropylene/ethylene-propylene-diene thermoplastic vulcanizates, Polym Degrad Stab 85 (2004) 831–839. https://doi.org/10.1016/J.POLYMDEGRADSTAB.2004.03.016.
[12] M.H. Zulkifli, M.S.M. Rasidi, N.A.M. Rahim, L. Musa, A.H. Masa, The Study of Recycled Ethylene Propylene Diene Monomer (EPDM-r)/Polypropylene (PP) Polymeric Blends, J Phys Conf Ser 2080 (2021) 012012. https://doi.org/10.1088/1742-6596/2080/1/012012.
[13] S. Kashi, R. Varley, M. De Souza, S. Al-Assafi, A. Di Pietro, C. de Lavigne, B. Fox, Mechanical, Thermal, and Morphological Behavior of Silicone Rubber during Accelerated Aging, Polym Plast Technol Eng 57 (2018) 1687–1696. https://doi.org/10.1080/03602559.2017.1419487.
[14] A. Boubakri, N. Haddar, K. Elleuch, Y. Bienvenu, Impact of aging conditions on mechanical properties of thermoplastic polyurethane, Mater Des 31 (2010) 4194–4201. https://doi.org/10.1016/j.matdes.2010.04.023.

[15] S. Pehlivan-Davis, J. Clarke, S. Armour, Comparison of accelerated aging of silicone rubber gasket material with aging in a fuel cell environment, J Appl Polym Sci 129 (2013) 1446–1454. https://doi.org/10.1002/APP.38837.
[16] J. Feng, Q. Zhang, Z. Tu, W. Tu, Z. Wan, M. Pan, H. Zhang, Degradation of silicone rubbers with different hardness in various aqueous solutions, Polym Degrad Stab 109 (2014) 122–128. https://doi.org/10.1016/J.POLYMDEGRADSTAB.2014.07.011.

IOP Conf. Series: Materials Science and Engineering **1319** (2024) 012023 doi:10.1088/1757-899X/1319/1/012023

Effects of Sulfuric Acid-Water Immersion on the Properties of Ethylene Propylene Diene Monomer, Thermoplastic Vulcanizates, and Eco-Friendly Thermoplastic Vulcanizates

Daniel Foltuţ[1*], Jan Pospisil[2] and Dragoş-Toader Pascal[3]

[1] Department of Materials and Manufacturing Engineering, Politehnica University of Timişoara, 1 Mihai Viteazul Blv., Timisoara 300222, Romania
[2] Vitesco Technologies, Na Rovince 879, 720 00 Ostrava 20, Czechia
[3] Vitesco Technologies Engineering Romania, Calea Buziaşului 11A, Timişoara 300699, Romania

*E-mail: daniel.foltut@student.upt.ro

Abstract. This research investigates the impact of sulfuric acid-water solutions on the physico-chemical characteristics of ethylene propylene diene monomer rubber (EPDM), thermoplastic vulcanizates (TPV), and ECO TPV materials. We evaluated the deterioration of these materials by subjecting them to acidic environments and using methods such as differential scanning calorimetry (DSC), volume and weight measurements, hardness measures, and scanning electron microscopy (SEM). The samples were submerged in solutions of H_2SO_4 with concentrations of 1M, 0.1M, and 0.001M at a temperature of 90°C for a duration of 1000 h. The results indicated that EPDM had superior chemical and thermal stability, accompanied by negligible changes in weight and volume. The TPV exhibited moderate stability, however the ECO TPV showed severe deterioration, as seen by major increases in weight and volume, as well as notable changes in DSC profiles. Surface degradation and cracking, especially in ECO TPV samples, were observed using SEM investigation. The hardness results indicated little changes in hardness after exposure to the various chemical aging techniques. The results indicate that EPDM and TPV are better suited for use in acidic conditions, such as fuel-cell systems. However, ECO TPV may need extra stabilization in order to enhance its endurance.

1. Introduction

The automotive industry is progressively prioritizing the reduction of its carbon emissions by using innovative methods such as incorporating fuel cells and embracing environmentally sustainable components. Fuel cells, specifically Polymer Electrolyte Membrane Fuel Cells (PEMFCs), are regarded as a viable energy source because of their notable efficiency and few emissions. Nevertheless, the longevity of PEMFCs is of utmost importance, since the effectiveness of their operation heavily relies on the performance of components like seals and gaskets. ethylene propylene diene monomer (EPDM) rubber is extensively used in seals because of its stability and resistance to deterioration, which makes it highly suitable for usage in PEMFCs. [1–6]

Simultaneously, the need for environmentally acceptable materials has resulted in the creation of thermoplastic vulcanizates (TPVs) and their environmentally friendly variations (Eco-TPVs). These materials possess the characteristics of both thermoplastics and elastomers, providing the advantages of being recyclable and having a reduced environmental footprint. Nevertheless, their behavior under different environmental circumstances, such as when subjected to sulfuric acid, remains little comprehended. Prior research has emphasized the significance of crosslinking systems in enhancing the mechanical characteristics of TPVs, such as their ability to withstand heat aging and their compatibility with polypropylene (PP). [7–9]

While there are options to predict the life of elastomers using numerical methods [10], thermal and chemical aging need to be performed to have accurate prediction on the aging of synthetic rubber materials.

Studies on the thermal aging of elastomers have shown that extended exposure to high temperatures may result in substantial changes in their mechanical characteristics and chemical composition. Tayefi et al. conducted a review on the process of accelerated aging and lifespan prediction of elastomers. They emphasized the significance of temperature-induced oxidative degradation, which leads to embrittlement and a decrease in elasticity [11]. Research conducted by Song et al. examined the impact of varying viscosities of EPDM on the crystallization behavior and mechanical characteristics of TPVs. The findings indicated that greater viscosities of EPDM improve thermal stability, leading to enhanced overall mechanical performance. [12]

The studies that were done until the time of the writing of this paper reveal that exposure to acidic solutions can lead to surface roughness, cracks, and weight changes over time [13, 14]. The degradation of EPDM in these conditions is primarily due to de-crosslinking and chain scission, which compromises the material's integrity and performance [14, 15].

Thermoplastic Vulcanizates (TPVs), such as Eco TPV, are being used as substitutes for EPDMs since they can be recycled and have a less environmental footprint. Research on the mechanical characteristics of TPVs has shown that various methods of crosslinking may have a substantial impact on their performance. Specifically, peroxide crosslinking has been found to provide high resistance to heat aging and mechanical stability [7,16]. However, the performance of these materials under acidic conditions, similar to those in fuel cells, remains less explored.

This study seeks to fill the gap in knowledge by examining the impact of sulfuric acid-water solutions on the physico-chemical characteristics of EPDM, TPV, and Eco TPV. We conduct a thorough investigation of material deterioration under acidic environments using a range of methodologies including volumetric, hardness, mass, thermal, and microscopic methods. The research evaluates the changes in weight and volume, hardness and micro-hardness, surface morphology, and thermal characteristics of various materials during extended exposure.

2. Experimental setup

2.1. Materials

Thermoplastic Vulcanizate (TPV) and Eco Thermoplastic Vulcanizate (Eco TPV) are sophisticated elastomeric materials designed to blend the flexibility of rubber with the capacity to be processed like thermoplastics. TPV is a composite material consisting of cross-linked EPDM rubber and polypropylene (PP), which imparts it with favorable mechanical qualities and thermal stability. The Eco TPV variation contains at least 15% recycled material from post-consumer sources, resulting in a lower carbon footprint compared to regular EPDM or the normal TPV version.

Ethylene Propylene Diene Monomer (EPDM) rubber, particularly the peroxidic cured type, is

Table 1. Material Properties[a].

Material	EPDM	TPV	ECO-TPV
Density	1.12 g/cm^3	0.92 g/cm^3	0.96 g/cm^3
Work Temperature	-40 to 150°C	-40 to 135 °C	-40 to 135 °C
Tensile Stress at 100% elongation	7.5 MPa	4.61 MPa	4.5 MPa
Shore A Hardness	77	87	83
Elongation at break	213%	526%	560%
Stress at break	19.2 MPa	10.4MPa	8.7 MPa

[a] Datasheet values.

known for its resistance to heat, oxidation, ozone, and weathering. EPDM's stable chemical structure and specific mechanical characteristics make it a suitable material for demanding applications such as seals and gaskets in fuel cells and automotive components [17–20]. The comparative properties of these materials are summarized in **Table 1**, highlighting their respective mechanical strengths and operational temperature ranges.

2.2. Hot water extraction test -Ion Leaching

The Hot Water Extraction Test (HWET) is specifically intended to assess the leachability and chemical stability of materials during hot water exposure. The test utilizes non-contaminating borosilicate glass bottles, deionized water with a conductivity of less than 1.0 μS/cm, and a portable conductivity tester to evaluate the water's conductivity before and after exposure. The use of borosilicate glass bottles guarantees the absence of contamination originating from the containers themselves, while the use of deionized water provides a controlled and uniform testing environment.

In order to prepare for the HWET, the samples are used in their original state. The specimens' weight is measured both before and after the test to determine any changes in mass. Subsequently, the specimens are put in receptacles that are filled with deionized water. During the testing step, the containers are subjected to agitation for a duration of one minute to guarantee thorough mixing, and the initial conductivity of the water is assessed at a temperature of 23°C, which is the standard room temperature. Subsequently, the containers are hermetically sealed and subjected to incubation at a temperature of 80°C. Conductivity measurements are conducted at several time intervals, such as 0 h, 24 h, and up to 168 h of soaking, in order to observe any changes in the electrical conductivity of the water. In the latter stages, the specimens are weighed after testing and it is ensured that the conductivity of the water is kept below 5 μS/cm in order to pass the test. This indicates the leachability and chemical stability of the materials under examination. [21]

2.3. Acidic Solution Aging Setup

Three aging media were considered for the accelerated aging of the samples: Fuel-Cell like media, three different solutions (1 mol/L solution of 96% H_2SO_4 and DI Water, pH≈ 0.1 (1M); 0.1 mol/L solution of 96% H_2SO_4 and DI Water (0.1M), pH≈ 0.84; 0.001 mol/L solution of 96% H_2SO_4 and DI Water, pH≈ 2.24 (0.001M)). The samples were soaked in the solutions detailed at 90°C for 1000h. This method was used to assess how different levels of acid affect the mechanical properties of materials over time. Similar studies have shown that acid environments have a significant impact on elastomer materials' durability and mechanical integrity, resulting in changes in properties such as stress and elongation at break [9, 22].

The aging process was designed to simulate the conditions encountered in fuel-cell systems, focusing on their durability and resistance to degradation.

2.4. Weight and volume measurements

In a first step, the samples are submerged in the sulfuric acid solutions (0.1M, 0.001M, and 1M) for a duration of 1000 h at a temperature of 90°C. Prior to immersion, the weight of each sample is measured both in air and in water to obtain baseline values. After being submerged, the weight measurements are taken again in both air and water to evaluate the alterations.
Subsequently, the gathered data is used to compute the percentage alteration in both weight and volume of the specimens. Formula 1 is used to calculate the percentage change in weight.

$$\Delta m_{100} = \left(\frac{m_i - m_0}{m_0}\right) \bullet 100 \tag{1}$$

where m_0 is the initial mass and m_i is the mass after immersion. Volume change is calculated with formula 2:

$$\Delta V_{100} = \left(\frac{m_i - m_{i,w} + m_{s,w}}{m_0 - m_{0,w} + m_{s,w}} - 1\right) \bullet 100 \tag{2}$$

where m_0 is the initial mass of the test piece, m_i is the mass of the test piece after immersion, $m_{0,w}$ is the initial mass of the test piece plus sinker in water, $m_{i,w}$ is the mass of the test piece plus sinker after immersion in water, and $m_{s,w}$ is the mass of the sinker in water.

2.5. Hardness and Micro-hardness method

The experimental procedure consisted of assessing the Shore A hardness and micro-hardness IRHD of all material samples in accordance with established standards. The Shore A hardness was evaluated using the ISO 48-4 standard, whereas the micro-hardness IRHD was calculated using the ISO 48-2 standard.

2.6. Differential Scanning Calorimetry

The DSC apparatus was set up with nitrogen gas as the purge gas, provided at a flow rate of 40 ml/min, to establish a non-reactive environment. The temperature program started at -60°C with a heating rate of 10 K/min, finally reaching a peak temperature of 200°C.

2.7. Scanning Electron Microscopy Method

In the Scanning Electron Microscopy (SEM) study, the samples were coated with a thin layer of gold to increase conductivity and boost picture resolution. The scanning electron microscope (SEM) used in this investigation was a VEGA TESCAN, operating under a high vacuum environment.

The equipment parameters were adjusted to enhance the examination of surface morphology with high precision. The voltage applied was constantly set at 20.0 kilovolts. The working distance ranged from 12.42 mm to 15.34 mm. Various levels of magnification were utilized to capture both macroscopic and microscopic attributes of the samples. Low magnification images (366x, 373x, 380x) were used to observe larger surface features and overall texture, while higher magnification images (791x, 778x, 738x, 806x) were employed to examine finer surface details and microstructural characteristics.

3. Results and discussion

3.1. Hot water extraction test results

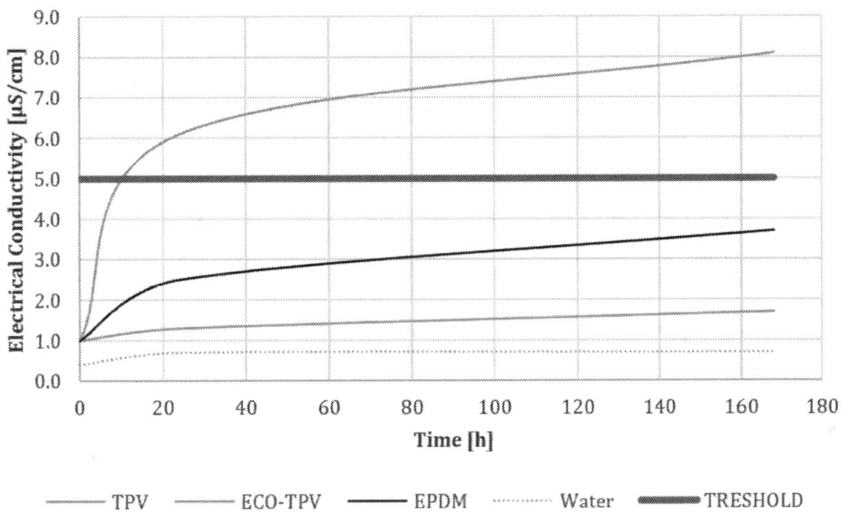

Figure 1 Electrical conductivity vs time for EPDM, TPV and ECO-TPV, measured at different time intervals of immersion in DI Water

Figure 1 depicts the variation in electrical conductivity (µS/cm) of TPV, ECO-TPV, EPDM, and water—over a period of 168 h. The DI water electrical conductivity is measured at regular intervals as a reference. Additionally, a threshold conductivity value is marked at 5.0 µS/cm. The ECO-TPV exhibits the highest electrical conductivity, surpassing the threshold within the first 20 hours and continuing to increase throughout the experiment, reaching approximately 8.0 µS/cm by the end. EPDM and TPV show moderate increases in conductivity, with EPDM surpassing the threshold around the 100-hour mark, indicating significant ionic mobility or degradation in the material leading to higher ion release. TPV remains below the threshold but shows a steady increase, reflecting some degree of ion release.

The differences in electrical conductivity values among the materials can be attributed to their varying compositions and structural properties. ECO-TPV's higher values suggest a greater susceptibility to ionic release or degradation under the test conditions, likely due to the presence of more polar or ionizable groups in its composition [23]. It is known that materials with higher levels of polar additives or fillers tend to exhibit higher conductivity when subjected to moisture or other aging conditions. [24]

3.2. Weight and volume change

Table 2 Volume change of the samples

	EPDM		TPV		ECO TPV	
Solution	**WEC [%] (σ)**	**VC [%] (σ)**	**WEC [%] (σ)**	**VC [%] (σ)**	**WEC [%] (σ)**	**VC [%] (σ)**
0.001M	0.533 (0.022)	-1.048 (0.232)	0.753 (0.071)	-0.298 (0.561)	0.948 (0.230)	0.205 (0.795)
0.1M	1.436 (0.035)	1.107 (0.062)	3.577 (0.065)	3.083 (0.250)	4.325 (0.037)	4.537 (0.236)
1M	1.882 (0.005)	1.637 (0.070)	5.224 (0.186)	4.866 (0.483)	6.145 (0.50)	6.501 (0.600)

after immersion in solution for 1000h at 90°C. The standard deviation σ is written in parentheses.

Table 2 presents the weight change (WEC) and volume change (VC) of EPDM, TPV, and ECO TPV samples after immersion in H_2SO_4 solutions of varying concentrations (0.001M, 0.1M, and 1M) for 1000 h at 90°C. The data reveals both positive and negative volume changes. Notably, EPDM shows a VC of -1.048% in the 0.001M solution, and TPV shows a VC of -0.298% in the same solution, indicating a reduction in volume. These negative values suggest that under low acid concentration, some materials may undergo shrinkage or densification, possibly due to crosslinking effects tightening the polymer network or loss of volatile components leading to a denser structure.

In contrast, as the acid concentration increases, all materials tend to exhibit positive volume changes, reflecting swelling and degradation. ECO TPV shows the highest WEC and VC values, reaching 6.145% and 6.501% respectively, in the 1M solution, indicating significant degradation. TPV also displays notable changes, with WEC and VC values peaking at 5.224% and 4.866% respectively in the 1M solution. EPDM, however, demonstrates better chemical stability with the lowest volumetric and weight changes, peaking at 1.882% WEC and 1.637% VC in the 1M solution.

3.3. Hardness and Micro-hardness change

Figure 2 illustrates the Shore A hardness (right) and micro-hardness mIRHD (left) of EPDM, TPV, and ECO TPV samples both as delivered (AD) and after immersion in rhw H_2SO_4 solutions for 1000 h at 90°C. TPV maintains the highest hardness values, with Shore A hardness showing less than a 5% decrease across all conditions, and mIRHD values remaining above 88, indicating superior resistance to chemical and thermal aging. EPDM exhibits moderate stability, with a Shore A hardness decrease of approximately 7-10% and mIRHD values consistently around 80-85. In contrast, ECO TPV shows a significant decrease in hardness, with Shore A values dropping by up to 15-20% and mIRHD values decreasing to around 78-80, particularly under higher acid concentrations.

IOP Conf. Series: Materials Science and Engineering **1319** (2024) 012023 doi:10.1088/1757-899X/1319/1/012023

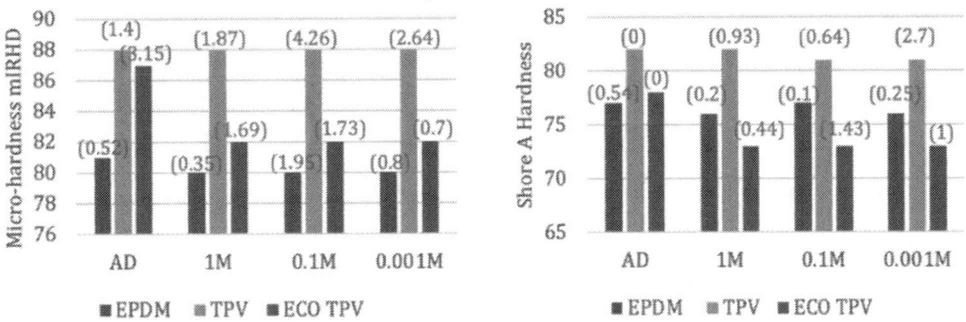

Figure 2 Shore A Hardness change of the samples (left) and micro-hardness mIRHD (right). The standard deviation σ is written in parentheses.

3.4. Differential Scanning calorimetry results

Figure 3 displays the Differential Scanning Calorimetry (DSC) curves for TPV (top left), ECO TPV (top right), and EPDM (bottom). The green curves represent samples in the As Delivered (AD) state, the blue curves correspond to samples aged in a 1M H_2SO_4 solution, and the red curves represent heat-aged samples, all aged for 1000 h at 90°C. For TPV, the AD state shows a distinct thermal transition, which shifts slightly upon aging in the acidic solution and more prominently after heat aging, indicating changes in the material's thermal stability and possible degradation. Similarly, the ECO TPV samples exhibit significant changes in their DSC profiles, with the most notable shifts observed in the heat-aged samples, suggesting substantial thermal degradation. EPDM shows relatively stable thermal behavior across all conditions, with minimal shifts in the thermal transitions, indicating better thermal stability and resistance to both chemical and thermal aging compared to TPV and ECO TPV.

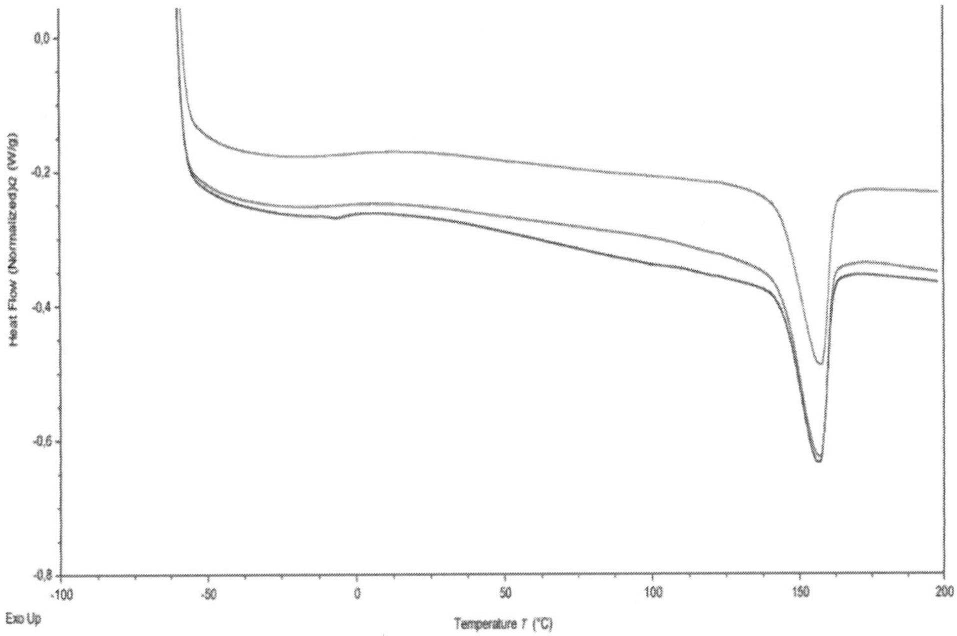

Figure 3a DSC curves for TPV

IOP Conf. Series: Materials Science and Engineering **1319** (2024) 012023 doi:10.1088/1757-899X/1319/1/012023

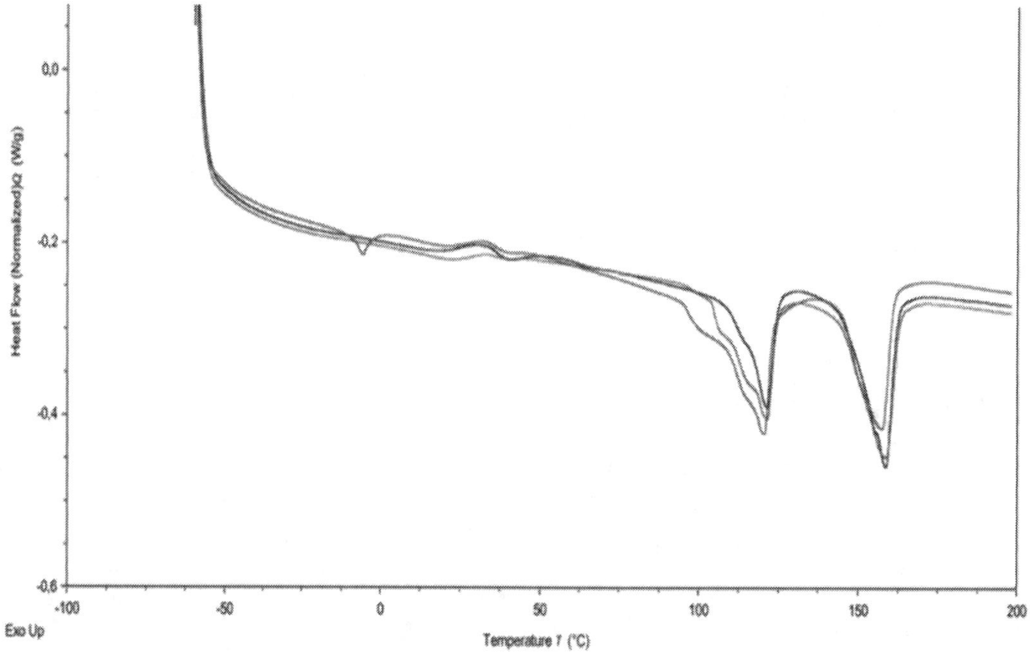

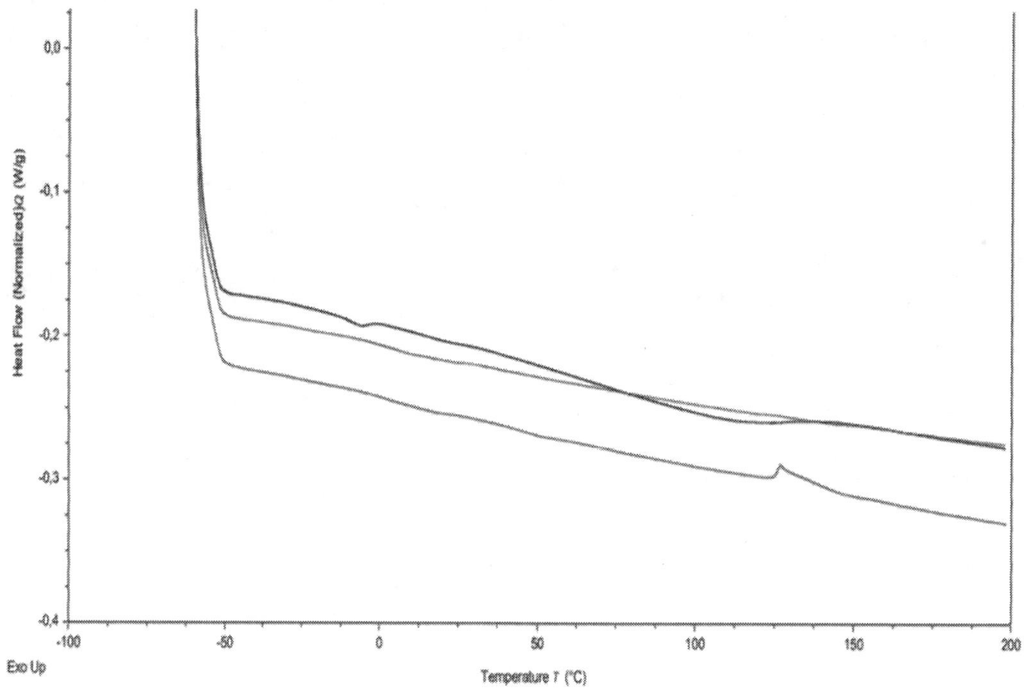

Figure 4b and 3c DSC curves for ECO TPV (b) and EPDM (c).

3.5. SEM Results

The Scanning Electron Microscopy (SEM) images (**Figure 4**) display the surface morphology of different samples under various conditions. All the images captured at the same magnification for consistency. The images reveal significant differences in surface texture and integrity among the samples. In the AD state, surfaces appear relatively smooth with minimal imperfections. However, after aging in the acidic solution, the surfaces exhibit increased roughness, micro-cracks,

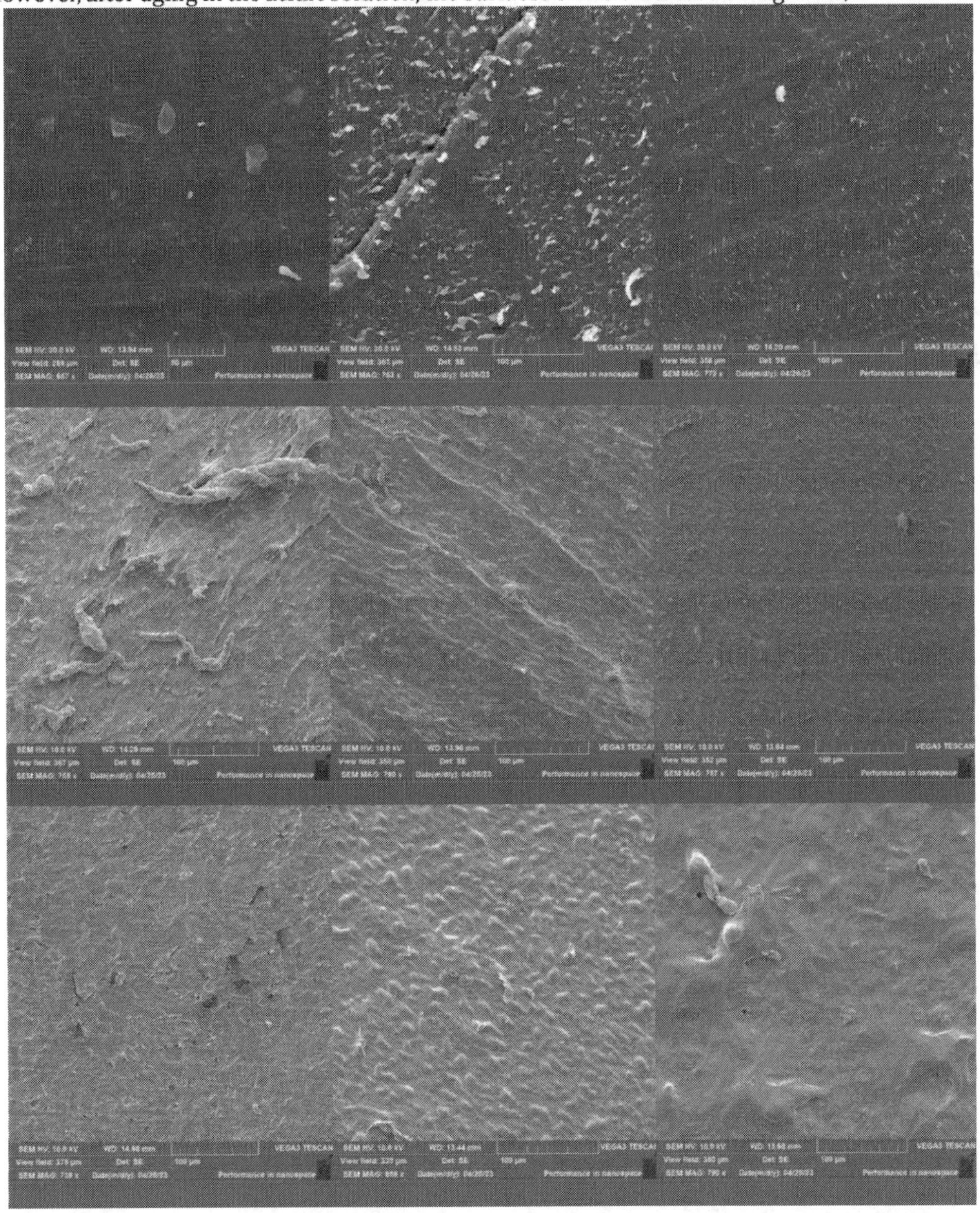

Figure 5 SEM images showing the surface of the studied samples. First row: EPDM AD (left), EPDM Heat Aged (center), EPDM aged in 1M (right); second row: TPV AD (left), TPV Heat Aged (center), TPV aged in 1M (right); third row: ECO TPV AD (left), ECO TPV

and erosion, particularly in the ECO TPV samples, indicating chemical degradation. Heat-aged samples also show considerable surface deterioration, with visible cracks and deformations, highlighting the impact of prolonged thermal exposure on material integrity.

4. Conclusions

This research examined the impact of immersing EPDM, TPV, and ECO TPV in sulfuric acid-water solutions on their physico-chemical characteristics. The investigation included volumetric, mass, thermal, and microscopic methods. The findings indicate that EPDM had the greatest resilience to chemical and thermal aging, with negligible changes in weight and volume. On the other hand, ECO TPV exhibited considerable deterioration, characterized by large increases in both weight and volume, as well as noticeable alterations in thermal properties as shown by DSC analysis. The TPV exhibited a state of intermediate stability, characterized by minor alterations in its characteristics. SEM research provided further evidence of material deterioration, namely demonstrating surface erosion and cracking, particularly in ECO TPV. The hardness measurement shown little change after being subjected to several chemical aging techniques. These data indicate that TPV and EPDM exhibit greater stability and suitability for uses in acidic conditions, although ECO TPV may need further stabilizing or preventive measures to improve its longevity.

Declaration

In preparing this paper, we utilized AI tools including ChatGPT (based on OpenAI's GPT-4 architecture) and QuillBot for paraphrasing, results interpretation, and drafting. We have reviewed, edited, and verified the content to ensure its accuracy and integrity.

Acknowledgments

We extend our special gratitude to Prof. Dr. ing. Viorel-Aurel ŞERBAN for serving as the PhD coordinator and to Vitesco Technologies Group AG for providing the study with the resources and support it needed.

References

[1] J. Liu, L.W. Zhang, B. Wang, Z.Q. Xu, W. Wang, Effect of Thermal Aging on Mechanical Properties of EPDM Rubber, Applied Mechanics and Materials 727–728 (2015) 318–321. https://doi.org/10.4028/WWW.SCIENTIFIC.NET/AMM.727-728.318.
[2] D. Yang, J. Ma, Q. Zhang, B. Li, P. Ming, C. Zhang, Accelerated Test of Silicone Rubbers Exposing to PEMFC environment, Progress in Natural Science: Materials International 30 (2020) 882–889. https://doi.org/10.1016/J.PNSC.2020.10.008.
[3] Q. Xu, J. Zhao, Y. Chen, S. Liu, Z. Wang, Effects of gas permeation on the sealing performance of PEMFC stacks, Int J Hydrogen Energy 46 (2021) 36424–36435. https://doi.org/10.1016/j.ijhydene.2021.08.137.
[4] X. Ling, Y.L. Hu, W.W. Lai, Y. Wang, Formulation of EPDM heat-resistant V-ribbed belt, IOP Conf Ser Mater Sci Eng 770 (2020) 012029. https://doi.org/10.1088/1757-899X/770/1/012029.
[5] W. Shuang-Hong, Z. Chang-Sheng, H. Chao-Fan, The aging property and storage life prediction of EPDM, J Phys Conf Ser 2478 (2023) 032046. https://doi.org/10.1088/1742-6596/2478/3/032046.
[6] J. Garche, L. Jörissen, Applications of fuel cell technology: Status and perspectives, Electrochemical Society Interface 24 (2015) 39–43. https://doi.org/10.1149/2.F02152IF/XML.
[7] K. Naskar, D. Kokot, J.W.M. Noordermeer, Influence of various stabilizers on ageing of dicumyl peroxide-cured polypropylene/ethylene-propylene-diene thermoplastic vulcanizates, Polym Degrad Stab 85 (2004) 831–839. https://doi.org/10.1016/J.POLYMDEGRADSTAB.2004.03.016.
[8] J. Yun, R. Patel, D.C. Worley, High-performance styrenic thermoplastic vulcanizates for long-term application, J Appl Polym Sci 105 (2007) 2996–3005. https://doi.org/10.1002/APP.24151.
[9] C.W. Lin, C.H. Chien, J. Tan, Y.J. Chao, J.W. Van Zee, Dynamic mechanical characteristics of five elastomeric gasket materials aged in a simulated and an accelerated PEM fuel cell environment, Int J Hydrogen Energy 36 (2011) 6756–6767. https://doi.org/10.1016/J.IJHYDENE.2011.02.112.

[10] L. Kunheng, C. Zhiyong, S. Wenku, Rubber aging life prediction based on interpolation and improved time-temperature superposition principle, Mater Res Express 9 (2022) 015301. https://doi.org/10.1088/2053-1591/AC45BA.
[11] M. Tayefi, M. Eesaee, M. Hassanipour, S. Elkoun, E. David, P. Nguyen-Tri, Recent progress in the accelerated aging and lifetime prediction of elastomers : A review, Polym Degrad Stab 214 (2023) 110379. https://doi.org/10.1016/J.POLYMDEGRADSTAB.2023.110379.
[12] L.F. Song, N. Bai, Y. Shi, Y.X. Wang, L.X. Song, L.Z. Liu, Effects of Ethylene-Propylene-Diene Monomers (EPDMs) with Different Moony Viscosity on Crystallization Behavior, Structure, and Mechanical Properties of Thermoplastic Vulcanizates (TPVs), Polymers 2023, Vol. 15, Page 642 15 (2023) 642. https://doi.org/10.3390/POLYM15030642.
[13] J. Tan, Y.J. Chao, J.W. Van Zee, W.K. Lee, Degradation of elastomeric gasket materials in PEM fuel cells, Materials Science and Engineering: A 445-446 (2007) 669-675. https://doi.org/10.1016/J.MSEA.2006.09.098.
[14] J. Tan, Y.J. Chao, H. Wang, J. Gong, J.W. Van Zee, Chemical and mechanical stability of EPDM in a PEM fuel cell environment, Polym Degrad Stab 94 (2009) 2072-2078. https://doi.org/10.1016/J.POLYMDEGRADSTAB.2009.07.009.
[15] D. McDonnell, N. Balfe, G.E. O'Donnell, Chemical ageing effects on the mechanical behaviour of ethylene-propylene diene monomer, Polym Test 64 (2017) 167-174. https://doi.org/10.1016/J.POLYMERTESTING.2017.10.001.
[16] M. Maiti, J. Patel, K. Naskar, A.K. Bhowmick, Influence of various crosslinking systems on the mechanical properties of gas phase EPDM/PP thermoplastic vulcanizates, J Appl Polym Sci 102 (2006) 5463-5471. https://doi.org/10.1002/APP.25106.
[17] R.Z. Abd Rashid, N. Johari, S.A. Mazlan, S.A. Abdul Aziz, N.A. Nordin, N. Nazmi, S.N. Aqida, M.A.F. Johari, Effects of silica on mechanical and rheological properties of EPDM-based magnetorheological elastomers, Smart Mater Struct 30 (2021) 105033. https://doi.org/10.1088/1361-665X/AC1F64.
[18] L.I. Murtazina, A.R. Akhmedgoraeva, R.Y. Galimzyanova, Y.N. Khakimullin, Construction Sealants Based on EPDM Modified with Silane-Terminated Urethane Prepolymers, IOP Conf Ser Mater Sci Eng 753 (2020) 052038. https://doi.org/10.1088/1757-899X/753/5/052038.
[19] M.H. Zulkifli, M.S.M. Rasidi, N.A.M. Rahim, L. Musa, A.H. Masa, The Study of Recycled Ethylene Propylene Diene Monomer (EPDM-r)/Polypropylene (PP) Polymeric Blends, J Phys Conf Ser 2080 (2021) 012012. https://doi.org/10.1088/1742-6596/2080/1/012012.
[20] R. Wang, Y. He, Z. Wang, H. Kang, J. Ma, H. Jin, Effect of Thermal Aging on Breakdown Strength of Epdm Rubber, IOP Conf Ser Mater Sci Eng 768 (2020) 062091. https://doi.org/10.1088/1757-899X/768/6/062091.
[21] S.A. Mueller, B.R. Kim, J.E. Anderson, M. Kumar, C. Huang, Leaching of Ions from Fuel Cell Vehicle Cooling System and Their Removal to Maintain Low Conductivity, SAE Technical Papers (2003). https://doi.org/10.4271/2003-01-0802.
[22] S. Kashi, R. Varley, M. De Souza, S. Al-Assafi, A. Di Pietro, C. de Lavigne, B. Fox, Mechanical, Thermal, and Morphological Behavior of Silicone Rubber during Accelerated Aging, Polym Plast Technol Eng 57 (2018) 1687-1696. https://doi.org/10.1080/03602559.2017.1419487.
[23] S.P. Ansari, F. Mohammad, Electrical Conductivity Studies of Conducting Polymer Nanocomposites in Ambient Conditions, Https://Doi.Org/10.1177/096739111602400406 24 (2016) 273-280. https://doi.org/10.1177/096739111602400406.
[24] G. Kraus, J.F. Svetlik, Electrical Conductivity of Carbon Black-Reinforced Elastomers, J Electrochem Soc 103 (1956) 337. https://doi.org/10.1149/1.2430324/XML.

IOP Conf. Series: Materials Science and Engineering **1319** (2024) 012024 doi:10.1088/1757-899X/1319/1/012024

Aluminium foams prepared using concentrated solar power at Synlight

Jaroslav Kováčik[1*], Peter Oslanec[1], Dmitrij Laaber[2], José Galindo[3], José Rodríguez[3] and Inmaculada Cañadas[3]

[1] Institute of Materials and Machine Mechanics, Slovak Academy of Sciences, Dúbravská cesta 9, 845 13 Bratislava, Slovakia
[2] DLR Institute of Future Fuels, Schneiderstraße 2, 52428 Jülich, Germany
[3] CIEMAT—Centro de Investigaciones Energéticas, Medioambientales y Tecnológicas, PSA–Plataforma Solar de Almería, Apartado 22, E-04200 Tabernas, Almería,Spain

*E-mail: ummsjk@savba.sk; jaroslav.kovacik@savba.sk

Abstract. Possibility to prepare AlSi10 aluminium foams and also gradient aluminium foams using concentrated solar energy was tested experimentally. Moreover, the experiments with incorporating of ductile Advanced Pore Morphology Al1050 foams elements into brittle AlSi10 foam was also investigated. It was possible after certain optimisation of the process to prepare sound AlSi10 foams using the artificial solar energy in Synlight. The prepared foams have the size of 100 mm x 100 mm x 10 mm. The densities of foams were in the range of 0.7 – 1.5 g x cm^{-3} depending on the density of used Advanced Pore Morphology ductile Al1050 foam elements. Finally, the homogeneity of chosen prepared foams were investigated using X-ray 3D tomography. In the case of composite foams, the local heterogeneities in the structure were observable, especially for Advanced Pore Morphology Al1050 foam elements with density above 1 g x cm^{-3}.

1. Introduction

The energy needs of buildings in Europe lead to the production of > 25% of greenhouse gases. Emissions are from the operation of buildings, and during renovation or construction of new ones [1]. 85-95% of the approximately 250 million buildings in Europe will still be in use in 2050 and 75% of these buildings have low energy efficiency. To reduce energy consumption, more than 90000 houses/per week ought to be renovated in EU. This is 3x more than the current rate [1, 2].

There is an unused alternative: two-thirds of the original industrial heat is wasted by pointlessly releasing it into the environment. If this heat could be captured, stored in a thermal battery and transferred to houses, where it would also be discharged into the surroundings, but this time purposefully through the building, heat for heating would not have to be produced at all, thus reducing not only CO2 emissions but also dust from combustion and emissions from building renovation. The storage PCM (phase changing materials) that melt or they solidify at the temperature we want to keep stable in the building could be used. During PCM phase transformation (melting / solidification) they can absorb/release a large amount of latent heat without a significant change in their temperature [3]. Their wider use runs into a fundamental problem associated with their low thermal conductivity in molten state, which prevents effective heat transfer to the entire volume of PCM. Currently, PCM is therefore mainly used either sealed in plastic capsules or as foils with a large surface area and small volume [4-7].

Since 1994 IMMM SAS, Bratislava developed, measured properties and structure, prototyped and industrially applied powder metallurgically prepared aluminium foams [8-12]: 2000 - the first industrial applications of Al foams - the door sill filled with Al-foam in Ferrari 360 and 430 Spider [8]. 2006 - crash absorber for Audi Q7 SUV safety net with over 120,000 parts per year. 2008 - crash absorption box for railway applications in the Sprinter Light Train (SLT) in Holland, (1000 pieces per year). 2010 - research activities related to reducing the energy demand of buildings using low-potential heating/cooling through ceiling panels made of aluminium foam – the first application so far at SAPA Profily in Žiar nad Hronom, Slovakia – 260 m^2. 2022 - serial production of aluminium foam chassis with cushioned heatsink, advanced thermal management and ergonomics for laser projectors Spectrum for Kvant Lasers s.r.o, Slovakia - 200 pcs / year.

The IMMM SAS, Slovakia is currently involved in projects concerning decarbonisation of buildings using developed heat storage battery based on aluminium foam filled with PCM. PCM is placed in the pores of AlSi10 aluminium foam, which form min. 80% of the total volume. It is possible to completely fill these pores with liquid PCM, using pressure infiltration. This solution eliminates partially the fundamental disadvantage of liquid PCM, i.e. low thermal conductivity. Further, it is technically possible - to store and transport residual industrial heat even with the temperature of the heating medium below 40°C. These batteries will be then transported to urban area in cost effective distance to heat the houses or commercial buildings.

When PCM is in fine pore structure of aluminium melting/solidification process takes significantly smaller time. AlSi10 aluminium foams are usually used now for this solution. However base thermal conductivity of this alloy (120 W/mK) is significantly smaller as conductivity of Al1050 alloy (200 W/mK). Ductile aluminium foam is a better solution, but it is more difficult to infiltrate its larger volumes necessary for the technical solutions with PCM due to thicker cell walls.

Plataforma Solar, Almeria in cooperation with Universidad Politécnica de Madrid, Spain developed and optimised preparation of various aluminium foams using concentrated solar energy (CSP) [13] and also the possibility to prepare Advanced Pore Morpohology (APM) foam parts. Technical solutions that they developed are unique in the world. Current research is oriented on the creation of APM aluminium foams and their industrial use inside of composite foams.

APM aluminium foams [14 – 15] have a wide range of possible applications due to their specific mechanical and thermal properties: for reinforcing hollow structures, deformation elements for absorbing impact energy, for vibration damping in dynamically excited components, and as an internal layer of composite materials [16-19]. Moreover, they could be also filled with phase change materials (PCM) to store the heat effectively and rapidly due to high thermal conductivity when Al1050 matrix material is used.

It is expected that by using of APM parts made from pure ductile aluminium foam that are placed inside AlSi10 aluminium alloy foams will shorten charging/discharging time of developed battery significantly and also improve the life cycle of battery. The aim of this work is to test the possibility to place inside AlSi10 foam ductile APM Al1050 foams. Further, in order to decrease energy cost of such production it is proposed to use concentrated solar energy for foaming of aluminium foams composed from ductile APM Al1050 foam and brittle AlSi10 foam.

2. Experimental

For all expriments powder metallurgically prepared [20-28]AlSi10 foaming precursors

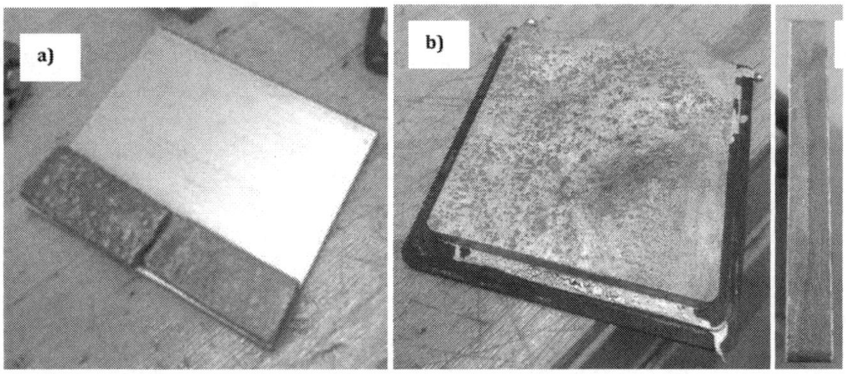

Figure 1. AlSi10 precursor of size 80 mm x 80 mm x 3mm, together with APM Al1050 foams approx. 40 mm x 20-30 mm x 8 mm: a). AlSi10 foam prepared using CSP: b) Side view on prepared foam: c).

containing 0.7 wt.% TiH_2 of the size 80 mm x 80 mm x 3 mm were used (see Fig. 1a). Further, 20 pcs of ductile APM Al1050 foams were foamed in electric furnace from Al1050 precursor containing 0.3 wt.% TiH_2 at 680 °C. Foaming time was 10 minutes. APM foams had various densities in the range of 0.7 – 1.5 g x cm^{-3}.

The foaming experiments were performed at Synlight facility, IFF, DLR, Julich, Germany [29]. Synlight is a large solar simulator, or the worlds largest artificial sun. Main advantage of Synlight is that its operation does not depend on fluctuating weather conditions.

The 149 modules are mounted on a 14m high platform in a beehive-shaped arrangement (Fig. 2a). Each module containins 2.7kWrad Xenon lamp with parabolic reflector, cooling unit, position system, power supply modules, and computer control. The key idea of Synlight is to

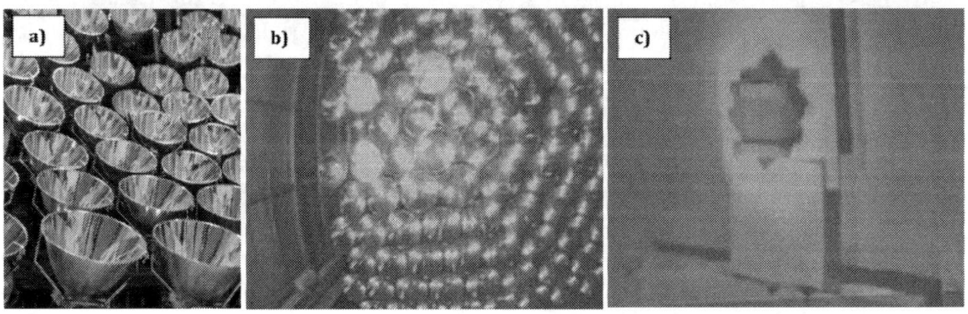

Figure 2. Synlight Xenon lamps mounted in beehive arrangement: a). Lamps switched on during foaming experiments: b). Foaming mould and solar spot during experiment: c).

share the facility resources by multiple experiments. Therefore the three test chambers opposite to the platform are placed behind fire and light proof roller shutters with a size of 4m x 4m [30].

Test chambers solar spots corresponds to solar energy of 315, 240 and 240 kW and peak flux of 12.5, 10 and 10 MW/m^2, respectively [31].

Test chamber 3 of Synlight facility was used for experiments. At the first it was necessary to create and develop the test set-up for aluminium foaming as Synlight has a different operation complexity than the usual Solar Furnace. There a knowledge obtained previously at PSA played a crucial role [13, 32]: the test setup consitent of mould holder, onto which the protective alumina ceramic plates were mounted to protect the test chamber and keep the mould warm from the back side. During solar foaming 3-4 lamps up to 8 kW power were used (Fig. 2b). The solar spot of each lamp is pointed approximatelly into the center of the foaming mould (Fig. 2c).

Primary 22 foaming tests of AlSi10 aluminium precursors in a mould have been carried out to optimise the foaming (see Fig. 1). At the first it was necessary to optimise the number of lamps necessary for foaming, then the position of the solar spots of each lamp, the way how to determine proper swith off time of the lamps and how to cool the mould to have sound foams.

Then solar foaming tests were undertaken at Synlight with the emphasis on preparation of composite foams of dual composition – already prepared APM Al1050 foams were placed inside mould together AlSi10 precursor (Fig. 1b) and such composition was foamed using CSP at Synlight. 32 foaming tests of APM Al1050 foams + AlSi10 aluminium precursors has been carried out, with various configurations due to complexity. The test bench have a possibility to store and later evaluate the temperature record during the CSP tests captured by 4 pieces of type K thermocouples. Cooling of the foam hot side after foaming was carried out by a fan.

Finally, X-ray micro tomography Phoenix Nanotom, IMMM SAS, Slovakia, for non destructive analysis of the internal structure of materials with with resolution better than 0.5 μm, was used to investigate the inner structure of prepared foams.

3. Results and discussion

It was possible to optimise solar foaming process to prepare sound AlSi10 foams using the solar energy in Synlight (Fig. 3). In the case of composite foams, the results were not so straightforward

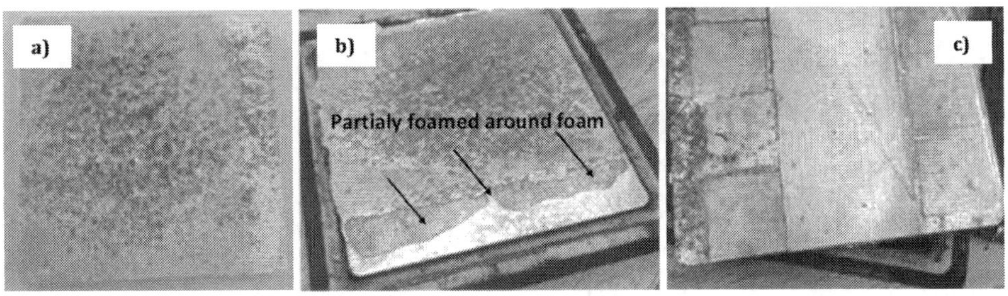

Figure 3. AlSi10 foam prepared using CSP: a). Partially foamed APM Al1050 foam inside AlSi10 foam: b). Fully foamed APM Al1050 foam into AlSi10 foam: c). Samples size 100 mm x 100 mm.

and the local heterogeneities were still observable (see Fig. 3b). The heterogeneity increased with increased density of APM Al1050 foams (density above 1 g x cm^{-3}). The prepared foams have size of 100 mm x 100 mm x 10 mm and the foaming takes place between 10 – 15 minutes depending on the number of used Xenon lamps. The densities of foams were in the range of 0.7 – 1.5 g x cm^{-3} depending on the density of used APM Al1050 ductile foams.

When mould, containing aluminium foam precursor, is heated using CSP there are following main factors influencing significantly foaming process: Size and shape of solar spot, value of

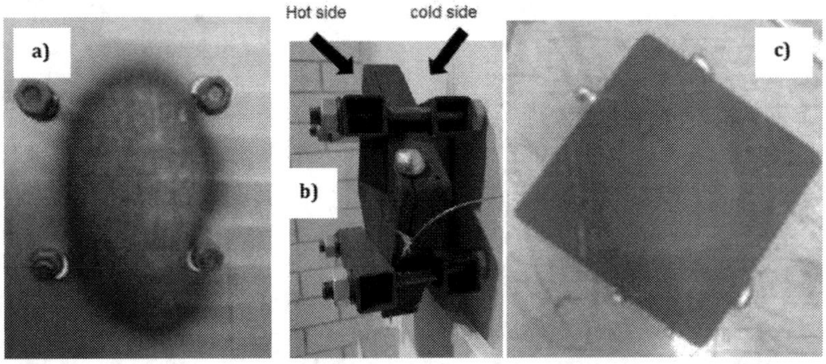

Figure 4. Size and shape of optimised solar spot during solar foaming: a). Hot and cold side of mould: b). Mould after foaming with leaking aluminium droplets: c).

focused solar energy per cm^2 and only one side of heat flow into the mould during CSP tests (Fig. 4). The main difficulty in the case of aluminium foaming is one side heat flow into the mould during CSP tests. This leads to significant temperature drop between the hot front and cold rear side of used mould.

The four thermocouples were used during the experiments to check the temperature of mould and to enable early detection of foaming of aluminium. Two of them (T1 and T2) were

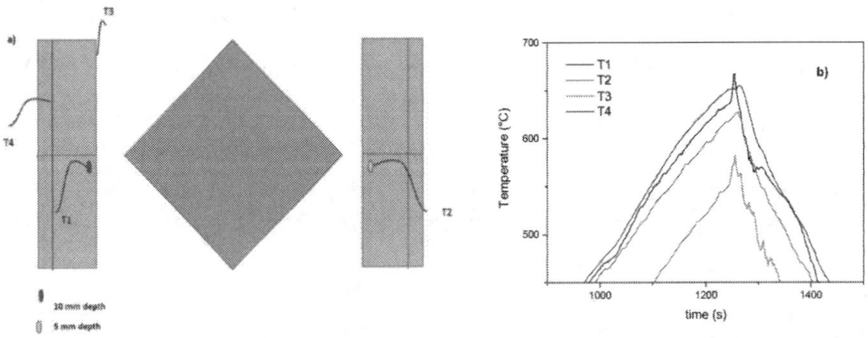

Figure 5. Placement of thermocouples in the mould: a). Temperature changes at the end of foaming at various positions inside and on the surface of mould: b) .

placed inside the holes drilled into the side of mould. They were in opposite corners of the mould (Fig. 5) in the depth of 5 and 10 mm, respectively. One thermocouple was outside on the side of mould (T4). It just touched the surface in the joint where bottom and top piece of mould attach to each other. The last thermocouple was placed on the rear side of the mould (T3).

As indicates Fig. 5b, thermocouples on hot side of mould reached around 650 °C and more. However, the temperature on the rear side scarcely reached 580 °C (T3). It is 70 °C temperature difference. Various test conditions were performed to decrease this drop, e.g., switching off the heating until hot side temperature starts to decrease and then again switch on, and again. The temperature drop is in this case a bit smaller, but it is still there. So the problem with cold rear side of mould remains in solar foaming and in the current experimental set up it could not be diminished. Possibly other methods need to be used, such as using kiln or even rotary kiln [33, 34] in solar spot for solar aluminium foaming.

From Fig. 5b it is evident that when foam expansion takes place, the temperature in some thermocouples show rapid increase of temperature in short period of time (T2, T3). This effect

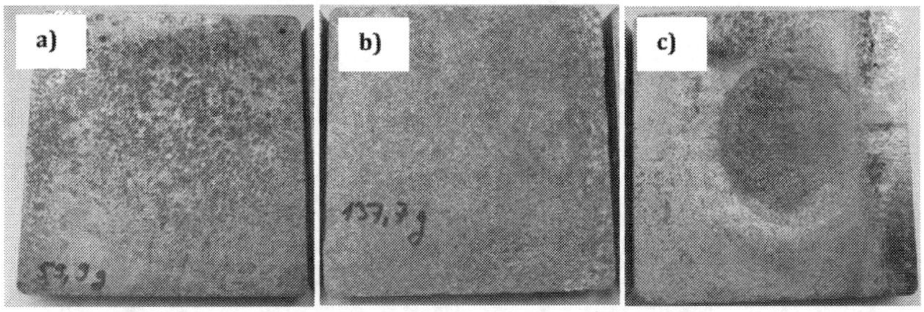

Figure 6. Foam samples investigated using XRD nanotomography: a) simple foam, b) gradient foam, c) composite foam with APM. Samples size 100 mm x 100 mm.

could be used in future automatisation of solar foaming.

The homogeneity of prepared foams structures was checked using X-ray 3D tomography with precision of 60 micro meters per pixel. Three basic types of AlSi10 foams were checked: simple foam, gradient foam and composite foam with APM (see Fig. 6).

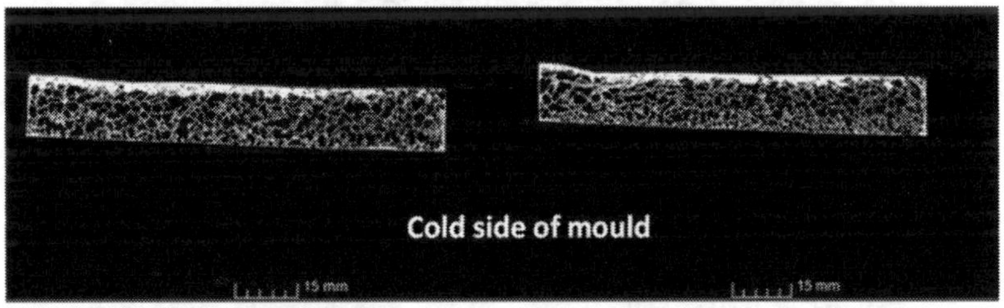

Figure 7. Inner structure of simple AlSi10 foam from Fig.6, density 0.6 g x cm^{-3} – two perpendicular cross sections in X-Y plane.

Simple foam structure is homogeneous with milimeter sized pores of irregular shape (Fig. 7). Cold side of foam has very thin surface skin. Hot foam side possesses folds due to foam collapse. Moreover, surface skin on this side is 2-3 mm thick. This is probably the result of pore collapse at hot side during foaming/cooling enhanced by pressure within pores inside the mould.

Gradient foam structure is non homogeneous with less than milimeter sized pores of irregular shape (Fig. 8). It has different thickness of surface skin. Thickness of the surface skin on the cold side of foam vary between 2 – 5 mm. Hot foam side possesses surface skin 1-3 mm thick. Density of foam is 1.4 g x cm^{-3}. This is the result of non-uniform heating of two pieces of aluminium foam precursor laid on top of each other [35] due to temperature drop between front and rear side of the mould.

Composite foam structure is very heterogeneous (Figs. 9-10) : While on hot side of the mould the collapsed foam is observed, AlSi10 foam possesses pores of almost simillar size. APM Al1050 foam has small porosity and higher density of 1.5 g x cm^{-3}. This is partially due to small amount of titanium hydride used in its production (0.3 wt.%) as well as due to slightly lower foaming temperature as is really necessary for sound foaming [25]. Fig. 10 confirms that ductile Al1050 foam is in full contact with brittle AlSi10 foam. Nevertheless, due to placement of APM

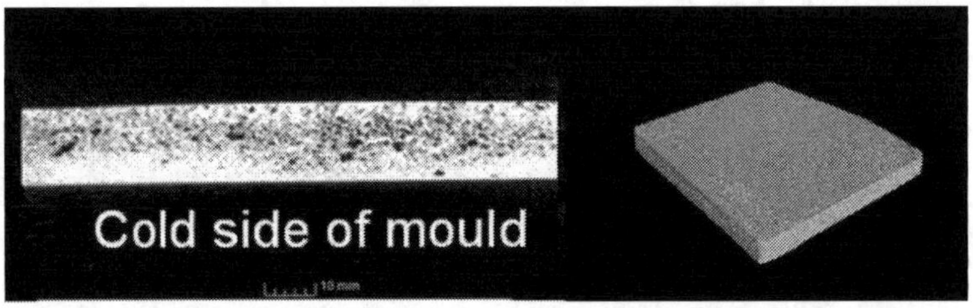

Figure 8. Inner structure of gradient AlSi10 foam from Fig.6, density 1.4 g x cm^{-3} –cross section in X-Y plane and overall picture with pores.

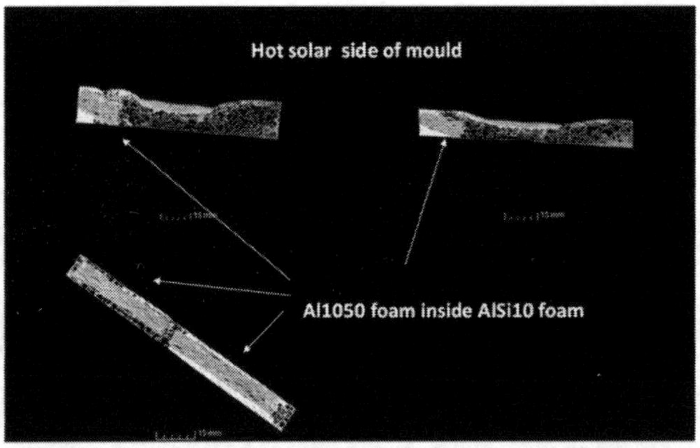

Figure 9. Inner structure of composite AlSi10 foam with APM Al1050 foam pieces from Fig.6, density 1.2 g x cm^{-3} - various cross sections in X-Y plane.

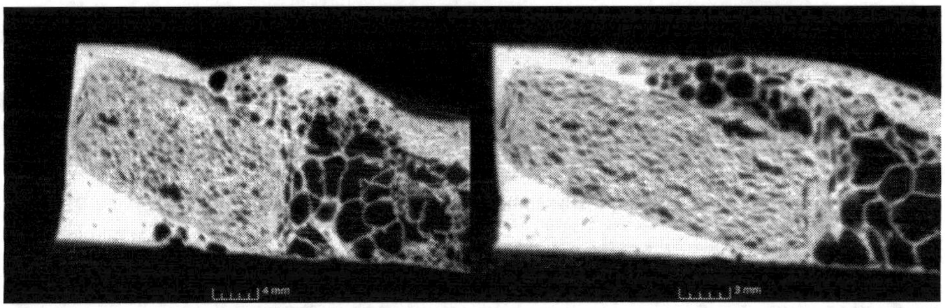

Figure 10. Detail of inner structure of composite AlSi10 foam with APM Al1050 foam pieces from Fig.6, density 1.2 g x cm^{-3}. Dense foam structure - APM Al1050 foam, density of 1.5 g x cm^{-3}.

Al1050 foam at the bottom of mould during foaming, the molten alumnium drained from AlSi10 foam and surrounds parially Al1050 foam from the left side of Fig. 10.

It is expected, that these results will enable further filling of both foams with PCM material, which is subject of future works.

4. Conclusions

The possibility to prepare simple AlSi10 foams, gradient AlSi10 aluminium foams and composite APM Al1050 foam + AlSi10 foam using concentrated solar energy (CSP) at Synlight facility was tested. Artificial sun was chosen, as it does not depend on the weather condition: When foaming is interrupted due to the clouds during the foam growth, it is not possible to continue the experiments.

After the creation of the test equipment, preliminary experiments and optimisation of the process it was possible to prepare sound AlSi10 foams and gradient AlSi10 foams using the solar energy in Synlight. The densities of simple foams, gradient foams and composite foams were in the range of 0.7 – 1.5 g x cm^{-3}. The density of used APM Al1050 ductile foams was between 1 – 1.5 cm^{-3}. In the case of composite foam, the local heterogeneities were caused by the presence of APM foams with density above 1 g x cm^{-3} inside AlSi10 foam.

The main difficulty in the case of using concentrated solar power for aluminium foaming is only one side heat flow into the mould during experiment. This leads to the significant temperature difference between the hot front and cold rear part of used mould. Various test conditions were performed to decrease this drop, but it is still present. It implies that the further optimisation of this process is necessary: The most probably new experimental set up with kiln or rotary kiln is necessary to develop.

In the future it is expected to fill composite foams with PCM material to improve heat flow into/from PCM material during heat storage/release in residual industrial heat batteries.

Acknowledgement

We thank the IFF, DLR for providing access to its installations, the support of its scientific and technical staff, and the financial support of the SFERA-III project (Grant Agreement No 823802). This research was also partially supported by VEGA project: Laser surface modification of Ti –TiB_2 bio-composites prepared by powder metallurgy process in order to increase their osseointegration (VEGA 2/0054/23). Partial financial support by Slovak government within the project No. 313011BWF3 under the Operational programme Integrated infrastructure (Project: Research on the possibilities of using waste heat from industry for heating in the public and commercial sector in Slovakia) is also gratefully acknowledged.

References

[1] EASAC. Available online: www.easac.eu (accessed on 19 June 2024).
[2] European Initiative for Energy-Efficient Buildings (E2B-EI). Available online: http://www.e2b-ei.eu/default.php (accessed on 19 June 2024).
[3] Kuznik, F.; David, D.; Johannes, K.; Roux, J.-J. A review on phase change materials integrated in building walls. Renew. Sustain. Energy Rev. 2011, 15, 379–391.

[4] Ahmad, M.; Bontemps, A.; Salle, H.; Quenard, D. Thermal testing and numerical simulation of a prototype cell using light wallboards coupling vacuum isolation panels and phase change material. Energy Build. 2006, 38, 673–681.

[5] Feldman, D.; Banu, D.; Hawes, D.W. Development and application of organic phase change mixtures in thermal storage gypsum wallboard. Sol. Energy Mater. Sol. Cells 1995, 36, 147–157.

[6] Kuznik, F.; Virgone, J.; Roux, J.-J. Energetic efficiency of room wall containing PCM wallboard: a full-scale experimental investigation. Energy Build. 2008, 40, 148–156.

[7] Sari, A. Form-stable paraffin/high density polyethylene composites as a solid–liquid phase change material for thermal energy storage: preparation and thermal properties. Energy Convers. Manag. 2004, 45, 2033–2042.

[8] García-Moreno, F. Commercial Applications of Metal Foams: Their Properties and Production. Materials 2016, 9, 85. https://doi.org/10.3390/ma9020085.

[9] Kováčik, J.; Simančík, F. Aluminium Foam - Modulus of Elasticity and Electrical Conductivity According to Percolation Theory. Scripta Mater. 1998, 39, 239–246.

[10] Simančík, F.; Kováčik, J. Electrical, thermal and acoustic properties of metallic foams. In Handbook of Cellular Metals: Production, Processing, Applications, Degischer, H.P., Kriszt, B., Eds.; Wiley-VCH: Weinheim, Germany, 2002; pp. 215–241.

[11] Sevostianov, I.; Kováčik, J.; Simančík, F. Elastic and electric properties of closed-cell aluminum foams cross-property connection. Mater. Sci. Eng. 2006, 420, 87–99.

[12] Gopinathan, A.; Jerz, J.; Kováčik, J.; Dvorák, T. Investigation of the Relationship between Morphology and Thermal Conductivity of Powder Metallurgically Prepared Aluminium Foams. Materials 2021, 14, 3623.

[13] Cambronero, L.E.G.; Cañadas, I.; Martínez, D.; Ruiz-Román, J.M. Foaming of aluminium–silicon alloy using concentrated solar energy. Sol. Energy 2010, 84, 879–887.

[14] Stöbener, K.; Baumeister, J.; Rausch, G.; Busse, M. Metal Foams with Advanced Pore Morphology (APM). High Temp. Mater. Process. 2007, 26, 231–238. https://doi.org/10.1515/HTMP.2007.26.4.231.

[15] Borovinsek, M.; Koudelka, P.; Sleichrt, J.; Vopalensky, M.; Kumpova, I.; Vesenjak, M.; Kytyr, D. Analysis of Advanced Pore Morphology (APM) Foam Elements Using Compressive Testing and Time-Lapse Computed Microtomography. Materials 2021, 14, 5897. https://doi.org/10.3390/ma14195897.

[16] Linul, E.; Galatanu, S.-V.; Marsavina, L.; Kováčik, J. Crushing behavior of closed-cell metallic foams: Anisotropy and temperature effects. J. Mater. Res. Technol. 2024, 30, 1436–1449. https://doi.org/10.1016/j.jmrt.2024.03.186.

[17] Galaţanu, S.-V.; Marşavina, L.; Kováčik, J.; Linul, E. Influence of density and loading speed on static and impact properties of closed-cell metallic foams. Eng. Fail. Anal. 2024, 161, 108297. https://doi.org/10.1016/j.engfailanal.2024.108297.

[18] Khezrzadeh, O.; Mirzaee, O.; Emadoddin, E.; Linul, E. Anisotropic Compressive Behavior of Metallic Foams under Extreme Temperature Conditions. Materials 2020, 13, 2329. https://doi.org/10.3390/ma13102329.

[19] Rajak, D.K.; Linul, E. Crushing response of Composite Metallic Foams: Density and High Strain Rate effects. J. Alloys Compd. 2021, 871, 159614. https://doi.org/10.1016/j.jallcom.2021.159614.

[20] Allen, B.C. U.S. Patent 3087807, 1963.

[21] Thornton, P.H.; Magee, C.L. The Deformation of Aluminum Foams. Metall. Mater. Trans. A 1975, 6, 1253–1263; doi:10.1007/BF02658535.

[22] Thornton, P.H.; Magee, C.L. Deformation Characteristics of Zinc Foam. Metall. Mater. Trans. A 1975, 6, 1801–1807; doi:10.1007/BF02642310.

[23] Baumeister, J. Method for Producing Porous Metal Bodies. German Patent 4018360, 29 May 1991.

[24] Banhart, J.; Baumeister, J. Deformation Characteristics of Metal Foams. J. Mater. Sci. 1998, 33, 1431–1440.

[25] Rodinger, T.; Ćorić, D.; Kováčik, J. The Effect of Foaming Agents on the Thermal Behavior of Aluminum Precursors. Materials 2024, 17, 710. https://doi.org/10.3390/ma17030710.

[26] Kovacik, J.; Orovčík, Ľ.; Jerz, J. High-temperature compression of closed cell aluminium foams. Kovove Mater. 2016, 54, 429–440. https://doi.org/10.4149/km_2016_6_429.

[27] Linul, E.; Marsavina, L.; Kováčik, J.; Sadowski, T. Dynamic and quasi-static compression tests of closed-cell aluminium alloy foams. Proc. Rom. Acad. Ser. A. 2017, 18, 361–369.

[28] Linul, E.; Pietras, D.; Sadowski, T.; Marsavina, L.; Rajak, D.K.; Kováčik, J. Crashworthiness performance of lightweight Composite Metallic Foams at high temperatures. Compos. Part A 2021, 149, 106516. https://doi.org/10.1016/j.compositesa.2021.106516.

[29] Wieghardt, K.; Funken, K.-H.; Dibowski, G.; et al. SynLight – the world's largest artificial sun. AIP Conf. Proc. 2016, 1734, 030038. https://doi.org/10.1063/1.4949090.

[30] Wieghardt, K.; Laaber, D.; Hilger, P.; et al. Engineering and erection of a 300kW highflux solar simulator. AIP Conf. Proc. 2017, 1850, 130013. https://doi.org/10.1063/1.4984507.

[31] Laaber, D.; von Storch, H.; Wieghardt, K.; et al. One year with synlight - Review of operating Experience. AIP Conf. Proc. 2019, 2126, 170007. https://doi.org/10.1063/1.5117677.

[32] Cambronero, L.E.G.; Cañadas, I.; Ruiz-Román, J.M.; Cisneros, M.; Corpas Iglesias, F.A. Weld structure of joined aluminium foams with concentrated solar energy. J. Mater. Process. Technol. 2014, 214, 2637–2643. https://doi.org/10.1016/j.jmatprotec.2014.05.032.

[33] Oliveira, F.A.C.; Fernandes, J.C.; Rodríguez, J.; Cañadas, I.; Galindo, J.; Rosa, L.G. Temperature uniformity improvement in a solar furnace by indirect heating. Sol. Energy 2016, 140, 141–150. https://doi.org/10.1016/j.solener.2016.11.004.

[34] Plaza, D.M.; Martinez, I.C.; Gasch, G.M.; Sufrategui, F.T.; García, J.R. A case study of the feasibility of using solar concentrating technologies for manufacturing ceramics. J. Clean. Prod. 2015, 87, 977–991. https://doi.org/10.1016/j.jclepro.2014.09.090.

[35] Nosko, M.; Simančík, F.; Florek, R. Reproducibility of aluminum foam properties: Effect of precursor distribution on the structural anisotropy and the collapse stress and its dispersion. Mater. Sci. Eng. A 2010, 527, 5900–5908. https://doi.org/10.1016/j.msea.2010.05.073.

Comprehensive Study of EPDM Rubber Properties: Temperature-Dependent Stiffness, Tensile Behavior and FEA Calibration

Daniel Foltuţ[1*], Estera Vălean[2], Georgiana-Iulia Şoşoi[3], Dragoş-Toader Pascal[4], Jan Pospisil[5] and Monica Buzdugan[4]

[1] Department of Materials and Manufacturing Engineering, Politehnica University of Timişoara, 1 Mihai Viteazul Blv., Timisoara 300222, Romania
[2] Department of Mechanics and Strength of Materials, Politehnica University of Timişoara, 1 Mihai Viteazul Blv., Timisoara 300222, Romania
[3] Department of Applied Chemistry, Organic and Natural Compounds Engineering, Politehnica University of Timişoara, Carol Telbisz 6, Timisoara 300001, Romania
[4] Vitesco Technologies Engineering Romania, Calea Buziaşului 11A, Timişoara 300699, Romania
[5] Vitesco Technologies, Na Rovince 879, 720 00 Ostrava 20, Czechia

*E-mail: daniel.foltut@student.upt.ro

Abstract. This work examines the temperature-dependent stiffness of Ethylene Propylene Diene Monomer (EPDM) rubber using tensile testing and Dynamic Mechanical Analysis (DMA). Additionally, it provides an appropriate model for Finite Element Analysis (FEA). The tensile tests indicate that EPDM has high rigidity and minimal elongation at -40°C, intermediate characteristics at 23°C, and a decrease in stiffness with rising temperature, up to 80°C. Nevertheless, a sudden rise in rigidity is seen at a temperature of 125°C, which may be ascribed to the strengthening influence of carbon black. The DMA findings confirm this pattern, demonstrating an upward trend in storage modulus and stiffness as temperature increases. At the same time, tan delta and loss modulus fall, suggesting a reduction in viscoelastic damping. The calibration of the Finite Element Analysis (FEA) model, using the Arruda-Boyce, Yeoh 3D, and Mooney-Rivlin 5-parameter models, determines that the Arruda-Boyce model is the most precise in accurately replicating the experimental data. The results indicate that the improved resistance to heat and increased rigidity of EPDM at high temperatures are affected by its material composition. However, more investigation is required to have a comprehensive understanding of the underlying processes.

1. Introduction

Elastomeric materials, such as EPDM rubber, are widely used in engineering applications because of their exceptional elasticity, weather resistance, and mechanical qualities [1,2]. EPDM's properties make it a preferred material in many areas like as automotive, construction, and aerospace. Gaining a comprehensive understanding of the mechanical properties of EPDM, particularly its response to different variables like as temperature, is essential for maximizing its design and performance in real-world applications. [3–5]

Prior research has thoroughly investigated the temperature-dependent mechanical characteristics of elastomers, emphasizing notable fluctuations in rigidity, tensile strength, and viscoelastic qualities in response to changes in thermal conditions. [6–11]. Wang et al. conducted a study on the aging behavior and storage life estimation of EPDM. They found that thermal oxidation aging has a substantial impact on the mechanical properties of EPDM. Their investigation shown that the overall crosslinking density and tensile permanent deformation progressively decline over time in the presence of thermal oxidative conditions, although the tensile strength and hardness exhibit an increase [12].

Similarly, Song et al. investigated the mechanical properties of aged EPDM insulation in high-voltage cable junctions under thermal-oxidative conditions. Their research revealed that the tensile strength and elongation of EPDM insulation diminished as it aged, particularly at elevated temperatures. The research also emphasized the development of pores and fractures inside the EPDM material as it aged, which led to the observed decrease in mechanical capabilities. [13]

Zhu and Li performed dynamic compression tests on silicone rubber, which is a kind of elastomer, at various temperatures to get insight into its mechanical properties when subjected to impact loading. Their findings demonstrated that the dynamic compressive strength of silicone rubber exhibited an increase in response to higher strain rates, indicating a phenomenon known as strain rate hardening. At lower temperatures, the material underwent a transition to a "glass state," which resulted in notable alterations in its stress-strain characteristics [5].

In addition, Zhang et al. (2010) conducted a study on the mechanical and thermal characteristics of silicone rubber/EPDM blends with the objective of improving the damping capabilities of silicone rubber. The researchers discovered that the addition of EPDM to silicone rubber enhanced its capacity to dampen vibrations at elevated temperatures. Nevertheless, the thermal stability of the mixes declined as the EPDM percentage increased, suggesting a compromise between enhanced damping and diminished thermal stability. [14,15]

The objective of this study is to examine the temperature-dependent rigidity of EPDM rubber. This study aims to gain comprehensive insights into the material's performance under different thermal conditions and how to utilize the determined properties in finite element analysis (FEA) simulations. This will be achieved by conducting tensile tests and dynamic mechanical analysis (DMA) at various temperatures, and calibrating FEA models.

2. Experimental setup

2.1. Studied material

The material used in the present study is an EPDM (Ethylene Propylene Diene Monomer) compound, known for its remarkable resistance to heat, ozone, and weather conditions. The EPDM compound has many essential constituents: The composition of the material consists of 60% EPDM, which contributes to its durability and resistance. Additionally, it contains 32% Carbon Black, which acts as a reinforcing agent to improve its tensile strength and stiffness. The material also includes 3.4% Zinc Oxide, which serves as a vulcanization activator to enhance cross-link density and improve its thermal stability. Lastly, 4.6% of miscellaneous additives and processing aids are added to customize the material's properties for specific applications. This composition is also enhanced by peroxidic curing.

2.2. Tensile testing at different temperatures

Tensile testing of EPDM rubber was performed using dumb-bell samples that satisfied the requirements of ISO 37 type 2 (**Figure 1**). The samples possess a total length of 75 mm, with a length of 25 mm for the narrow section, and a width of 4 mm for the narrow portion (cross section of 8 mm^2). The testing was conducted in a controlled environment with a humidity level of 53%. The temperature settings were adjusted to evaluate the material's performance under several situations, especially at -40°C, 23°C, 80°C, and 125°C. The experiments were carried out with an Instron Tensile Testing Machine, with a test velocity adjusted to 500 mm/min. The video extensometer was used to record the strain measurements.

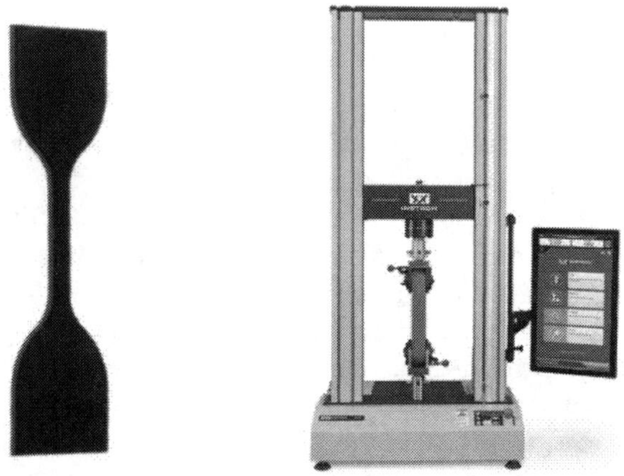

Figure 1 Example of ISO 37 type 2 rubber sample

2.3. Dynamic Mechanical Analisys (DMA)

A single cantilever test setup was used to perform the Dynamic Mechanical Analysis (DMA) of the EPDM rubber. The experiments were conducted at a frequency of 1 Hz, with an amplitude of 50 µm. In order to comprehend the viscoelastic characteristics of the substance, the rate of heating was established at 5 K/min, and the temperature was altered within the range of 20°C to 160°C.

3. Results and discussion

3.1. Temperature dependent tensile behavior

EPDM tensile test results are displayed in **Figure 2**. The results are not as one might expect, especially with the rising temperature. Starting at -40°C, the material exhibited the highest stiffness and the lowest elongation, indicating a brittle behavior. At room temperature (23°C), the EPDM displayed moderate stiffness with balanced mechanical properties, demonstrating both good strength and elasticity. With the raising temperature up to 80°C results in material softening. However, an unusual trend was noted at 125°C, where the stiffness was higher than expected, surpassing the stiffness measured at 80°C. This specific behavior suggests complex interactions within the material composition under specific thermal conditions.

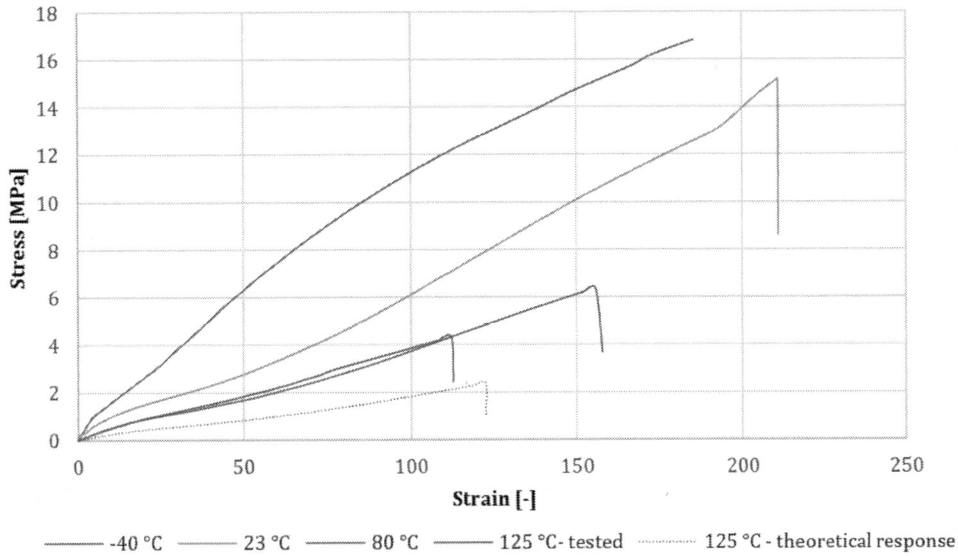

Figure 2 Tensile Test Response of EPDM at Various Temperatures

The unexpected increase in rigidity at 125°C, as opposed to 80°C, may be ascribed to the elevated concentration of carbon black in the EPDM composition. Carbon black is a well recognized reinforcing filler that greatly improves the mechanical characteristics of rubber compounds, such as rigidity, tensile strength, and resistance to wear and tear [16]. By including a 30% concentration of carbon black, the EPDM compound has enhanced thermal conductivity, resulting in increased heat dissipation. As a result, the network of polymer chains becomes more stable at higher temperatures, which decreases the material's propensity to become soft. In addition, the inclusion of 5% zinc oxide serves as both an activator in the vulcanization process and aids in the formation of cross-links between polymer chains [17], but to a lesser extent than carbon black. The combination of these factors results in the observed increase in stiffness at 125°C, highlighting the importance of carbon black in maintaining the rigidity and mechanical stability of the EPDM compound under thermal stress.

IOP Conf. Series: Materials Science and Engineering **1319** (2024) 012025 doi:10.1088/1757-899X/1319/1/012025

3.2. Dynamic Mechanical Analysis results

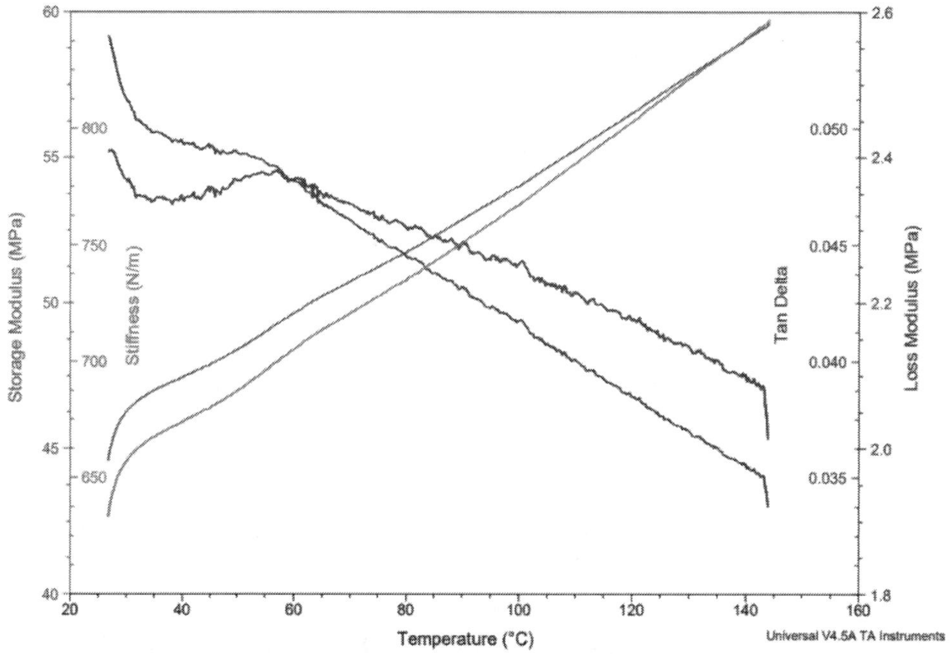

Figure 3 DMA result of the studied EPDM

Figure 3 shows the findings of DMA analysis conducted on the examined EPDM material. The findings demonstrate the fluctuation of storage modulus, loss modulus, stiffness, and tan delta in relation to temperature. As the temperature rises from 20°C to 160°C, the storage modulus (the green curve), which indicates the material's capacity to store elastic energy, also increases. This suggests that the material's rigidity increases as the temperature increases, which goes against the usual behavior of most elastomers.

At the same time, the stiffness, which is measured in N/m and shown by the pink curve, also rises as the temperature increases. This response implies that at higher temperatures, there can be extra formation of connections or strengthening of the structure inside the EPDM material. This is probably driven by the significant amount of carbon black present. Carbon black improves the thermal stability and mechanical qualities of EPDM by enhancing its thermal conductivity and strengthening the polymer network. As a result, EPDM becomes stiffer at higher temperatures.

The tan delta curve, shown as the brown curve, exhibits a progressive drop in the damping factor as the temperature rises, suggesting a reduction in energy dissipation. This tendency is essential for applications that need less damping at higher temperatures. On the other hand, the loss modulus (shown by the blue curve) drops as the temperature rises. This indicates a decrease in the material's capacity to dissipate energy as heat, which in turn reflects a decrease in its viscoelastic damping capabilities.

IOP Conf. Series: Materials Science and Engineering **1319** (2024) 012025 doi:10.1088/1757-899X/1319/1/012025

3.3. FEA material modelling

Three calibration models were used in the material modeling process: the Arruda-Boyce (AB) model, the Yeoh 3D model, and the Mooney-Rivlin (MR) 5-parameter model. Usually, when there is just experimental data for tensile testing, the AB or Yeoh models are used since they have more accuracy in modeling real-world behavior. The Arruda-Boyce model was used in this work to calibrate the stress-strain (S-S) curve. While the other models may be helpful in some situations, they exhibited an excessive level of inflexibility in the resultant biaxial curves, rendering them less appropriate for this particular application.

The top row of Figure 4 shows the calibration of the stress-strain (S-S) curves using three different models: Arruda-Boyce (left), Yeoh 3D (middle), and Mooney-Rivlin 5-parameter (right). The blue dotted lines represent the experimental data, while the solid lines represent the model predictions. The bottom row illustrates the biaxial stress-strain response curves for the same models. The Arruda-Boyce model (bottom left) closely matches the experimental data with moderate rigidity. The Yeoh 3D model (bottom middle) and the Mooney-Rivlin 5-parameter model (bottom right) show higher rigidity in the biaxial response modelling, deviating from the known behavior.

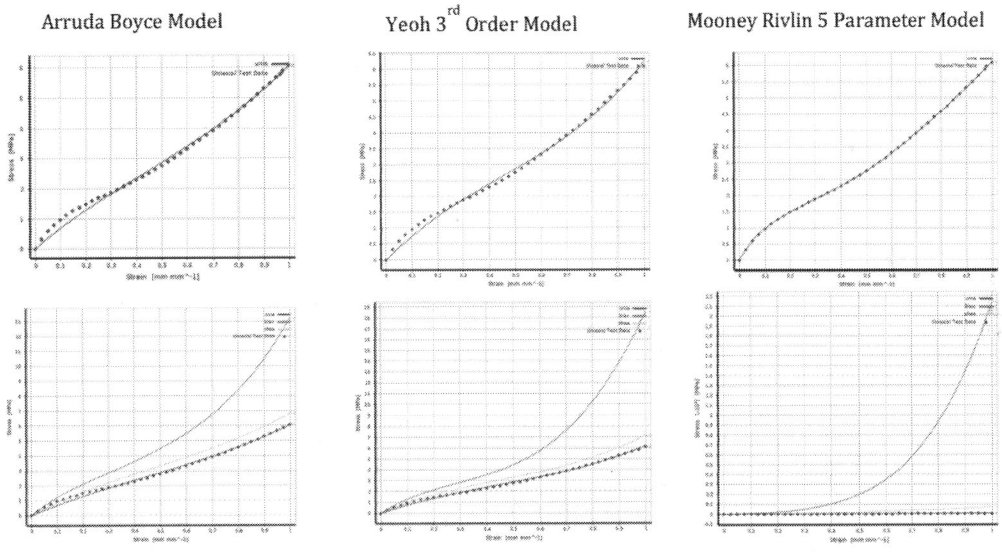

Figure 4 Curve fitting with different material models

4. Conclusions

The findings of this research suggest that the mechanical properties of EPDM rubber, which are influenced by temperature, show both anticipated and unforeseen patterns. At lower temperatures, such as -40°C, the material exhibits a high level of rigidity and a low degree of elongation, which suggests that it behaves in a brittle manner. At temperatures of 23°C and 80°C, the EPDM exhibits the characteristic behavior of an elastomer, with a moderate level of rigidity and enhanced ability to stretch. Surprisingly, the material exhibits increased rigidity at 125°C compared to 80°C. The anomaly may be ascribed to the reinforcing effects of carbon black, which improves the thermal stability and mechanical qualities of the EPDM composite.

Dynamic Mechanical Analysis (DMA) corroborates these results, since the storage modulus and stiffness exhibit an upward trend as the temperature rises. The tan delta and loss modulus measurements demonstrate that when temperature increases, there is a decrease in the ability to dissipate energy, indicating a reduction in viscoelastic damping. These studies suggest that EPDM maintains and maybe enhances its structural integrity at high temperatures, possibly owing to increased cross-linking or structural strengthening taking place inside the material, impacted by the high concentration of carbon black.

This research emphasizes the need for more inquiry in order to comprehend the fundamental processes that cause the heightened rigidity at higher temperatures. The calibration of the FEA model using three material models (Arruda-Boyce, Yeoh 3D, and Mooney-Rivlin 5-parameter) shows variable levels of accuracy, with the Arruda-Boyce model being the closest the experimental data.

Declaration

In preparing this paper, we utilized AI tools including ChatGPT (based on OpenAI's GPT-4 architecture) and QuillBot for paraphrasing, results interpretation, and drafting. We have reviewed, edited, and verified the content to ensure its accuracy and integrity.

Acknowledgments

We extend our special gratitude to Prof. Dr. ing. Viorel-Aurel ŞERBAN for serving as the PhD coordinator and to Vitesco Technologies Group AG for providing the study with the resources and support it needed.

References

[1] X. Mao, S. Xu, C. Wu, Dynamic Mechanical Properties of EPDM Rubber Blends, Polym Plast Technol Eng 47 (2008) 209–214. https://doi.org/10.1080/03602550701817611.

[2] T. Šarac, N. Quiévy, A. Gusarov, M.J. Konstantinović, Influence of γ-irradiation and temperature on the mechanical properties of EPDM cable insulation, Radiation Physics and Chemistry 125 (2016) 151–155. https://doi.org/10.1016/J.RADPHYSCHEM.2016.03.024.

[3] B. Shen, D. Zhang, S. Li, C. Wang, Experimental study of cooling water temperature on properties of short fiber and rubber composites, IOP Conf Ser Earth Environ Sci 189 (2018) 032019. https://doi.org/10.1088/1755-1315/189/3/032019.

[4] H. Arsyad, L.H. Arma, M. Syahid, M. Khalid, An experimental study of tensile properties and vibration absorption characteristic of ground tire rubber (GTR)/HDPE waste: Effect of temperature and heating time., IOP Conf Ser Mater Sci Eng 1034 (2021) 012159. https://doi.org/10.1088/1757-899X/1034/1/012159.

[5] X. Zhu, W. Li, Research on Dynamic Compression Testing of Silicone Rubber under Different Temperatures, J Phys Conf Ser 2002 (2021) 012047. https://doi.org/10.1088/1742-6596/2002/1/012047.

[6] M. Lu, J. Yun, M.H. Chen, L. Sun, An experimental investigation of lateral pressure effects on stress relaxation of elastomers, Mater Res Express 6 (2019) 085313. https://doi.org/10.1088/2053-1591/AB1EFA.

[7] D. Sangian, Improving the performance of dielectric-elastomer actuators at elevated operating temperatures by means of thermal softening, Smart Mater Struct 29 (2020) 025013. https://doi.org/10.1088/1361-665X/AB6079.

[8] M. Hemmatian, R. Sedaghati, S. Rakheja, Characterization and modeling of temperature effect on the shear mode properties of magnetorheological elastomers, Smart Mater Struct 29 (2020) 115001. https://doi.org/10.1088/1361-665X/ABB359.

[9] W. Zhang, X. Gong, S. Xuan, W. Jiang, Temperature-dependent mechanical properties and model of magnetorheological elastomers, Ind. Eng. Chem. Res. 50 (2011) 6704–6712. https://doi.org/10.1021/ie200386x.

[10] Y. Wan, Y. Xiong, S. Zhang, Temperature effect on viscoelastic properties of anisotropic magnetorheological elastomers under compression, Smart Mater Struct 28 (2018) 015005. https://doi.org/10.1088/1361-665X/AAEAF8.

[11] J. Zhang, J. Sheng, X. Liu, L. Liu, J. Zhao, H. Chen, Temperature effect on electromechanical properties of polyacrylic dielectric elastomer: an experimental study, Smart Mater Struct 29 (2020) 047002. https://doi.org/10.1088/1361-665X/AB79B7.

[12] W. Shuang-Hong, Z. Chang-Sheng, H. Chao-Fan, The aging property and storage life prediction of EPDM, J Phys Conf Ser 2478 (2023) 032046. https://doi.org/10.1088/1742-6596/2478/3/032046.

[13] J. Song, Z. Wang, H. Ji, R. Wang, Y. Zhou, Y. He, H. Jin, Mechanical Behavior of Aged EPDM Insulation of High-Voltage Cable Joints in Thermal-Oxidative Environment, IOP Conf Ser Mater Sci Eng 768 (2020) 062088. https://doi.org/10.1088/1757-899X/768/6/062088.
[14] C. Zhang, K. Pal, J.U. Byeon, S.M. Han, J.K. Kim, A study on mechanical and thermal properties of silicone rubber/EPDM damping materials, J Appl Polym Sci 119 (2011) 2737–2741. https://doi.org/10.1002/APP.31697.
[15] S. Çavdar, T. Özdemir, A. Usanmaz, Comparative study on mechanical, thermal, viscoelastic and rheological properties of vulcanised EPDM rubber, Https://Doi.Org/10.1179/174328910X12647080902970 39 (2010) 277–282. https://doi.org/10.1179/174328910X12647080902970.
[16] M.E. Spahr, R. Rothon, Carbon Black as a Polymer Filler, Polymers and Polymeric Composites: A Reference Series (2016) 1–31. https://doi.org/10.1007/978-3-642-37179-0_36-2.
[17] M. Maciejewska, The effect of zinc oxide nanoparticle morphology on activity in crosslinking of carboxylated nitrile elastomer, (2009). https://doi.org/10.3144/expresspolymlett.2009.68.

IOP Conf. Series: Materials Science and Engineering **1319** (2024) 012026 doi:10.1088/1757-899X/1319/1/012026

Statistical method of extracting assembly stresses in over-constrained solder-less press-fit pin technology for vibration reliability analysis

Mihai DRIENOVSKY [1*], Daniel Marius POP [2], Cristian Sorin NES [3] and Arjana DAVIDESCU [4]

[1, 2, 4] Mechanic, "Politehnica" University of Timisoara, Pta. Victoriei Nr.2, Timisoara, Romania
[3] Powertrain Solution, Vitesco Technologies Group AG, Siemensstraße 12, Regensburg, Germany

*E-mail: mihai.drienovsky2@student.upt.ro

Abstract. Solderless press-fit technology is increasingly favored over lead-free solder mountings due to cost-effective design benefits amid lead-free regulations. However, its over-constrained nature causes assembly issues and preload stress in pins, leading to fatigue failure under vibration. This paper extends previous research by presenting a method to evaluate the robustness of these assemblies under vibration-induced fatigue, considering assembly-induced stresses as static preload stress. The proposed statistical method uses Monte Carlo simulations to compute the distribution of pin failures due to fatigue, providing a comprehensive analysis that traditional worst-case FEA simulations cannot. This approach aids in identifying worst-case scenarios, predicting failure percentages, ranking designs, and improving design robustness through iterative feedback during the design phase.

1. Introduction

Since the 2006 implementation of the Restriction of Hazardous Substances (RoHS) Directive, which mandates the exclusion of lead from electronic assemblies, the industry has been compelled to explore alternative technologies that comply with environmental standards while ensuring reliability and cost-effectiveness [1]. This directive significantly impacted traditional soldering practices by raising the melting points of lead-free solders, introducing new challenges in the mounting process of electronic components.

One promising alternative that has emerged is solderless press-fit pin technology. This method, which eliminates the need for lead-based solder, offers a robust solution to the increased complexity and costs associated with higher melting point solders. Press-fit technology uses mechanical interconnections, avoiding thermal stress during assembly and making it particularly suitable for sensitive electronic components [2]. This shift is crucial for industries like automotive electronics, where reliability under extreme operating conditions is paramount [3] especially as electronic components must endure rigorous vibration conditions.

IOP Conf. Series: Materials Science and Engineering **1319** (2024) 012026 doi:10.1088/1757-899X/1319/1/012026

However, despite its advantages, press-fit technology is not without its challenges. Research has shown that failures in electronic connectors often result from a complex interplay of factors including oxidation, corrosion, fretting wear, mechanical vibrations, and assembly preloads, all of which contribute to the degradation of connector performance [4,5]. Notably, assembly preloads combined with vibration are frequently implicated in the failure of press-fit assemblies [6].

Building on this foundation, this paper extends our previous research [7] by introducing a statistical method to analyze the lifetime reliability of press-fit pin technology under vibration conditions. It incorporates the impact of manufacturing defects on individual components, offering a comprehensive reliability assessment beyond the traditional "Worst-Case tolerance calculation." [8, 9]. By employing Monte Carlo simulations, the proposed method evaluates the distribution of pin failures due to fatigue, providing insights into the reliability and robustness of press-fit assemblies under operational stresses.

2. Model for extracting assembly resulted stress.

2.1. Case study and boundary conditions

To better describe the method and the assumptions considered, the following 8 electrical pin case study is proposed with a 5g electrical component attached. The electrical pins and corresponding holes are considered to have the same position tolerance of ±0.03 mm shown in Figure 1. For analyzing the multiple pin design, a single pin is simulated in Ansys, a finite element analysis tool, the force and stress are extracted as a function of displacement results shown in Figure 2.

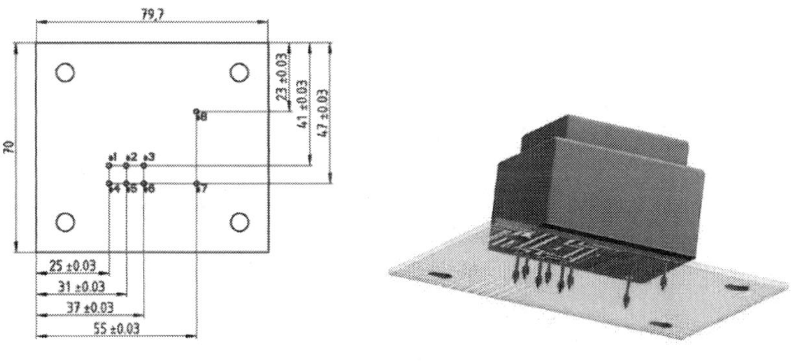

Figure 1. Design with 8 press-fit electrical pins and corresponding PCB holes with the proposed position tolerances (pins and holes are considered to have the same position tolerances relative to the component they belong to)

For demonstrative purposes and due to a vast range of press-fit electrical pins design solutions, each with different pin-hole contact behaviors, pin design and simulation was done closer to the model limitation than to any particular solution, resulting in axis-symmetric pins with the same diameter as the holes. Pin material is CuSn6 R560 according to datasheet [10], being a common material found in automotive electrical press-fit pins.

IOP Conf. Series: Materials Science and Engineering **1319** (2024) 012026 doi:10.1088/1757-899X/1319/1/012026

2.2. Model

For the multiple press-fit problem, a 2D model is proposed in which the electric press-fit pins are considered as virtual springs with a frictionless connection between two perfectly rigid bodies. The virtual springs behave according to the inputted force and stress curves shown in Figure 2.

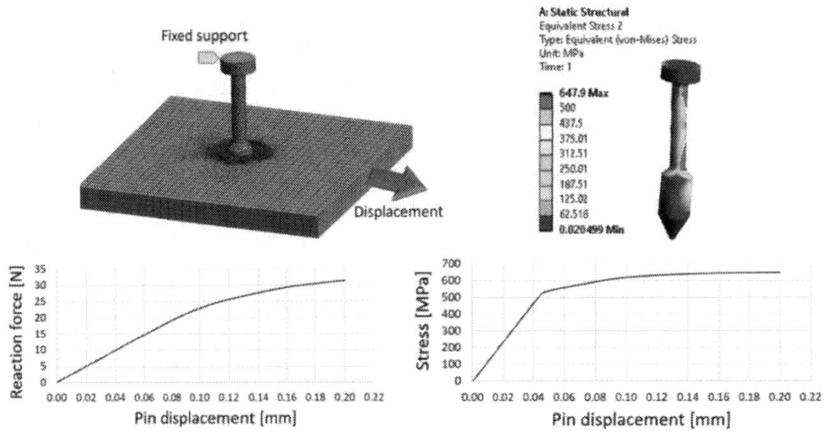

Figure 2. Reaction force and stress curve as a function of pin displacement extracted from Ansys simulation.

A simulation program was developed to determine the electrical pin stress resulting from the assembly between the electronic component and the PCB. For a randomly generated case, the program searches for the position and orientation of each component where the equilibrium Equation 1 are satisfied. The resulting force and torque, which are the vector sum of all the forces and torques acting on an object, should be zero. However, due to computer rounding errors, a perfect zero value may not be achieved in all cases. Therefore, it is necessary to define the smallest value to be considered as zero, referred to as the "*allowed_deviation*".

$$\overrightarrow{resultingForce} = \sum \vec{F} = \vec{0} + \overrightarrow{allowed_{deviaiton}} \tag{1}$$

$$\overrightarrow{resultingTorque} = \sum \vec{\tau} = \vec{0} + \overrightarrow{allowed_{deviation}}$$

The solving algorithm is detailed in previously published work [7]. The flow diagram in Figure 3 describes the main steps involved in finding the static equilibrium position. First, the program minimizes the resulting force by moving the object in the direction of the resulting force until it meets the conditions specified in Equation 1, referred to as the 'Translation Loop.' When the resulting force is 'zero,' the 'Rotation Loop' starts, rotating the object in the direction of the resulting torque until it also reaches 'zero.' Finally, the program checks if the full static equilibrium condition Equation 1 is satisfied. If it is, the program saves the data and stops the case; if not, it loops back to the 'Translation Loop.' The current program can simulate only isotropic pins, and the pin-hole press-fit connection can be addressed only through the provided stress and force curves. The model does not analyze the dimensional variation of the press-fit and its impact on the force and stress curves.

IOP Conf. Series: Materials Science and Engineering **1319** (2024) 012026 doi:10.1088/1757-899X/1319/1/012026

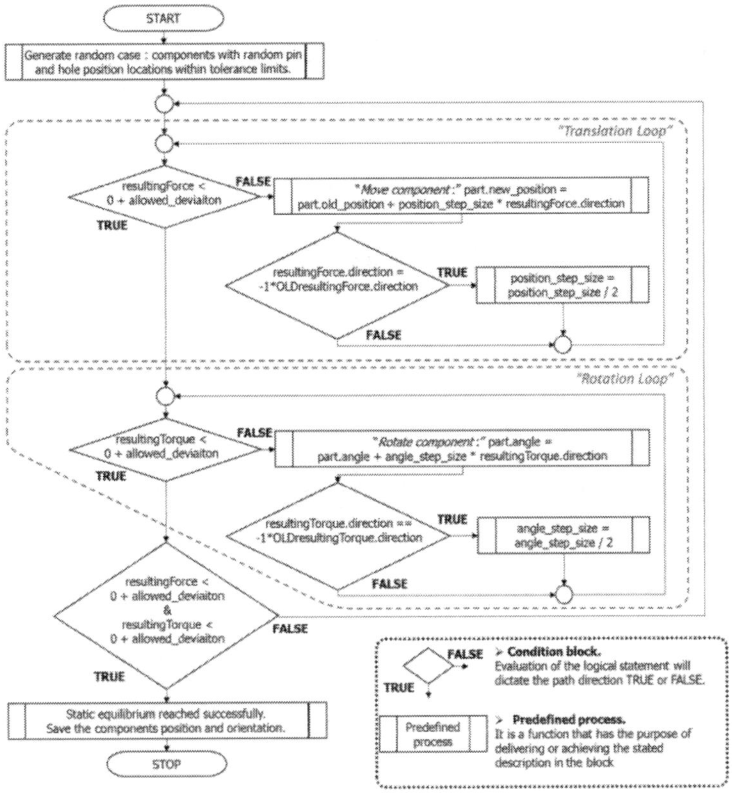

Figure 3. Algorithm for finding static equilibrium position for a given multiple-pin assembly.

2.3. Monte Carlo

The Monte-Carlo simulation method is widely used in the scientific domain, mainly due to its ease of use. The method consists of performing statistical analysis on many samples, generated by passing randomly generated values through predefined mathematical models [11, 12]. In the presented case, we study the probability of occurrence of specific mechanical stress values for each pin by analyzing the stress frequency occurrence in 10,000 randomly generated cases. For each case a random pin and hole position is generated according to a uniform distribution, within specified tolerance ranges and then the static equilibrium condition is solved, resulting in a certain position and orientation of the parts, from where the pin deformation and automatically pin stress is extracted.

Example of resulting data is depicted in Figure 4 where the stress values for Pin1 are shown in a histogram, because all of the pins follow a similar pattern regarding stress values, maximum and minimum values for all pins are shown in Table 1. The "odd shape" of the histogram (two local maximums) from Figure 4, occurs due the bilinear shape of the inputted stress curve.

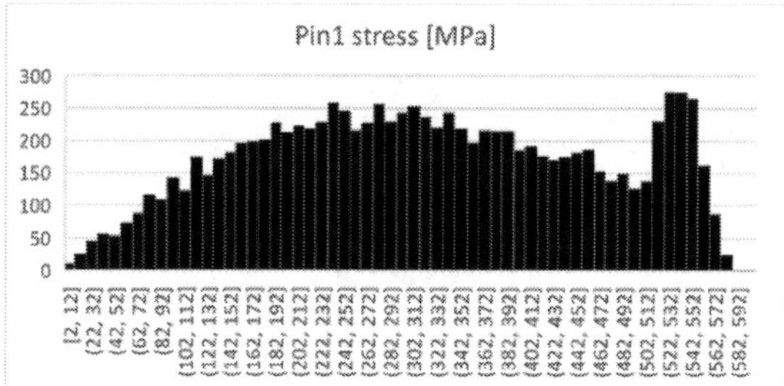

Figure 4. Histogram showing frequency occurrence of stress values for Pin1.

Table 1. The most extreme cases of stress from 10 000 from Monte-Carlo simulations.

	Pin1	Pin2	Pin3	Pin4	Pin5	Pin6	Pin7	Pin8
Max Stress [MPa]	599.3	590.4	596.1	597.2	598.9	593.3	593.9	587.2
Min Stress [MPa]	1.7	4.2	3.4	4.7	0.7	2.5	2.2	1.3

Table 2. Vibration power spectral density according to ISO16750-3

Frequency	Column heading
10	10
100	10
300	0.51
500	20
2000	20

3. Random vibration analysis

During normal operation, random vibrations occur in vehicles that travers uneven terrain. For the current study the random vibration was performed using ANSYS 2021R1, applying the power spectral density shown in Table 2 according to the ISO16750-3 "Road vehicles - Environmental conditions and testing for electrical and electronic equipment – Part 3 Mechanical Loads" standard [13].

IOP Conf. Series: Materials Science and Engineering **1319** (2024) 012026 doi:10.1088/1757-899X/1319/1/012026

Fatigue calculation in random vibration is a complex task because the response will be random. In engineering, Steinberg's three interval method [14] is widely used to address this issue, where the Von Mises stress response obeys Gauss distribution in terms of occurrence frequency. ANSYS results are summarized in Table 3 containing expected frequency and stresses for each pin.

Table 3. ANSYS results for each pin (Expected frequency and Von Mises stress according to Steinberg's three interval method).

Column heading	Von Mises p[%] Occurrence	Pin1	Pin2	Pin3	Pin4	Pin5	Pin6	Pin7	Pin8
f[Hz] Expected frequency	-	1552	1537	1757	1606	1607	1616	1576	1563
1σ Von Mises [MPa]	68.3%	11.6	8.5	8.3	20.5	9.1	8.1	28.9	25.8
2σ Von Mises [MPa]	27.2%	23.2	17.0	16.6	41.0	18.2	16.2	57.8	51.6
3σ Von Mises [MPa]	4.3%	34.8	25.5	24.9	61.5	27.3	24.3	86.7	78.4

4. Extracting failure rate

To determine the failure rate for a given test duration, for the present case 22 hours according to ISO16750-3 standard [13], and for the given vibration profile shown in Table 2, the stress values from the simulations should be passed through material S-N fatigue curve to determine the maximum number of allowed cycles denoted "*N_xσ*", which then are to be compared with the cycles executed during the test denoted "*n*" from Equation 2.

$$n_{x\sigma} = T * f * p_{x\sigma} \tag{2}$$

$n_{x\sigma}$ = *number of stress cycles executed*
$T\ [s]$ = *test duration*
$f\ [Hz]$ = *expectedfrequency (Table 2)*
$p[\%]$ = *% of occurance for a given stress*

Because the proposed study considers two different sources that generate mechanical stress in the pins, the assembly induced stresses which is static, and the vibration oscillation induced stresses which is dynamic. The two stress sources generate an oscillating stress profile with non-zero mean, which can't be directly analyzed through S-N fatigue curve of the material. To overcome this issue, the Goodman equivalent mean stress correction method [15, 16], Equation 3, is proposed for generating an equivalent stress with zero mean stress.

$$\sigma_{a,eq} = \frac{\sigma_a}{1 - \frac{\sigma_m}{R_m}} \tag{3}$$

$\sigma_{a,eq}[MPa] = zero\ mean\ stress\ equivalent\ amplitude$
$\sigma_a[MPa] = oscilation\ stress\ amplitude$
$\sigma_m[MPa] = non - zero\ mean\ stress$
$R_m[MPa] = material\ tensile\ yield\ stress$

Using Miner's cumulative fatigue Equation 4, it is possible to extract a total damage index R out of Steinberg's three interval method [14,17].

Failure probabilities, Equation 5 can be extracted by calculating index R for each Monte-Carlo case and determine the percentage of cases that fail.

$$R = \frac{n_{1\sigma}}{N_{1\sigma}} + \frac{n_{2\sigma}}{N_{2\sigma}} + \frac{n_{3\sigma}}{N_{3\sigma}} \tag{4}$$

$R = Miner's\ cumulative\ damage\ index$
$n_{x\sigma} = number\ of\ stress\ cycles\ executed$
$N_{x\sigma} = number\ of\ stress\ cycles\ allowed$

$$if\ R < 1\ then\ "survived"\ full\ test\ duration \tag{5}$$

Table 4. Scrap rates for each pin and for different test durations.

Test Duration	Pin1 Scrap %	Pin2 Scrap %	Pin3 Scrap %	Pin4 Scrap %	Pin5 Scrap %	Pin6 Scrap %	Pin7 Scrap %	Pin8 Scrap %
100%	13.1 %	9.2 %	9.3 %	22.8 %	9.1 %	8.4 %	33.6 %	21.7 %
50%	12.8 %	8.5 %	8.9 %	21.9 %	8.6 %	7.9 %	32.1 %	20.7 %
25%	11.9 %	7.2 %	7.6 %	20.6 %	7.7 %	6.3 %	29.5 %	19.2 %

Results from the simulation can be seen in Table 4 where scrap rate is shown for each pin and Table 5 where the results from using current statistical method are compared with results from using Worst-Case tolerance calculation. To better illustrate the benefits of current method, percentage of scrap is evaluated for 3 cases, where duration of the vibration test would be changed to 100%, 50% and 25% of original test duration. Scrape rates from Table 5 are not the equivalent to sum of scrape rates of individual pins shown in Table 4, because an assembly is considered scraped if one or more of its pins are broken.

Table 5. Comparison of results: Statistical Method vs. Worst-Case Scenario Tolerance Calculations.

Test Duration	Scrap % Statistic method	Scrap % Worst Case method
100%	74.2 %	100 %
50%	72.4 %	100 %
25%	68.5 %	100 %

5. Conclusion

This paper proposes a statistical method for analyzing and extracting vibration reliability information for an over-constrained multiple electric pin assembly, given assembly induced stress by components individual tolerances.

The Monte-Carlo type of analysis offers reliability information on individual pins shown in Table 4 or entire assembly in Table 5, where one or more defects are considered as one assembly defect.

When comparing the presented method with worst-case tolerance calculation, the benefits are clearly shown, where Worst-Case calculation returns a resulting cumulative damage *R=2,117,534>1* (detailed in Appendix A), current method allows for a percentage type of view how the assembly behaves at different stages of the vibration test (25%, 50%, 100%).

Current algorithm performs a simulation in 20s whiles in Ansys the same simulation would take roughly 2h, making this method useful for Monte-Carlo simulations. The process can be accelerated, by opening several instances of the same program and running the simulations in parallel and collecting the data from all the instances.

Method presented in this paper, allows for comparing different designs proposals, with new pin-hole nominal positions or different pin shapes, allowing selection of designs with the highest chance of passing vibration requirements. The view on percentage of passing or failing the vibration requirements, provides for some degree of insight into cost considerations.

6. Acknowledgements

This work was partially supported by Vitesco Technologies Romania SRL.

Appendix A. Calculation steps for worst case method

To better illustrate the difficulty of interpreting reliability by using worst case method, Pin 6 from Table 3 was selected, having the lowest vibration stress.

- Worst case tolerance calculation would result in a $pin\ deformation = 0.6\ mm$
- From pin stress-deformation curves (Figure 2) results in a Pin Static stress $\sigma_m = 558.56\ MPa$
- Considering material tensile yield stress [10] $R_m[MPa] = 600MPa$ and Table 3 data from Equation 3 result zero mean stress equivalent amplitudes:
  - $\sigma_{a,eq,1\sigma} = 373\ MPa$
  - $\sigma_{a,eq,2\sigma} = 746\ MPa$
  - $\sigma_{a,eq,3\sigma} = 1119\ MPa$
- For a $T = 22\ hours$ test and expected frequency $f = 1616\ Hz$ results in following number of stress cycles executed:
  - $n_{1\sigma} = 87{,}415{,}258$
  - $n_{2\sigma} = 34{,}684{,}531$
  - $n_{3\sigma} = 5{,}541{,}846$
- From materials S-N curve [10] allowed number of stress cycles:
  - $N_{1\sigma} = 246$
  - $N_{2\sigma} = 19$
  - $N_{3\sigma} = 19$
- Resulting in cumulative damage $R = 2{,}117{,}534 > 1$

References

[1] Directive, R. "Directive 2002/95/EC of the European Parliament and of the Council of 27 January 2003 on the restriction of the use of certain hazardous substances in electrical and electronic equipment." Official Journal of the European Union 13 (2003): L37

[2] Ubando, A. T., et al. "Analysis of the effects of geometry on the press fit application in automotive power modules." IOP Conference Series: Materials Science and Engineering. Vol. 1109. No. 1. IOP Publishing, 2021.

[3] Ubando, A. T., et al. "Sliding speed effect in press-fit connection using Finite Element Analysis." IOP Conference Series: Materials Science and Engineering. Vol. 1109. No. 1. IOP Publishing, 2021

[4] Munukutla, Aravind, and Anil Kurella. "INTERMITTENT CONNECTOR FAILURES IN ELECTRONIC ASSEMBLIES."

[5] Tohmyoh, Hironori, et al. "Analysis of solderless press-fit interconnections during the assembly process." (2008): 031007.

[6] Wang, Ruoyu, and Liangjun Xu. "Mechanical behavior investigation of press-fit connector based on finite element simulation and its reliability evaluation." Microelectronics Reliability 116 (2021): 114010

[7] Drienovsky M., Arjana D., Dmytro R. "Tolerance Analysis of an Over-Constrained Assembly with Press-Fit Solderless Electric Contact Pins."Joint International Conference of the International Conference on Mechanisms and Mechanical Transmissions and the International Conference on Robotics. Springer, Cham, 2020.

[8] ASME publication Y14.41-2003, Digital Product Definition Data Practices.

[9] Bryan R. Fischer „ Mechanical Tolerance Stackup and Analysis" 2nd Edition, ISBN 1439815720, CRC Press, 2011

[10] CuSn6 datasheet available at: https://www.kupferinstitut.de/mediathek/datenblaetter/ (Jun 2022)

[11] Metropolis, N., & Ulam, S. "The Monte Carlo Method", Journal of the American Statistical Association, 44(247), 335–341, 1949

[12] George S. Fishman, „Monte Carlo. Concepts, Algorithms, and Applications", ISBN 978-0-387-94527-9, Springer, New York, 1996

[13] Technical Committee ISO/TC 22 / SC 32. ISO 16750-3(E), Road Vehicles—Environmental Conditions and Testing for Electrical and Electronic Equipment—Part 3: Mechanical Loads; ISO copyright office: Geneva, Switzerland, 2012.

[14] Liu Z., Caifu Q, Zuzhi C. "PSD Random Vibration Strength and Fatigue Analysis of a CNG Tube Trailer."Mechanical Engineering and Materials. Springer, Cham, 2021. 173-182.

[15] Nicholas T. High cycle fatigue: a mechanics of materials perspective. Elsevier, 2006.

[16] Goodman, J., "Mechanics Applied to Engineering", Longman, Green & Company, London, 1914, accessed at: Mechanics applied to engineering : Goodman, John, 1862- : Free Download, Borrow, and Streaming : Internet Archive

[17] Irvine T. "Miner's Cumulative Damage via Rainflow Cycle Counting."Rn (theory)1 (2013): 26.

IOP Conf. Series: Materials Science and Engineering **1319** (2024) 012027 doi:10.1088/1757-899X/1319/1/012027

Vibration behaviour of additively manufactured AlSi10Mg specimens

Christian Öhl, Emanoil Linul* and Sergiu Valentin Galatanu

Politehnica University Timisoara, Department of Mechanics and Strength of Materials, 1 Mihai Viteazu Blvd., 300222-Timisoara, Romania

*E-mail: emanoil.linul@upt.ro

Abstract. Additive manufacturing is in continuous development, and the obtained components still have some shortcomings. The vast majority of mechanical characterizations are performed on polymer materials, using in particular the material extrusion process, while metallic components face not negligible deficiencies. This paper presents an experimental investigation of the vibration behaviour of selective laser melting (SLM) 3D printed specimens. The specimens were printed "on-edge" from AlSi10Mg aluminium alloys and subjected to fatigue tests. By using an electrodynamic shaker, controller, acceleration transducer, and laser vibrometer, the tests have been performed and the data analysed. After the tests, a correlation was obtained by using a finite element analysis (FEA). The damping coefficients (alpha and beta) are obtained from the experimental data. These coefficients are then used to check the accuracy of each FEA model. The results reveal the correlation between structural damping and anisotropy of additive manufacturing specimens, and their impact on high-cycle fatigue performance.

1. Introduction

The industry has developed a lot due to new materials and technologies [1-3]. Additive manufacturing (AM) is one of the current technologies that attracts the most interest worldwide in terms of the creation of new prototypes in the industry [4-6]. 3D printing technology allows the use of various materials, from polymers to metals [7-9]. The major advantage of AM is that prototypes for the desired parts can be quickly produced in production or in the laboratory [10-12].

Selective laser melting (SLM) is one of the most attractive additive technologies. This technology is similar to Laser Powder Bed Fusion (LPBD) and Selective Laser Sintering (SLS). The difference between SLM, LPBD and SLS is that both SLM and LPBD are specifically used in the production of metal objects compared to SLS, which is used for polymer objects [13].

The process involves heating the metal powder using a high-power laser to completely melt the material and fuse it to the final product. The most common metals used in this process are: aluminium, copper, steel, titanium and some precious metals [14].

The mechanical properties of the hollow-strut lattice metamaterials of AlSi10Mg produced by LPBF present excellent yield strengths even with much lower absolute densities [13].

The influence of heat treatment on the SLM produced alloy specimens of AlSi10Mg were observed by Jian et al. [15]. If the heat treatment temperature increases, the strength and hardness of the alloy decrease, and the elongation increases.

Regarding the mechanical behaviour of alloy structures, in the literature, the focus is on tensile, compression and fatigue tests. There are no investigations about the vibration behaviour of the parts produced by SLM [16-20]. However, for other additive manufacturing technologies, the vibration behaviour was highlighted [21].

At the same time, investigations that analyse the mechanical performances of the products using numerical simulations were observed, but as in the case presented above, the emphasis is on the static simulations, respectively fatigue simulations [13, 22]

This work investigates the effect of anisotropy on the vibration behaviour of additively manufactured AlSi10Mg specimens by using an accelerated fatigue-testing methodology [23, 24]. The printing orientation of the specimens has been performed vertically and by using an electrodynamic shaker, controller, acceleration transducer, and laser Doppler vibrometer, the tests have been performed and the data analysed.

2. Materials, samples and methods

Aluminium alloy AlSi10Mg was used to manufacture the samples. The chemical composition of AlSi10Mg material is presented in Table 1.

Table 1. Chemical composition [%]

Si	Mn	Fe	Mg	Cu	Zn	Ni	Pb	Al
10	< 0.45	< 0.55	0.3	< 0.05	< 0.1	< 0.05	< 0.05	Balance

In this study, specimen geometry was used based on ISO 1099 standard [25] with both extremities having modified lengths, one long-length of 18 mm and short-length of 10 mm, as shown in Fig. 1a. All specimens have been 3D printed vertically (see Fig. 1b).

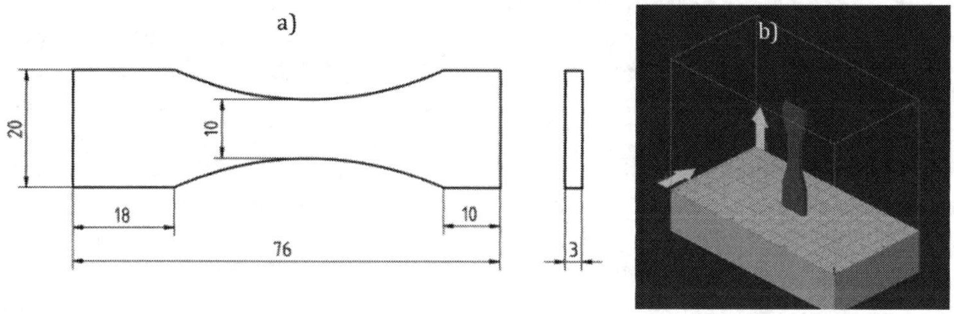

Figure 1. Specimen dimensions based on the ISO 1099 standard, in mm (a) and their orientation (b)

Three specimens have been tested (Fig. 2a), one by one clamped on a fixture that is mounted and screwed on the electrodynamic shaker shown in Figs. 2b and 2c. The fixture is designed to not allow the specimen to move, with a fit tolerance from the top to the bottom. An acceleration transducer is mounted on the middle of the fixture, keeping the test in range of the set profile. A fluorescent adhesive tape has been placed on both extreme sides of each specimen (see Fig. 2a), which allows a Doppler vibrometer to measure the natural vibration frequencies from specimens in real time. Specimens were fixed on the short-length in the first setup and on the long-length in the second setup to vibrate in a vertical direction.

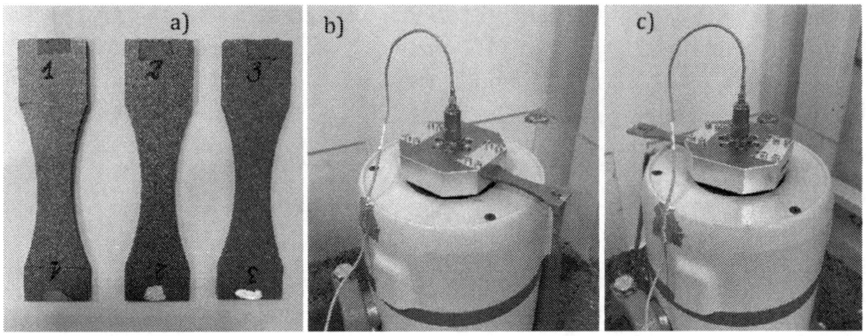

Figure 2. Tested specimens (a) together with long-length (b) and short-length (c) vibration setups

A linear sinusoidal sweep rate of 50 Hz was used from a frequency bandwidth of 20 to 1000 Hz with 1g constant acceleration on the entire spectrum. The Q-factor is defined as the ratio of the centre frequency of the resonance to half-power bandwidth of the resonance [26, 27]. The Q factor represents the sharpness or steepness of a resonance. It is calculated using Eq. (1). Damping is an energy dissipation mechanism that causes vibrations to gradually decrease over time until they stop, as described by Silas [28]. The vibrational energy is transformed into sound or heat. The amount of damping depends on the velocity of motion, the material and the frequency of vibration. Structural damping (ξ) was determined using Eq. (2), taking into account the relationship between the damping and the Q-factor.

$$Q_{factor} = \frac{\text{Resonant frequency}}{\text{-3db bandwidth (BW)}} = \frac{f_0\ [\text{Hz}]}{f_2\ [\text{Hz}] - f_1\ [\text{Hz}]} \tag{1}$$

$$\xi = \frac{1}{2Q} \tag{2}$$

3. Numerical analysis

A commercial programme, Ansys Workbench 2023 R2 has been used to perform numerical analysis and correlate the results. To perform the analysis, a mesh with element size 1 mm was chosen and the automatic method Fig. 3c with aluminium as the material property (Density of 2250 kg/m^{-3}, Young`s modulus of 59100 MPa and Poisson ratio of 0.33 as value). These properties correlate with the investigated material (AlSi10Mg) by modifying the Young`s modulus in Ansys to find the peak in frequency scale and damping ratio to find the peak on acceleration scale. The specimens were fixed support on each extreme side of the specimen one by one and an acceleration of 9.806 m/s^2 was applied (Figs. 3a and 3b).

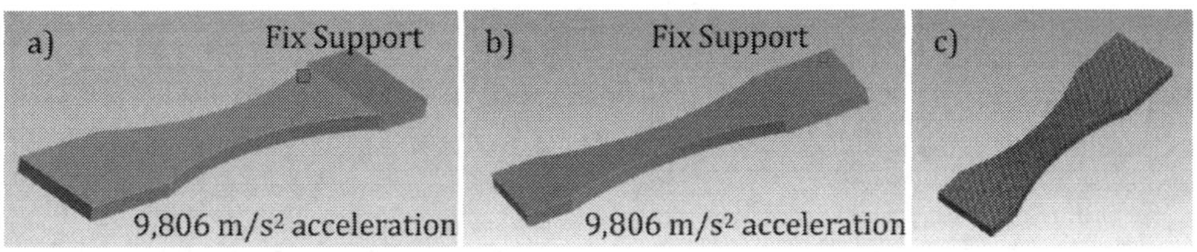

Figure 3. Short-length (a) and long-length (b) boundary conditions, and Mesh Model (c)

In numerical models, damping can be defined by viscous damping (forces proportional to the fundamental frequency of the structure) and material damping (forces balanced to strains), or a combination of both. In this, Rayleigh damping was used, which inputs the mass (M) and stiffness (K) matrix within the damping matrix [C], Eq. (3).

$$[C] = \alpha\,[M] + \beta\,[K] \tag{3}$$

where α and β are Rayleigh coefficients (proportional damping coefficients).

By extracting the natural frequency ω_i and the damping ratio ξ from the experimental test, Eq. (4) allows the calculation of the Rayleigh coefficients α and β.

$$\xi_i = \frac{\alpha}{2\omega_i} + \frac{\beta\omega_i}{2} \tag{4}$$

4. Results and discussions

The experimental results of the present investigation are shown in Fig. 4. The frequency response function in the report with acceleration has been plotted for each extreme side of the specimen, more exactly for long and short-length.

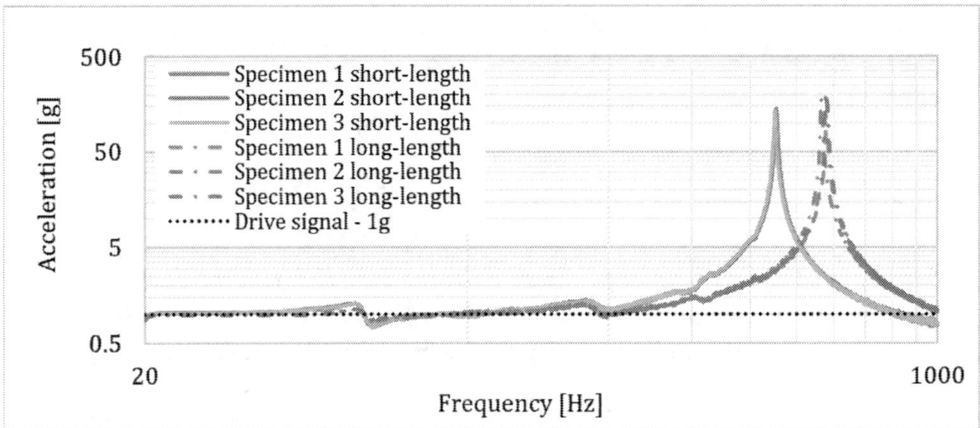

Figure 4. Frequency response functions for short-length and long-length fixed sides

It can be observed that all three specimens with long-length fixed have almost the same amplitude at the same frequency scale, this applies to the other side, short-length fixed. For the long-length it can be observed that acceleration is between 168-213 g and for the fixed short-length is between 132-140 g. On the frequency scale short-length specimens have a maximum resonance peak at 452 Hz and long-length at 574 Hz, with a difference between them of 120 Hz.

Long-length specimens have a larger clamping area and less free mass when compared to the short-length specimens, have on frequency scale a resonance frequency between 563-580 Hz, it also seems that long-length variation results are larger than short-length specimens. It can be concluded that the long-length has a resonance frequency at a higher frequency with a difference of 120 Hz than short-length and also a higher amplitude on the acceleration scale by 50 g.

An average on frequency and acceleration scale of all 3 specimens for short-length and long-length fixed, have been plotted in Fig. 5. In frequency histogram (Fig. 5a), short-length presents a resonance frequency of 452 Hz with a small deviation and long-length at 574 Hz in frequency range with a greater deviation than short-length. In the acceleration histogram (Fig. 5b), short-length has an amplitude of 136 g and long-length 187 g, long-length also has a larger variation on the acceleration scale than short-length specimens.

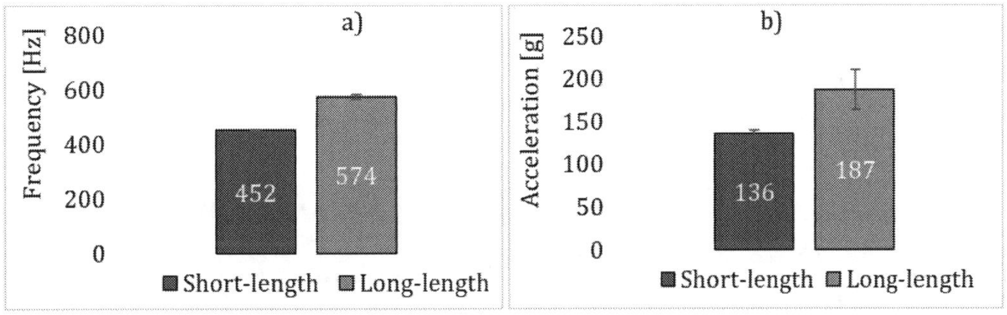

Figure 5. Resonance frequency (a) and mean amplitude (b)

In the damping ratio average graph from Fig. 6 it can be observed that they have almost the same damping (about 9% difference), just the long-length it's the smaller ones.

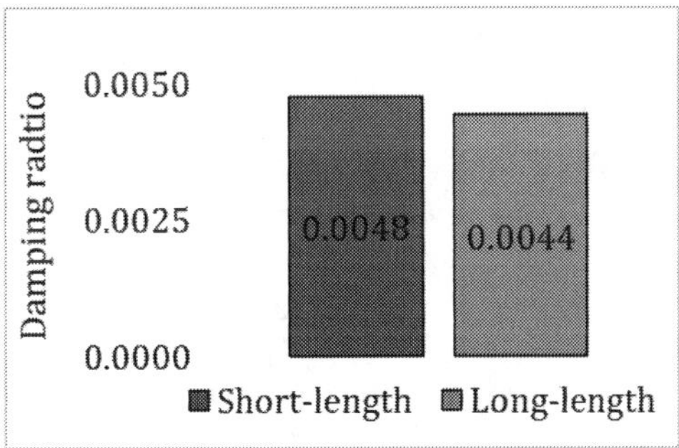

Figure 6. Damping ratio

Figure 7 shows an experimental-numerical comparison between the responses of the long-length and short-length specimens. A good correlation between test and simulation can be observed. The curves from pre-pulse and post-pulse with large differences between simulation and test are because simulation software enables getting only 21 data values for frequency and acceleration scale, while test software enables 1000 data recording values, more accurately and simulation software automatically puts random points from the beginning until the resonance peak and after, until the end.

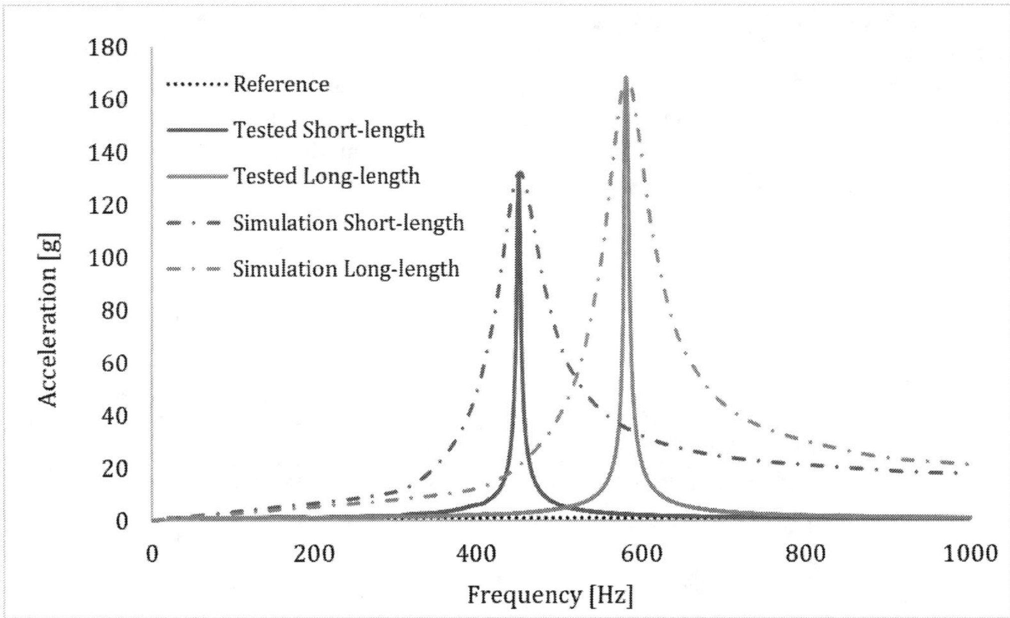

Figure 7. Frequency response functions: experimental results vs numerical

5. Conclusions

The paper investigates the vibration behaviour of AlSi10Mg samples additively manufactured by the selective laser melting (SLM) process. The mechanical tests are performed on two fixing configurations of the SLM-printed samples (short-length <SL> and long-length <LL>), their comparison being highlighted. The following were found:

- Depending on the fixed type of the samples (SL-F and LL-F), the acceleration-frequency curves show an almost perfect overlap.
- The LL fixation shows higher values of the resonance frequency (574 Hz) compared to the SL fixation (452 Hz).
- The average amplitude of the LL fixation is 187 g, 37.5% higher than that of the SL fixation (136 g).
- The damping ratio is almost the same for both configurations, however, it is slightly higher for the SL fixation (about 9%).
- A good correlation of the experimental and numerical results was obtained for both fixations of the samples.

References

[1] C. Jianu, S. I. Spiridon, G. Ionita, 2020, Buletinul AGIR, 3-4, 102–107.
[2] C. Vălean, I.N. Orbulov, A. Kemény, et al., 2024, Engineering Failure Analysis, 161, 108335.
[3] D.A. Şerban, A.V. Coşa, G. Belgiu, R. Negru, 2022, Polymers, 14(18), 3822.
[4] C. Vălean, L. Marşavina, M. Mărghitaş, et al., 2020, Procedia Structural Integrity, 28, 1134-1139.
[5] N. Shahrubudin, T.C. Lee, R. Ramlan, 2019, Procedia Manufacturing, 35, 1286-1296.
[6] D.I. Stoia, L. Marsavina, E. Linul, 2020, Polymers, 12 (3), 640.
[7] R.F. Fernandes, J.S. Jesus, L.P. Borrego, et al., 2023, Procedia Structural Integrity, 54, 300-306.
[8] C. Vălean, M. Baban, D.K. Rajak, et al., 2024, Construction and Building Materials, 414, 135015.
[9] D.G. Zisopol, M. Minescu, D.V. Iacob, 2024, Technology and Applied Science Research, 14(2), 13592-13597.

[10] D.I. Stoia, E. Linul, 2024, Theoretical and Applied Fracture Mechanics, 132, 104478.
[11] D.A. Şerban, C. Marşavina, A.V. Coşa, et al., 2021, Mathematics, 9(13), 1495.
[12] C. Vălean, E. Linul, G. Palomba, G. Epasto, 2024, Journal of Materials Research and Technology, 30, 1470-1481.
[13] X. Wang, Q. Lu, P. Zhang, et al., 2024, Optics & Laser Technology, 174, 110500.
[14] R. Kimberley, Y. Dong, A. Pramanik, et al., 2023, Journal of Manufacturing and Materials Processing, 7(5), 168.
[15] H. Jian, M. Yang, W. Fang, et al., 2024, https://doi.org/10.1007/s11665-024-09271-3.
[16] D.A. Lados, D. Apelian, J.K. Donald, 2006, Acta Materialia, 54, 1475-1486.
[17] D. Buchbinder, H. Schleifenbaum, S. Heidrich, et al., 2011, Physics Procedia, 12, 271-278.
[18] F. Trevisan, F. Calignano, M. Lorusso, et al., 2017, Materials, 10, 76.
[19] S. Siddique, M. Imran, F. Walther, 2017, International Journal of Fatigue, 94, 246-254.
[20] P. Wei, Z. Wei, Z. Chen, et al., 2017, Applied Surface Science, 408, 38-50.
[21] I.I. Ailinei, S.V. Galaţanu, L. Marşavina, 2021, Structural Integrity and Life, 22, 25–28.
[22] M. Costas, D. Morin, M. de Lucio, et al., 2020, European Journal of Mechanics - A/Solids, 81, 103966.
[23] M. Cesnik, J. Slavic, M. Boltezar, 2012, Journal of Sound and Vibration, 331, 5370-5382.
[24] X. Ling, Y. Shixi, G. Chunbiao, 2012, Optics and Lasers in Engineering, 50, 1596-1601.
[25] ISO 1099, 2017, Metallic materials — Fatigue testing — Axial force-controlled method.
[26] I.I. Ailinei, S.V. Galaţanu, L. Marşavina, 2022, Engineering Failure Analysis, 137, 106206.
[27] I.I. Ailinei, S.V. Galaţanu, D.F. Mădroane, L. Marşavina, 2022, Procedia Structural Integrity, 42, 1422-1427.
[28] G. Silas, Mechanics. Mechanical Vibrations, Ed. Didactica si Pedagogica, 1968 (in Romanian).

IOP Conf. Series: Materials Science and Engineering **1319** (2024) 012028 doi:10.1088/1757-899X/1319/1/012028

THE INFLUENCE OF CITRIC ACID ON THE CORROSION PROCESS OF DIFFERENT METALLIC MATERIALS

Nataliia RUDENKO [1,a], George-Daniel DIMA [2,b], Mircea Laurențiu DAN[2,c*]

[1]University Politehnica Timisoara, Innovation and Technology Transfer Center, 2 V. Pârvan, 300223 Timisoara, Romania

[2]University Politehnica Timisoara, Faculty of Industrial Chemistry and Environmental Engineering,, 6 V. Pârvan, 300223 Timisoara, Romania

[a]nataliia.rudenko@upt.ro, [b]george.dima@student.upt.ro, [c]mircea.dan@upt.ro,

Keywords: corrosion behaviour, citric acid, corrosion techniques, chronoamperometric studies

Abstract

The paper presents the results from a study of the corrosion behaviour of different metallic materials, such as copper, aluminium and OL 52 steel in the presences of various concentrations of citric acid added in neutral solution (Na_2SO_4 0.25 M).

Citric acid (2-hydroxy-1,2,3-propanetricarboxylic acid) is among the most commonly utilized compounds in the food and beverage sectors. This non-toxic chemical can be derived from natural, renewable sources such as citrus fruits, pineapples, kiwis, raspberries, strawberries, and both black and red currants. Corrosion studies carried out previously demonstrated a mixed effect of citric acid, it can be a good inhibitor for aluminum alloys in a strongly acidic or strongly basic media as well as in neutral NaCl media, for low carbon-steel and copper in an acidic environment, but also, a good, accelerated corrosion agent in an alkaline solution for metals such as iron, copper, zinc and tin.

Potentiodynamic polarization and chronoamperometry techniques were used in the experimental part of this paper to evaluate the inhibition efficiency of the citric acid in test solutions. Also, Tafel method for the kinetic parameters determination for studied corrosion process.

1. Introduction

Carbon steel finds wide application in many fields of industry due to its availability, facility of production, relative low cost and good mechanical properties having at the same time other important properties. It corrodes when it interacts with acid solutions during different technological processes, such as: cleaning with acid agents, transportation of solution containing acid. Also, corrosion is the main reason for the destruction of metal at the bottom of ships, pipelines, power lines (transmission tower) and other objects [1-3].

Aluminum alloys are commonly used in industry due to their unique properties, such as high hardness, but low density. At the same time during the operating process a hard-to-soluble oxide film is formed on the metal surface, protecting the surface from aggressive external influence. Due to its extensive use, aluminum alloys often come in contact with natural or technological environments, resulting in their destruction [4].

Copper is used in various industrial sectors, as a base for the manufacturing of elements and parts of electrical appliances. Also, the material is widely used in the installation and maintenance of water supply systems, heating systems and gas pipelines. Copper corrosion appears when copper reacts to its media for a long time. This can lead to structural damage and an increased risk for fire hazards if left unchecked [5].

IOP Conf. Series: Materials Science and Engineering **1319** (2024) 012028 doi:10.1088/1757-899X/1319/1/012028

To study corrosion process of steel, copper and aluminum citric acid was chosen. Citric acid is an organic compound that acts as an antioxidant, conservating agent and purifier, widely used in the food industry, cosmetic industry, medicine metallurgy and other related fields. Citric acid is found in many fruits and vegetables, but it is produced on an industrial scale from various starch hydrolysates, cane or beet molasses or sugar [6-7].

Citric acid solutions, alongside hydrochloric and sulfuric acid solutions, are process liquids for removal of mineral impurities and corrosion products from metal surfaces. An advantage of citric acid ($C_6H_8O_7$), in comparison with HCL and H_2SO_4, its corrosivity, towards to the most important engineering materials such as steel, is low. This allows to use it at industry, as opposed inorganic acids, without additional inhibitor protection [8-9].

The aim of this work is to study the corrosion behavior of different metallic materials, such as copper, aluminum and OL 52 steel in the presences of various concentrations of citric acid added in neutral solution – blank solution (Na_2SO_4 0.25 mol L^{-1}).

2. Materials and methods

2.1 Materials

In order to study the metallic materials corrosion behaviour in neutral media (0.25 mol L^{-1} Na_2SO_4 solution- blank solution) and the inhibitory or acceleratory effect of citric acid, different concentrations of citric acid between 0.25; 0.5; 1; 2.5; 5; 7.5; 10 g /L solution have been used. Citric acid ($C_6H_8O_7$) has IUPAC name 2-hydroxy-1,2,3-propanetricarboxylic acid and its chemical structure is shown in Figure 1.

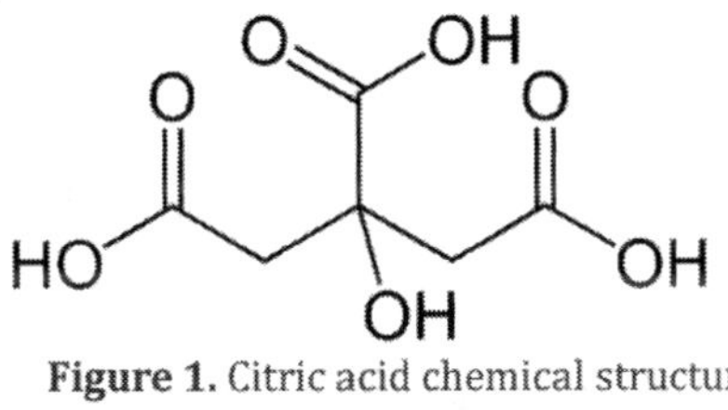

Figure 1. Citric acid chemical structure

The corrosion test electrode consists of a carbon steel bar made from OL 52, with its chemical composition detailed in Table 1. Also, as working electrodes, a bar of Cu of 99% purity and Al of 99% purity were used during the experimental determinations.

Table 1. Chemical composition of OL 52 electrode

Fe	C	Si	Mn	Ni	P	S	Cr	Mo	V	N	Nb	Ti	Al	Cu
max 96.865	max 0.2	max 0.5-1.65	max 0.9	max 0.5	max 0.035	max 0.03	max 0.3	max 0.1	max 0.12	max 0.015	max 0.05	max 0.03	max 0.02	max 0.35

2.2. Methods

The electrochemical behavior of citric acid in a neutral environment has been investigated using cyclic voltammetry on a platinum electrode with an active surface area of 0.5 cm^2 [10-12]. Various techniques, including linear polarization (Tafel method) and chronoamperometry, were employed to study the effects of citric acid on the corrosion processes of carbon steel, copper, and aluminum. The experimental setup comprised a conventional 100 mL three-electrode glass cell and a Biologic SP150 potentiostat/galvanostat [10]. The equipment was connected to two graphite counter electrodes, a working electrode (carbon steel, copper and aluminum) with a 1 cm^2 surface area, and an Ag/AgCl reference electrode. All potentials were referenced to this electrode ($E_{Ag/AgCl}$ = 0.197 V/NHE). Prior to each experiment, the working electrode surface was polished with abrasive papers of varying roughness, then cleaned in an ultrasonic bath and rinsed with distilled water. The electrode potential was allowed to stabilize for 60 minutes before each measurement.

The cyclic voltammograms (CVs) were recorded at various scan rates ranging from 5 to 500 mV s^{-1}, whereas linear voltammograms (LVs) were recorded only at a low scan rate of 1 mV s^{-1} to provide quasi-stationary conditions [10-12].

3. Results and discussion

3.1 Cyclic voltammetry

In order to observe the influence of citric acid added in the supporting electrolyte (0 Na_2SO_4 0.25 mol L^{-1}) onto the electrochemical behavior of the tested materials, voltametric studies were carried out in a wide potential range -1.75 V and +1.25 V.

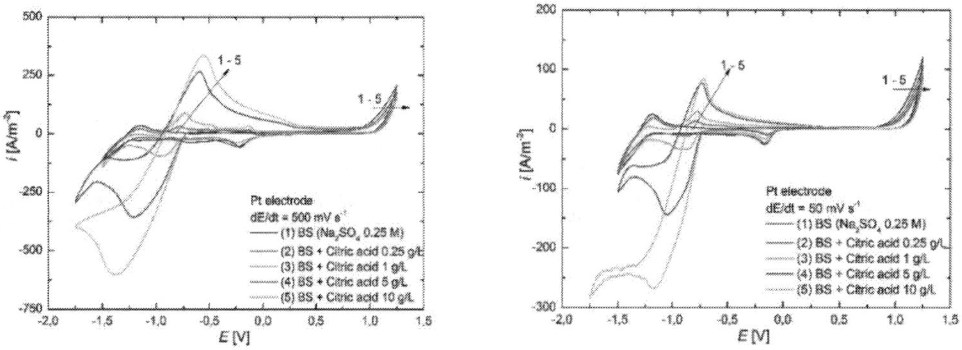

a b

Figure 2. Cyclic voltamograms recorded on Pt in Na_2SO_4 0.25M in the absence and presence of different concentrations of citric acid, at polarization rate of: 500 mV/s (a), 50 mV/s (b)

The cyclic curves recorded on platinum electrode in the absence and presence of citric acid at the polarization speed 500 mV/s and 50 mV/s respectively, were shown in figures 2a, b.

In the CVs recorded in a sulfate electrolyte solution without citric acid, starting anodic polarization from the open circuit potential (OCP) reveals only the curve corresponding to the oxygen evolution reaction (OER). During the reverse scan, a cathodic peak appears around -0.25 V, which is associated with the reduction of superficial platinum oxides formed on the electrode surface, as well as the reduction of superficial remanent oxygen.

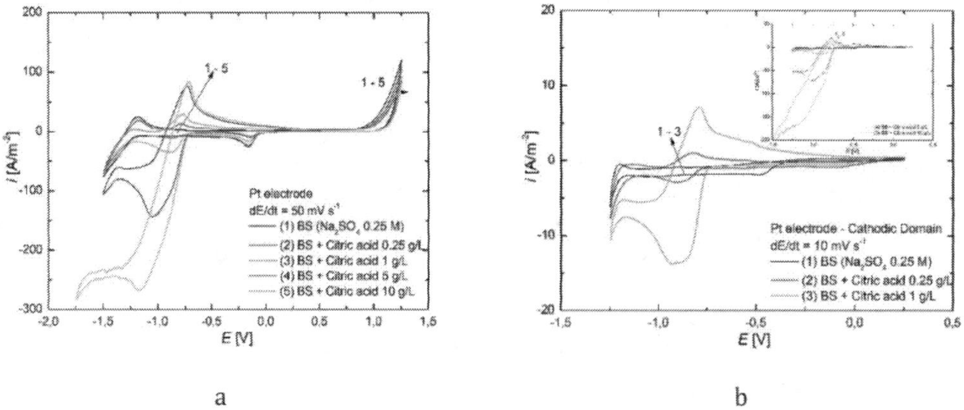

a b

Figure 3. Cyclic voltammograms traced on Pt in solutions of 0.25M Na_2SO_4 in the absence and presence of different citric acid concentrations in: anodic domain (a), cathodic domain (b), at scanning rate 10 mV

At more negative potentials (-1.25 V), the hydrogen evolution reaction (HER) becomes apparent. Based on the analysis of the voltammograms, it can be seen, in the presence of citric acid, two additional peaks are recorded anodic at approximately -0.65 V and cathodic at approximately 1.4 V associated with a reversible oxidation-reduction process.

To describe electrochemical process that take place on the anodic domain, respectively on the cathodic domain, cyclic voltammograms were performed in these two domains. The results are shown in figures 3 a, b.

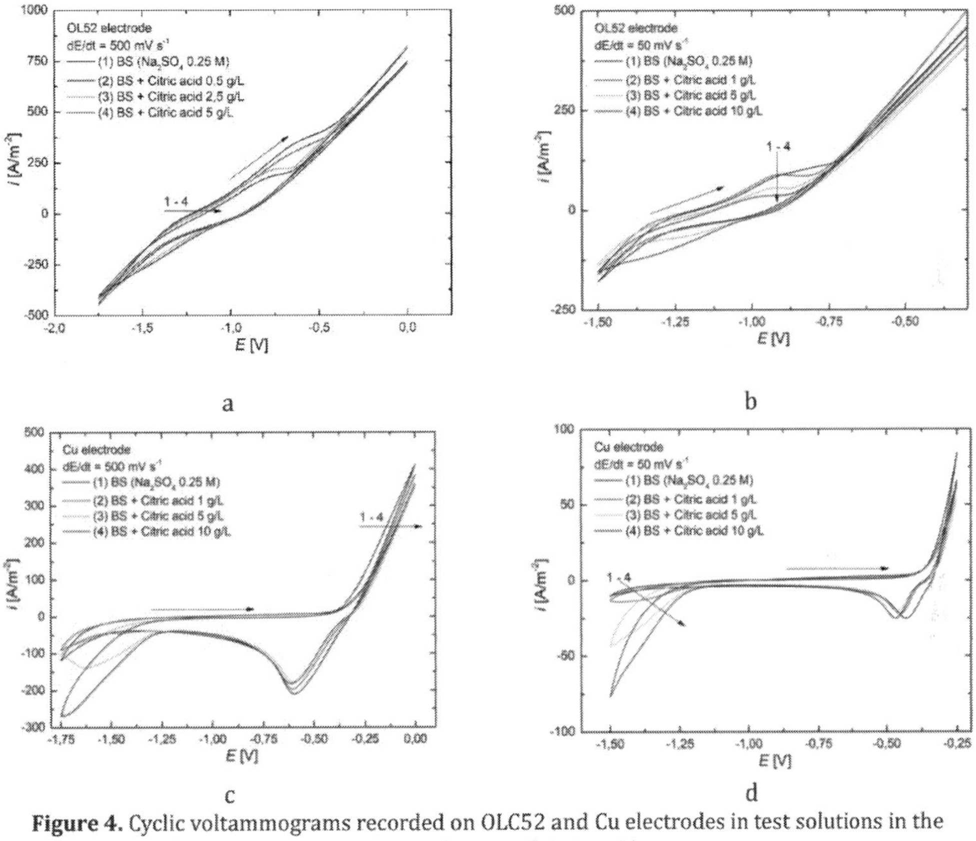

Figure 4. Cyclic voltammograms recorded on OLC52 and Cu electrodes in test solutions in the presence and absence of citric acid

Cyclic voltammetry also studied the electrochemical behavior of the tested materials in solutions used in experimental determinations. The figure 4 shows the cyclic curves presented on steel OLC 52 and copper respectively. The shape of the cyclic voltammograms plotted on aluminum is similar to that on OL 52, considering that only the curves for copper and OL 52 were included in the study. Analyzing these curves, we can state that on these metals don't appear specific peaks during oxidation or reduction process.

3.2 Chronopotentiometry studies

The initial step in the corrosion studies involved measuring the open circuit potential (OCP) for the metals in the testing solutions. Additionally, the potential values of the three metal electrodes in all test solutions were recorded over time. The time duration of hour is considered sufficient to register a state of equilibrium of metal-solution system.

Figures 5 a, b, c show OCP variations for three electrodes. It can be seen, that in case of steel and aluminum (fig. 5 a, c), the increase of amount of citric acid added to supporting electrolyte, led to a shift towards more positive values of the electrolyte potential. This indicates an increase in the metal's corrosion resistance.

For copper electrode (Fig. 5b), the OCP values shift towards negative values, signifying an enhancement in the metal's corrosion resistance. It can be considered that citric acid acts as corrosion inhibitor in the case of copper.

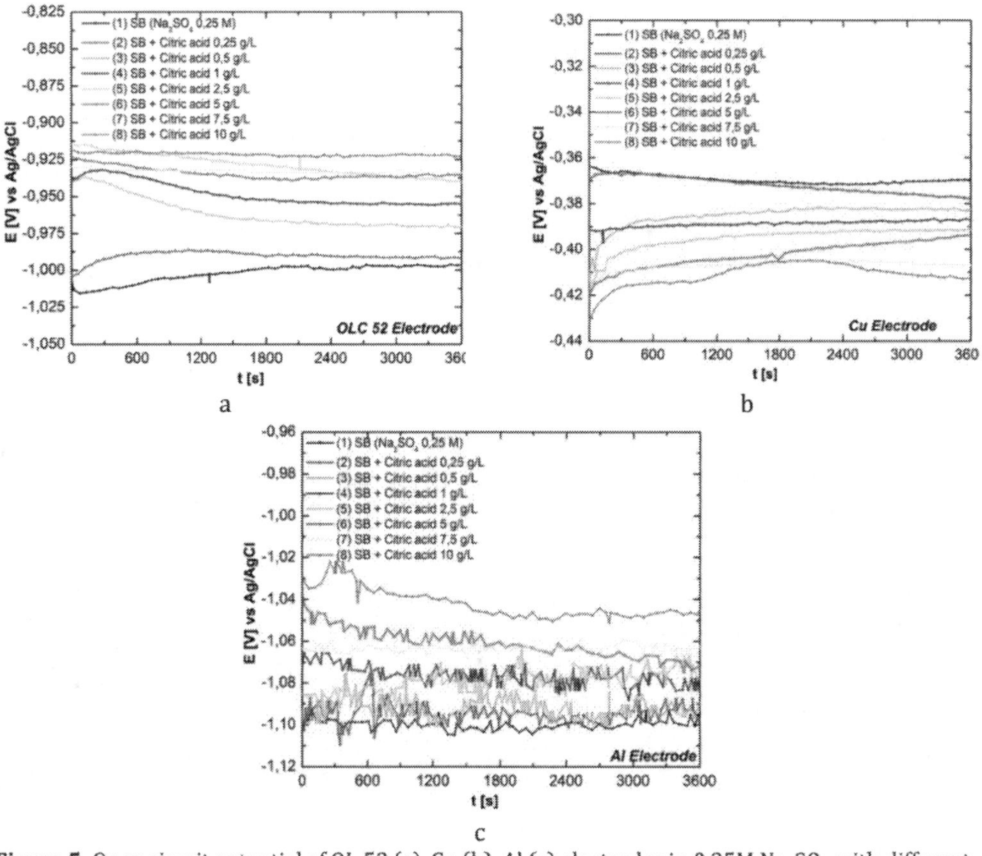

Figure 5. Open circuit potential of OL 52 (a), Cu (b), Al (c) electrodes in 0.25M Na_2SO_4 with different citric acid concentration

3.3 Corrosion studies by linear polarization method (Tafel method)

Parameters of corrosion process by linear polarization method or the Tafel method were determinate for each studied metal. Logarithmic Tafel curves obtained on the OL 52 electrode are shown in figure 6.

The polarization parameters for the corrosion processes of OL 52, copper, and aluminum in a neutral solution, including the corrosion potential (E_{corr}), corrosion current density (i_{corr}), cathodic and anodic Tafel slopes (b_c and b_a), and corrosion rate (v_{corr}), were determined by extrapolating the potentiodynamic curves.

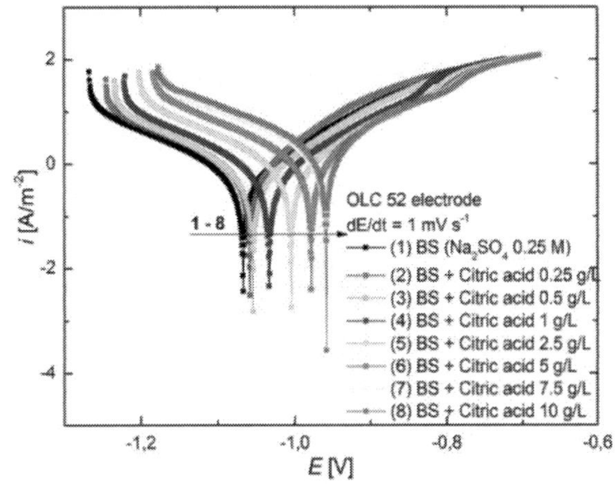

Figure 6. Tafel polarization diagrams drawn on OL 52 electrode in 0.25M Na_2SO_4 solution in the absence/presence of different concentrations of Citric Acid at 25°C

Table 2 presents specific parameters of the corrosion process obtained by modelling the Tafel curves using SP 150 biological potentiostat software. It is observed that corrosion rate increases with the concentration of citric acid added in experimental solution. Analyzing the range of values, the tested steel is not suitable for the mentioned applications since the corrosion rate is greater than 0.1 mm/y.

Table 2. Electrochemical parameters determined from linear polarization curves

Electrolyte	E_{corr}, mV	i_{corr}, µA/cm²	β_a, mV/dec	$\lvert\beta_c\rvert$, mV/dec	v_{corr}, mm/y
BS	1.103	41.52	118	146	0.710
BS+0.25 g/L CA	1.096	55.82	121	150	0.734
BS+0.5 g/L CA	1.063	60.11	124	148	0.785
BS+1 g/L CA	1.032	63.83	127	154	0.849
BS+2.5 g/L CA	1.006	72.82	134	160	1.082
BS+5 g/L CA	0.982	72.78	134	167	1.483
BS+7.5 g/L CA	0.964	134.7	135	172	2.002
BS+10 g/L CA	0.959	192.7	145	177	2.488

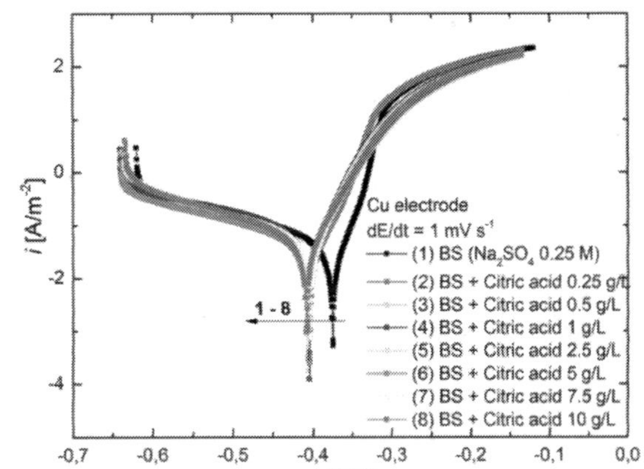

Figure 7. Tafel polarization diagrams drawn on Cu electrode in 0.25M Na_2SO_4 solution in the absence/presence of different concentrations of Citric Acid at 25°C

Also by using the method of linear polarization, the inhibitory character of citric acid for the corrosion process of copper in neutral solution was demonstrated in figure 7.

Table 3. Electrochemical parameters determined from linear polarization curves

Electrolite	E_{corr}, mV	I_{corr}, μA/cm²	βa, mV/dec	βc, mV/dec	V_{corr}, mm/y
BS	375.5	8.98	41.1	349	0.106
BS+0.25 g/L CA	405.1	7.18	65.6	340	0.085
BS+0.5 g/L CA	404.9	5.56	53.2	332	0.067
BS+1 g/L CA	404	4.34	37.1	270	0.050
BS+2.5 g/L CA	406	3.96	40.4	221	0.046
BS+5 g/L CA	403.7	3.61	37.5	192	0.041
BS+7.5 g/L CA	403.9	3.33	39.2	164	0.038
BS+10 g/L CA	410	2.63	39.8	151	0.030

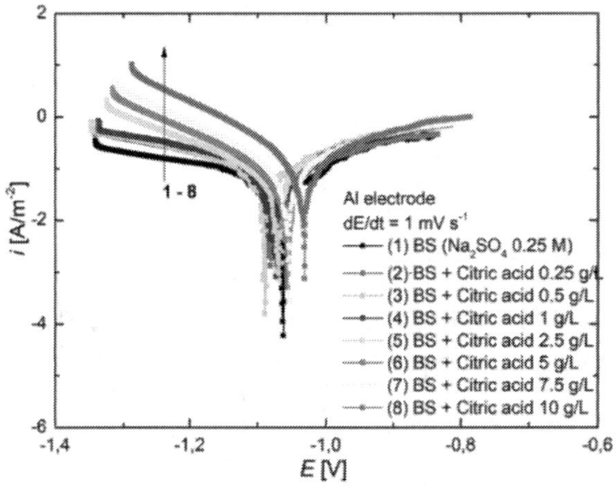

Figure 8. Tafel polarization diagrams drawn on Al electrode in 0.25M Na_2SO_4 solution in the absence/presence of different concentrations of Citric Acid at 25°C

Values obtained for aluminum are presented in figure 8 and table 4. Corrosion rate increases similarly with the increase of citric acid concentration.

Table 4. Electrochemical parameters determined from linear polarization curves

Electrolite	E_{corr}, mV	I_{corr}, μA/cm²	βa, mV/dec	βc, mV/dec	v_{corr}, mm/y
SB	1.063	7.31	219	413	0.079
SB+0.25 g/L CA	1.079	8.41	318	351	0.103
SB+0.5 g/L CA	1.090	9.39	362	482	0.117
SB+1 g/L CA	1.096	9.54	345	472	0.128
SB+2.5 g/L CA	1.129	10.51	170	199	0.137
SB+5 g/L CA	1.137	12.21	177	182	0.148
SB+7.5 g/L CA	1.155	13.57	215	168	0.160
SB+10 g/L CA	1.181	15.71	257	154	0.171

IOP Conf. Series: Materials Science and Engineering **1319** (2024) 012028 doi:10.1088/1757-899X/1319/1/012028

3.4 Chronoampherometric studies

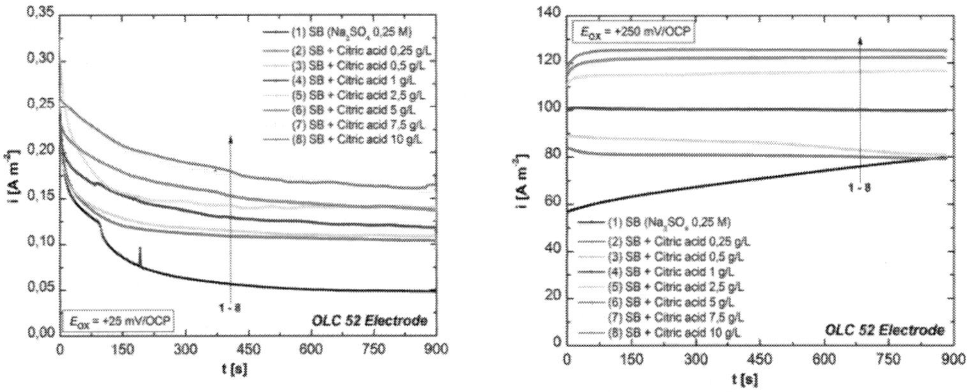

a b

Figure 9. Chronoamperometric measurements recorded on OL 52 electrode in test solutions, in the presence of various citric acid, at E_{Ox} = +25 mV/E_{cor} (a), respective E_{Ox} = +250 mV/E_{corr} (b) for 15 min.

To confirm the data obtained by Tafel method, amperometric studies were also performed. These are based on the oxidation at constant value of potential of studied metal in the test solutions and recording of current density values over time. Current values and corrosion rate increase proportionally to increasing citric acid concentration.

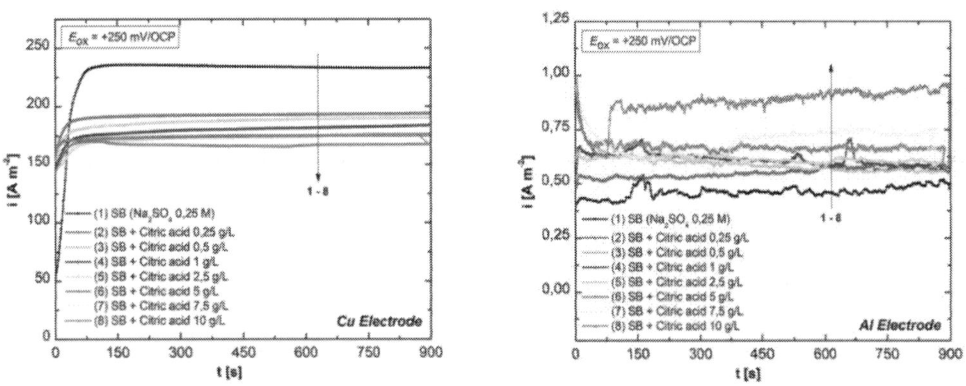

a b

Figure 10. Chronoamperometric measurements performed on Cu and Al electrodes in test solutions, in the presence of various citric acid, at E_{Ox} = +250 mV/E_{corr}

It is observed again, in the case of determinations for metallic copper, the highest values of current density are recorded in the supporting electrolyte, following their decrease with increasing of added citric acid concentration.

Conclusions

The study has been carried out by using electrochemical techniques such as cyclic voltammetry, linear polarization (Tafel method), cronopotentiometry and chronoamperometry

For metallic copper, the inhibition efficiency increases with the concentration of citric acid.

This is likely due to the adsorption of citric acid on the metal surface. Citric acid forms a stable copper-citrate complex with copper, whereas with aluminum and OL 52 it leads to the formation of corrosive products on the surface, preventing the adsorption of citric acid on these two metals. The study demonstrates that citric acid can serve as an eco-friendly inhibitor to prevent copper corrosion in a neutral Na_2SO_4 solution. However, citric acid increases the corrosion rate for OL 52 and aluminum in a neutral solution.

References

[1] J. Wysocka, S. Krakowiak, J. Ryl, 2017, Electrochim. Acta, 258, 1463-1475
[2] K. Kamarska, 2023, J. Ecol. Eng. 24(1), 307–311
[3] Ya.G. Avdeev, K.L. Anfilov, E.P. Rukhlenko, Yu.I. Kuznetsov, 2021, Int. J. Corros. Scale Inhib., 10(3), 911–923
[4] P. Matheswaran, A.K. Ramasamy, 2012, E-Journal of Chemistry, 9(1), 75-78
[5] Y. Avdeev, M. Tyurina, V. Rabinkov, A. Luchkin, 2017, XXVI R-S-P Seminar 2017, Theoretical Foundation of Civil Engineering
[6] M. Cui, Y. Yu, Y. Zheng, 2021, Polimers, 13, 1923
[7] A. Abbas, A. Y. Adesina, R. K. Suleiman, 2023, Metals, 13(8), 1479
[8] S. Suwarno, H. Muhtadi, 2020, Key Engineering Materials, 867, 224-228
[9] N. Mazinanian, I. O. Wallinder, Y. S. Hedberg, 2015, Corrosion Science and Technology, 14 (4), 166-171
[10] N. Vaszilcsin, A. Kellenberger, M.L. Dan, D.A. Duca, V.L. Ordodi, 2023, Materials, 16, 5555
[11] A. Kellenberger, D.A. Duca, M.L. Dan, M. Medeleanu, 2022. Materials, 15, 2918
[12] D.A. Duca, M.L. Dan, N. Vaszilcsin, 2018, IOP Conf. Ser.: Mater. Sci. Eng. 416 012043

Design and structural optimization of an assembled re-entrant auxetic structure

Dan-Andrei ȘERBAN[1*]

[1] Department of Mechatronics, University Politehnica Timişoara, Timisoara, Romania

*E-mail: dan.serban@upt.ro

Abstract. Auxetics represent a class of metamaterials that possess the unique quality of exhibiting a negative Poisson ratio. Considering the fact that the cross-section is reduced during compressive loadings, this class of materials and structures exhibit superior energy absorption capabilities in impact loadings, as the density of the material increases near the contact region. In this study, a novel approach in obtaining re-entrant auxetic structures is presented, consisting of the assembly and joining of components with simple geometries, that can be manufactured through conventional means (as opposed to additive manufacturing). The dimensions and aspect ratios of the structural members was evaluated numerically in order to obtain the optimal stiffness and strength values for a given mass. The design aspect also included special features that allows the assembly through shape, which facilitates the subsequent joining operation (welding or brazing).

1. Introduction

Auxetic structures represent a peculiar class of metamaterials with their main feature being a negative Poisson ratio, which implies a reduction of the cross-section during compression and an increase of the cross-section in tension [1, 2, 3]. This property can be technologically obtained from conventional foams (i.e. material subjected to triaxial compression whilst heated beyond the softening point of the base material and subsequently cooled [4, 5]) or they can be engineered (design and manufacturing from bulk materials [6, 7]).

Considering this aspect that grants auxetic structures unique features, the main applications of such materials are as energy absorption components (as the material contracts during compression, it increases its density and, consequently, its mechanical properties) for the 3D structures and as shape following meshes (plane structures that can follow rounded contours without wrinkling) for the 2D structures [8, 9, 10]. Thus, the first goal of the study is to determine the optimal geometrical parameters of the re-entrant auxetic structures investigated in this study in order to maximize the rigidity to weight ratio.

Given the complexity of the cellular structures, 3D auxetic materials are generally manufactured through additive manufacturing [6, 7, 11], which can be a disadvantage in terms of the low productivity. One of the goals of this study is to develop a re-entrant auxetic structure that can be assembled from parts with relatively simple geometries that can be manufactured using conventional technologies (machining, water jet / laser / plasma / electrical discharge cutting, etc.).

2. Design of the re-entrant structure

The structure investigated in this study consisted of cubic cells based on two hexagons with four angles $\alpha < 90°$ (called re-entrance angles) each, offset by 90° around the y-axis, with an additional two vertical struts (Figure 1 a and b). The structure was obtained by performing a tessellation where the adjacent cells were offset by half the width/length of the circumscribed cube along the x and z directions (Figure 1 c). The resulting structure exhibits negative Poisson ratio when subjected to loadings about all three Cartesian axes.

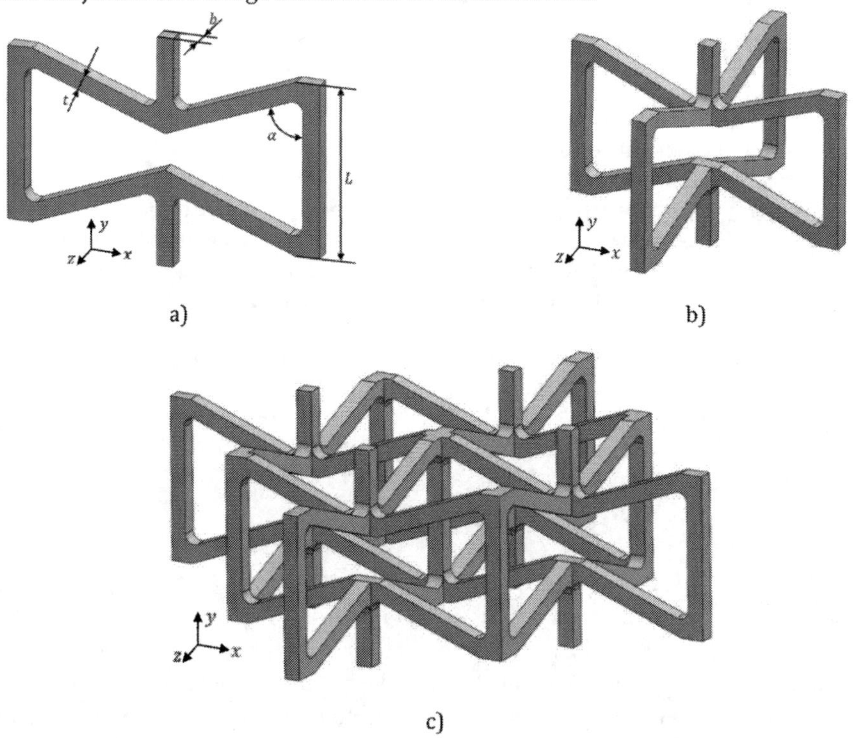

Figure 1. Re-entrant auxetic structure design

The structures had three variable parameters, namely the strut thickness t, the strut depth b and the strut length L. For the first part of the study, the thickness and the depth of the strut were linked through a parameter $k = t/b$. The influence of the re-entrance angle on the mechanical properties was investigated in a prior study [12] and was set for a value of 70°.

3. Structure optimization

The optimization process consisted of two stages: in the first part, a fixed value for the volume of the structure was chosen, and the influence of the parameter k on the stiffness of the structure was investigated. After the optimal value was obtained, the second part of the process consisted in the determination of the optimal strut thickness to length ratio, considering the stiffness normalized to the volume of the structure. In both cases, the stiffness of the structures was investigated for compressive and flexural loads.

3.1. Optimal thickness to depth ratio

The first step in the determination of the optimal thickness to depth ratio was the evaluation of the volume variation with strut thickness t and the parameter k. In consequence, the values for the thickness and depth of the strut were varied in a given range and the volume was evaluated using the CAD software for each pair, maintaining the L value constant. An equation was fitted to the obtained surface (Figure 2 a), which was solved for a given value of the volume, resulting in a variation of the strut thickness t with the parameter k (Figure 2 b).

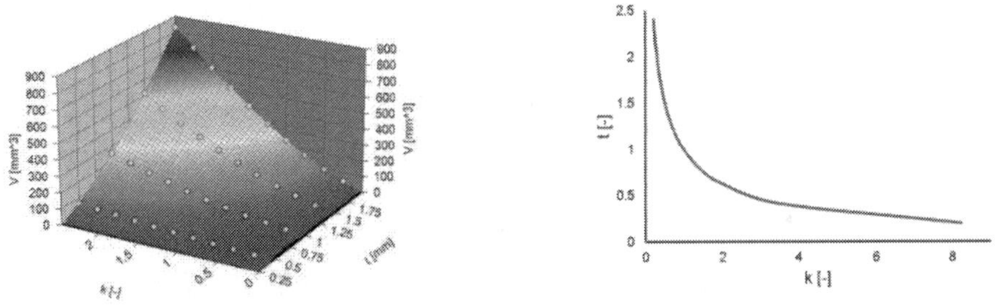

a) b)

Figure 2. Structure volume variation with the parameters t and k (a) and t variation with k for constant volume (b)

For this study, the k parameter was varied between 0.2 (Figure 3 a) and 8.2 (Figure 3b) in 10 increments.

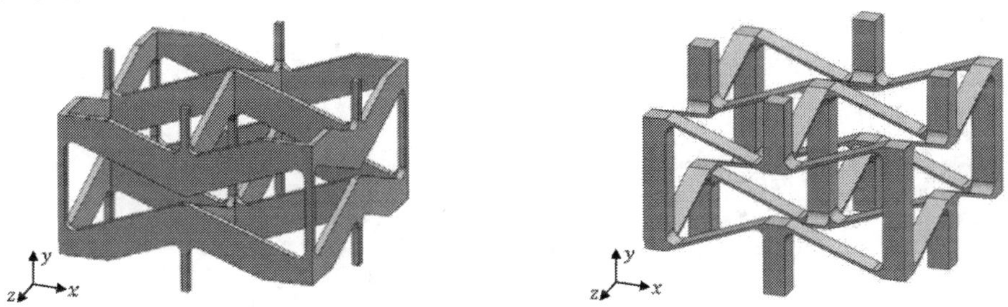

a) b)

Figure 3. Auxetic cells for $k = 0.2$ (a) and for $k = 8.2$ (b)

The resulting structures were subsequently tessellated to a 5 x 1 x 1 pattern, in order to obtain auxetic beams, destined for the flexural simulations (Figure 4). A $0.02 \cdot L$ thick face was added to the upper and lower regions of the beams to obtain a more uniform stress distribution during the numerical analyses.

The compression simulations used the unit cells presented in Figure 3, meshed using second order tetrahedral elements (C3D10) of 0.5mm average size (Figure 5 a). The material model consisted of a linear elastic formulation with Young's modulus of $2 \cdot 10^5$ and a Poisson ratio of 0.3. As boundary conditions, the outer surfaces of the four lower vertical struts were subjected to y-axis symmetry while a y-axis displacement was imposed to the free surface of the upper vertical struts. As field output, the reaction forces from the mesh nodes subjected to the y -axis symmetry R_j and the displacement of the top surfaces u were recorded and used to evaluate the compressive stiffness E_c of the structure.

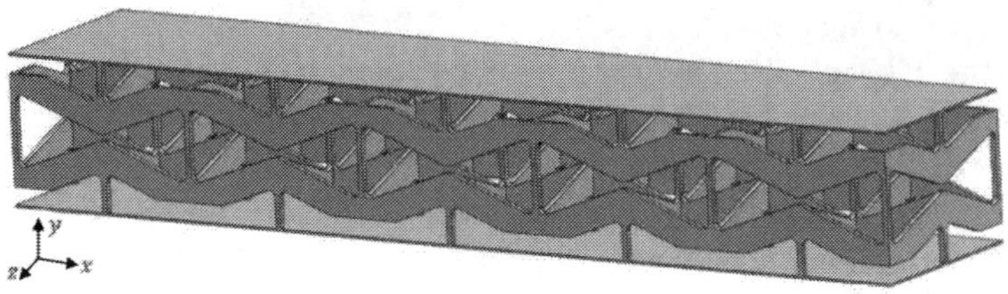

a)

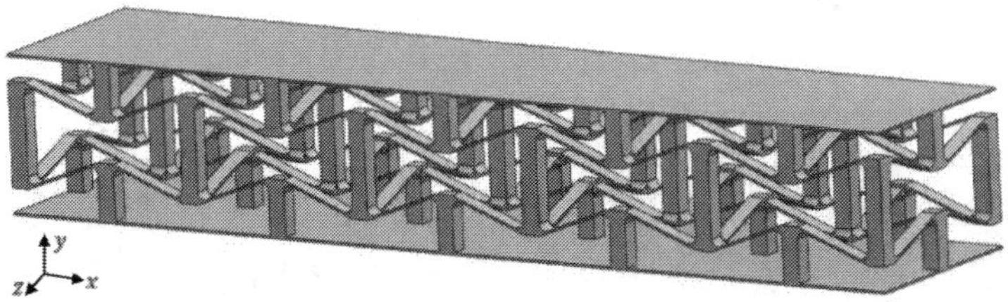

b)

Figure 4. Auxetic beam for $k = 0.2$ (a) and for $k = 8.2$ (b)

$$E_c = \frac{\sigma}{\varepsilon} = \frac{h \cdot \left(\sum_{j=1}^{n} R_j\right)}{u \cdot A} \tag{1}$$

where σ is the nominal stress, ε is the nominal strain, h is the initial height of the structure and A is the cross-section area of the circumscribed cube, normal to the y-axis.

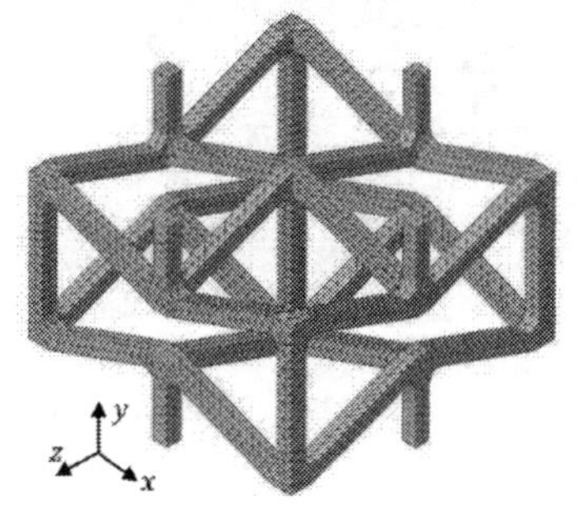

a) b)

Figure 5. Meshed model for the compression (a) and flexural (b) simulations

For the flexural simulations, apart from the beam models presented in Figure 4, half-cylinder discreet rigid bodies were used for the supports (positioned exactly underneath the furthest most strut for each structure) and the indentor. The deformable models used second order tetrahedral elements of 0.5 mm average size while the discrete rigid bodies used rectangular elements of 0.2 mm average size (Figure 5 b). A surface-to-surface interaction was defined between the rigid bodies and the adjacent surfaces of the deformable body, using a frictionless tangential behaviour and a "hard contact" normal behaviour. As boundary conditions, the rigid bodies corresponding to the support were fixed while a displacement along the y-axis was imposed to the indentor. The reaction force R_I and the displacement u_I in the indentor were recorded as field output, and the flexural stiffness E_f was evaluated using the relation

$$E_f = \frac{\sigma}{\varepsilon} = \frac{R_I \cdot S^3}{4 \cdot u_I \cdot w \cdot H^3} \tag{2}$$

where S is the support span, w is the depth of the beam and H is the height of the beam.

Figure 6 presents the variation of the relative stiffness (actual stiffness value for a given structure divided by the maximal stiffness value recorded) with the parameter k for the compressive and flexural simulations. It can be observed that, for both loading cases, the optimal value for the k parameter is 1 (the thickness of the strut equals its depth).

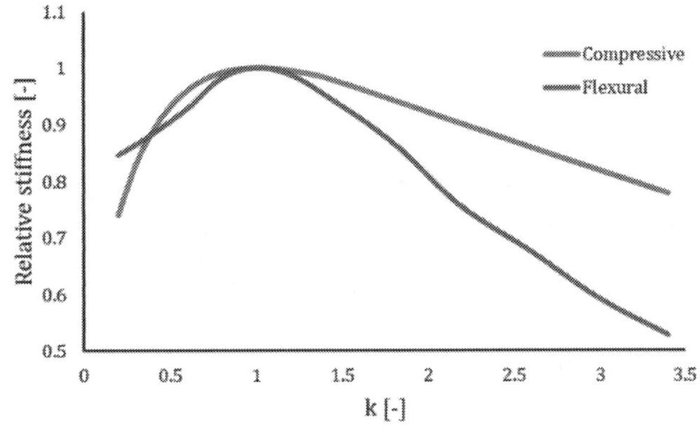

Figure 6. Variation of relative stiffness with the k ratio for the compressive and flexural tests

3.2. Optimal thickness to length ratio

In order to determine the optimal thickness to length ratio, the strut length L was maintained constant and the strut thickness t was varied from an arbitrary small value (equivalent to $t/L = 0.025$, Figure 7 a) to the largest value that determined valid (non-intersecting) structures (equivalent to $t/L = 0.275$, Figure 7 b).

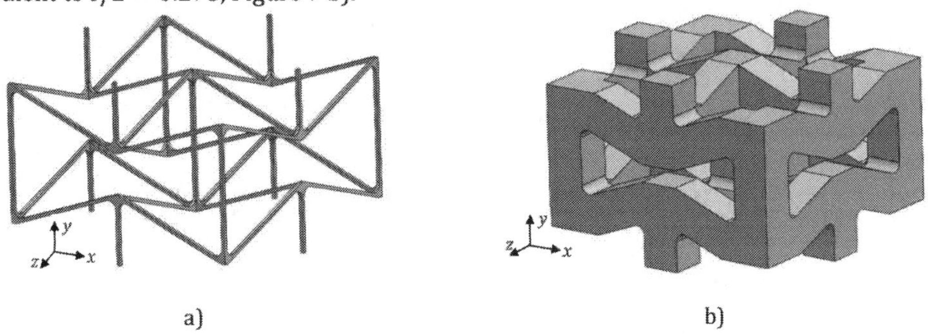

Figure 7. Auxetic cells for $t/L = 0.025$ (a) and for $t/L = 0.275$ (b)

In a fashion similar to the previous part of the study, the unit cells were tessellated and $0.02 \cdot L$ thick faces were added to the resulting beams (Figure 8).

IOP Conf. Series: Materials Science and Engineering **1319** (2024) 012029 doi:10.1088/1757-899X/1319/1/012029

The compressive and flexural simulation setups were identical with the ones presented previously, with the sole difference being the element size, whose size was chosen so that the mesh generated at least two elements per strut thickness.

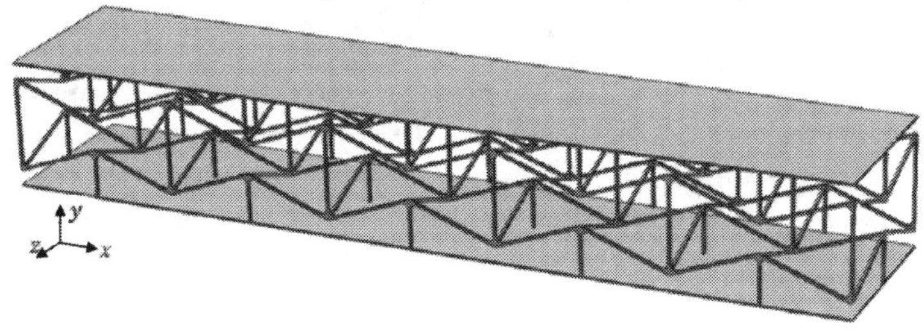

a)

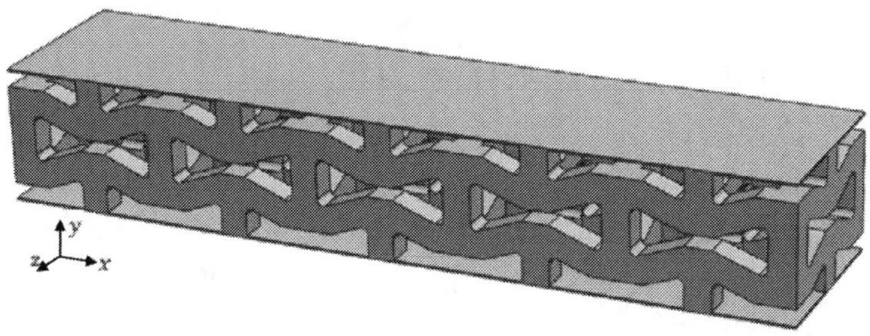

b)

Figure 8. Auxetic beam for $t/L = 0.025$ (a) and for $t/L = 0.275$ (b)

In order to obtain a valid comparison between the results, the evaluated stiffness of the structures was divided by the structure volume so that, for a constant base material density, the optimal structure can be considered as the one that determines the highest value for a fixed structural mass. The variation of the normalized stiffness to volume ratio with the t/L ratio is presented in Figure 9 for the compressive and flexural simulations. As with the previous investigations, both cases determined the optimal value for the largest t/L ratio evaluated.

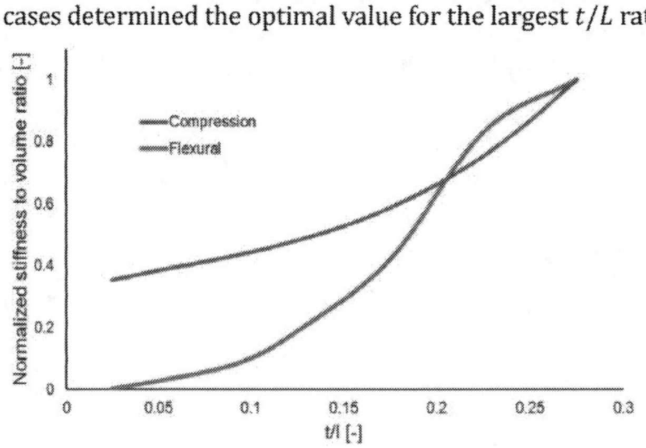

Figure 9. Variation of relative stiffness with the k ratio for the compressive and flexural tests

4. Design for manufacturing

For the investigated auxetic structure with a re-entrance angle value of 70°, the optimal aspect ratio consisted of a strut thickness to depth ratio of 1 and a strut thickness to length ratio of 0.275. Considering these structural parameters, the resulting unit cell was divided into planar parts that can be manufactured from sheet metal, metal plates, planar composite layups etc. using conventional cutting/machining technologies. The chosen division consisted of three types of parts: one type corresponding to the outer faces of the cell (Figure 10 a), one type corresponding to the mid-plane cell region (Figure 10 b) and one type that accounted for the connecting struts (Figure 10 c). In order to facilitate with the structural assembly and temporarily align the components before permanent fixing (welding or brazing), rectangular orifices and slots were introduced, as well as corresponding bosses to the assembly counterparts. In addition, the outer vertical struts of the parts from Figure 10 (a) and (b) were designed with a reduced thickness of $t/2$. The assembled unit cell (Figure 10 c) contains two instances of the first type of part, one instance of the second type of part and 12 instances of the third type of part. An example of a 3 x 3 x 2 assembled structure is presented in Figure 10 (e).

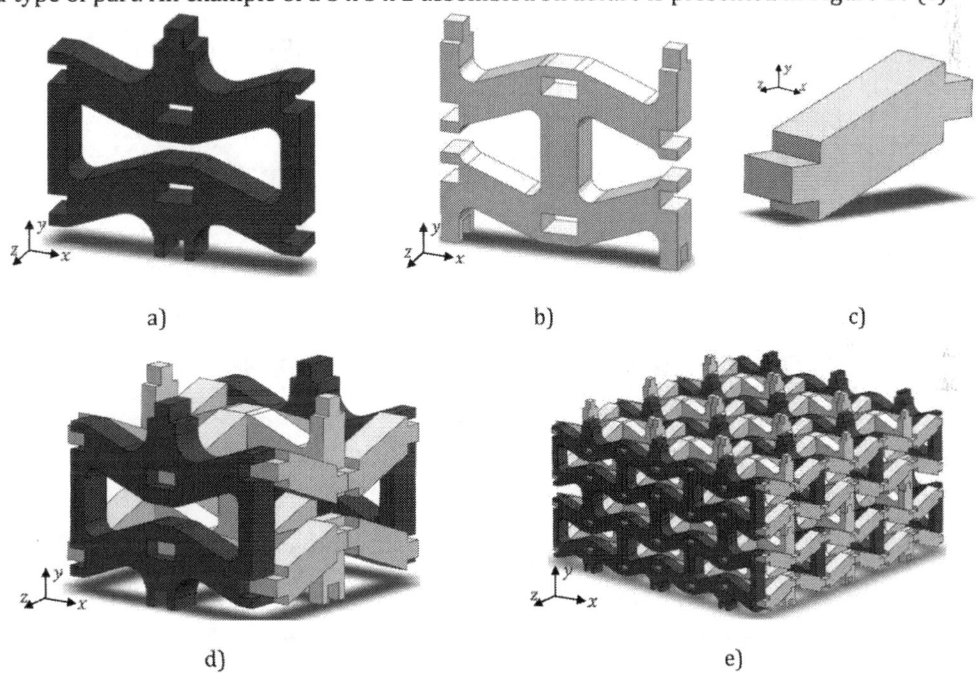

Figure 10. Structure components (a), (b) and (c), assembled unit cell (d) and assembled 3 x 3 x 2 structure (e)

5. Conclusions

In this study, the optimal structural parameters of a re-entrant auxetic structure were determined for compressive and three-point bending loadings. The first step focused on the evaluation of the influence that the strut thickness to depth ratio has on the stiffness of the structure, as smaller ratios determine high section moduli values for the horizontal struts but low cross-section areas for the vertical struts while larger ratios determine the exact opposite, when maintain a constant volume. From the numerical analyses it was concluded that the optimal structure has equal values for the thickness and the depth of the struts.

The second step investigated the variation of the stiffness normalized to the structural volume for variations in the strut thickness to length ratio, considering the optimal strut cross-section aspect ratio determined previously. In this case, smaller values for the given ratio determine lower stiffness at low volumes, while the increase in the structures' stiffness due to the augmentation of the t/L ratio comes at the cost of an increased volume. The simulation results show that for both the compression and flexural tests, the optimal results were obtained for the highest t/L ratio.

Having obtained the optimal structural parameters in terms of stiffness, the resulting structure was divided in order to allow the manufacturing of this type of auxetic structure using conventional machining or cutting operations. Consequently, three types of planar components were generated, which were provided with slots and bosses in order to facilitate the temporary assembly before the permanent solidarization of the structure, which could be obtained through welding or brazing.

6. Acknowledgements

This work was funded by the Research Grant PN-III-P2-2.1-PED-2021-1134, contract number 586PED/2022.

7. References

[1] A. Kolpakov, "Determination of the average characteristics of elastic frameworks," *Journal of Applied Mathematics and Mechanics,* vol. 49, no. 6, pp. 739-745, 1985.

[2] R. Almgren, "An isotropic three-dimensional structure with Poisson's ratio =−1," *Journal of Elasticity,* vol. 15, pp. 427-430, 1985.

[3] K. Evans, "Auxetic polymers: a new range of materials," *Endeavour,* vol. 15, no. 4, pp. 170-174, 1991.

[4] R. Lakes, "Foam Structures with a Negative Poisson's Ratio," *Science,* vol. 235, no. 4792, pp. 1038-1040, 1987.

[5] J. Cho and R. Lakes, "Fracture toughness of re-entrant foam materials with a negative Poisson's ratio: experiment and analysis," *International Journal of Fracture,* vol. 80, no. 1, pp. 73-83, 1996.

[6] M. Grujicic, R. Galgalikar, J. Snipes, R. Yavari and S. Ramaswami, "Multi-physics modeling of the fabrication and dynamic performance of all-metal auxetic-hexagonal sandwich-structures," *Materials and Design,* vol. 51, p. 113–130, 2013.

[7] A. Spagnioli, R. Brighenti, M. Lanfranchi and F. Soncini, "On the auxetic behaviour of metamaterials with re-entrant cell structurs," *Procedia Engineering,* vol. 109, pp. 410-417, 2015.

[8] S. Mohsenizadeh, R. Alipour, M. F. N. A. Shokri Rad and Z. Ahmad, "Crashworthiness assessment of auxetic foam-filled tube under quasi-static axial loading," *Materials and Design,* vol. 88, p. 258–268, 2015.

[9] W. Liu, N. Wang, T. Luo and Z. Lin, "In-plane dynamic crushing of re-entrant auxetic cellular structure," *Materials and Design,* vol. 100, p. 84–91, 2016.

[10] L. Yang, O. Harrysson, H. West and D. Cormier, "Mechanical properties of 3D re-entrant honeycomb auxetic structures realized via additive manufacturing," *International Journal of Solids and Structures,* Vols. 69-70, p. 475–490, 2015.

[11] A. Indreș, D. Constantinescu and O. Mocian, "Bending behavior of 3D printed sandwich beams with different core topologies," *Material Design & Processing Communications,* vol. 3, no. 4, p. e252, 2021.

[12] D. Șerban, "Variation of Poisson's Ratio with Axial Strain for Three-," *Material Design & Processing Communications,* vol. 2022, p. 2623601, 2022.

Exploring Antifungic Properties of Low Vacuum Dealloyed Recycled Brass Nanoporous Powder

Gyorgy Thalmaier[1*], Niculina Argentina Sechel[1], Ioan Vida-Simiti [1,2] Andreea Hegyi [3,4]

[1] Materials Science and Engineering Department, Technical University of Cluj-Napoca, Cluj-Napoca, Romania
[2] Technical Science Academy of Romania, Bucharest, Romania
[3] Cluj-Napoca Branch, NIRD URBAN-INCERC, Cluj-Napoca, Romania
[2] Department of Environmental Engineering and Sustainable Development Entrepreneurship, Technical University of Cluj-Napoca, Cluj-Napoca, Romania

*E-mail: Gyorgy.Thalmaier@sim.utcluj.ro

Abstract. Fungal infections pose a significant health concern globally, due to the evermore common resistance to classical antifungal therapies. One possible way to overcome this tendency is to use new pathways offered by the nanotechnology. In this respect it has been studied the possibility of utilizing recycled waste in the form of brass machining chips that are transformed into micron-porous copper (particles) by vacuum dealloying. In a subsequent step it was evidenced the potential antifungal properties. To manufacture the samples powder metallurgy was combined with vacuum dealloying, and a complex characterization to study the possible utilization of these powders at a laboratory level. The machined precursor powders were low vacuum dealloyed (P~10^{-1} torr) at 500 °C, 550 °C, 600 °C and 650 °C, and the dealloying time was set between 1-4 hours. The porous particles were characterized by SEM-EDS analysis. The antifungal property was evaluated by exposing the samples on a medium of Potato Dextrose Agar (PDA) previously contaminated with a solution of Aspergillus Niger spores. The growth of fungi was analysed daily up to seven days using optical microscopy.

1. Introduction

Fungal infections pose a significant health problem worldwide especially if combined with increasing resistance to known antifungal therapies. Nanoporous materials, characterized by their high surface area and unique physicochemical properties, have emerged as potential candidates for combating fungal infections. Copper is known to be effective for fungal diseases. Copper ions kill by modifying proteins and enzymes at cellular levels cells of pathogens they contact. By maximizing it's specific surface area these antifungal properties may be enhanced. To increase the specific surface area, the eco-friendly process of vacuum dealloying can be used.

Vacuum dealloying uses the different evaporation rates specific for each material to selectively remove one or more metals from the initial alloy. The aim of this selective evaporation is to induce an extremely high number of vacancies, that in their turn, can link together forming porosity. Depending on the evaporated species volume fraction and the efficiency of the dealloying process, this porosity can be varied in a wide range. The parameters of the dealloying process (the most important being time and temperature) influence the pores size permitting a

IOP Conf. Series: Materials Science and Engineering **1319** (2024) 012030 doi:10.1088/1757-899X/1319/1/012030

precise control over the final porous structure [1-3]. The theoretical backbone of this process is the Kirkendall effect [4] and the working temperature is below the melting point of the precursor [5].

The dealloying method used does not use pore formers specific for the porous materials made by powder metallurgy, so it's incomplete removal can be prevented. The main advantage of the vacuum dealloying process is it's simplicity, and straightforward nature. The scientific interest on this method is generated by a relatively simple way of achieving a nanoporous structure starting from a master alloy that contains an element in high concentration that can be easily sublimated [6]. The evaporation of this element is easily controlled so a good repeatability of the results is obtainable.

Compared to other dealloying methods (chemical, electrochemical or liquid metal), this method eliminates the need to use hazardous materials and permits the recovery of the evaporated metals becoming an environmentally friendly process [7-12].

In this studied we used the brass chips, a secondary raw material in the circular economy, for evidencing the possibility of utilizing recycled waste in the form of brass machining chips that are transformed into micron-porous copper (particles). We also aimed to investigate the potential for improved antifungal properties. Due to the high vapor pressure between the alloying elements, low vacuum dealloying can be used to fabricate nanoporous copper particles from recycled brass turning chips by selectively evaporating the other components from the alloy [7].

2. Materials and methods

The chips were machined from a commercial brass piece without the use of cooling liquids to minimize the surface contamination of the precursor material.

The as obtained precursors were subjected to a low vacuum dealloying process (~10^{-1} torr) at 500 °C, 550 °C, 600 °C and 650 °C, for 1-3 hours. After dealloying the powders were characterized by scanning electron microscopy (JEOL JSM 5600 LV). The obtained samples pores size was estimated from SEM images using the ImageJ software. The EDS probe was utilized for local chemical composition measurements in order to calculate the porosity.

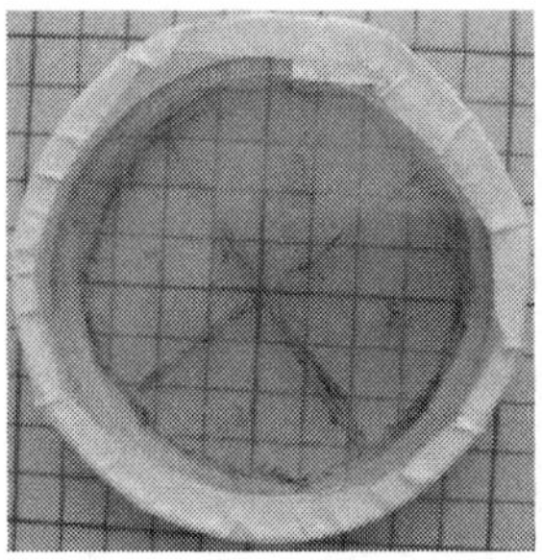

Figure 1. Contamination of the test specimens

The coarsening of the nanopores can be described by the equation [9] since it depends on the time and temperature:

$$d(t)^n = k_0 t e^{\frac{-E}{RT}} \quad (1)$$

Where Ds is the diffusivity; d(t) is the pore size at time t; k_0, K, and D_0 are constants and $k_0 = KD_0$; n is the coarsening exponent; R is the universal gas constant; T is the temperature; and E is the activation energy [9].

The antifungal property was evaluated using an aqueous solution of 39 g/l Potato Dextrose Agar (PDA) nutrient substrate contaminated with Aspergillus niger - FAg18003 spores. 0.005 g of porous copper were placed in 4 diametrically opposite points of each Petri dish. For each test a sterile 1 µl loop containing biological material taken directly from the growth vessels of pure cultures and the nutrient medium was contaminated by tracing two diameters and the circumference of the vessel, without touching the test material (Cu), so as shown in Fig. 1.

After contamination, the systems were sealed and kept in the incubator at 22°C±1°C, for 7 days. At predetermined time intervals (1, 2, 3 and 7 days after contamination) the viability of mold spores, the fungal development on the surface of the nutrient medium and on the surface of the granules of the tested material and the appearance/maintenance of the halo of inhibition was evaluated using optical microscopy.

3. Results and discussions

3.1 Low vacuum dealloying

The starting brass powders have a chemical composition of approx. 56.2 % copper, 42.2 % zinc and some lead (1.8 %). It is known that zinc evaporates in vacuum easily even at low temperatures and is one of the easiest elements to evaporate. In contrast copper evaporates considerably slower, so, due to the large difference in the saturated vapor pressures between them is possible to selectively evaporate zin from the brass alloy. A competition arises between the outward diffusion of the Zn atoms and their evaporation from the chips surface, the latter being the limiting factor during low vacuum dealloying, as the EDS analysis presents an increase of the Zn content in the samples surface at the beginning of the dealloying process (figure 2).

Initial chips

Composition
Cu: 56.2 wt. %
Zn: 42.2 wt. %
Pb: 1.8 wt. %

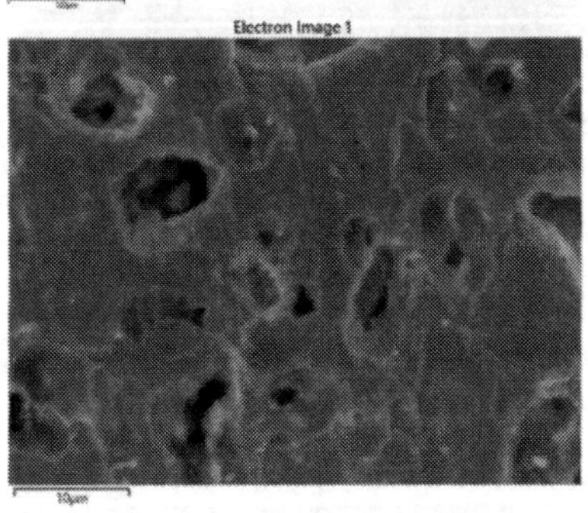

Partially dealloyed
(600 °C /0.25h)

Composition
Cu: 28.8 wt. %
Zn: 57.7 wt. %
Pb: 13.5 wt. %

Dealloyed chips
(600 °C /1h)

Composition
Cu 98.2 %
Zn 1.8 %

Figure 2. Local chemical composition variation during the dealloying process

In figure 3 the results of the image analysis on the SEM images are presented. In the obtained images the pores size was measured using the ImageJ software on two directions. The presented results are the median values on several pores. One can observe that both temperature and time increases the obtained pores size since it permits more atomic movement. By further evaluating the results presented (figure 4) one can determine the corresponding kinetic parameters active during these tests.

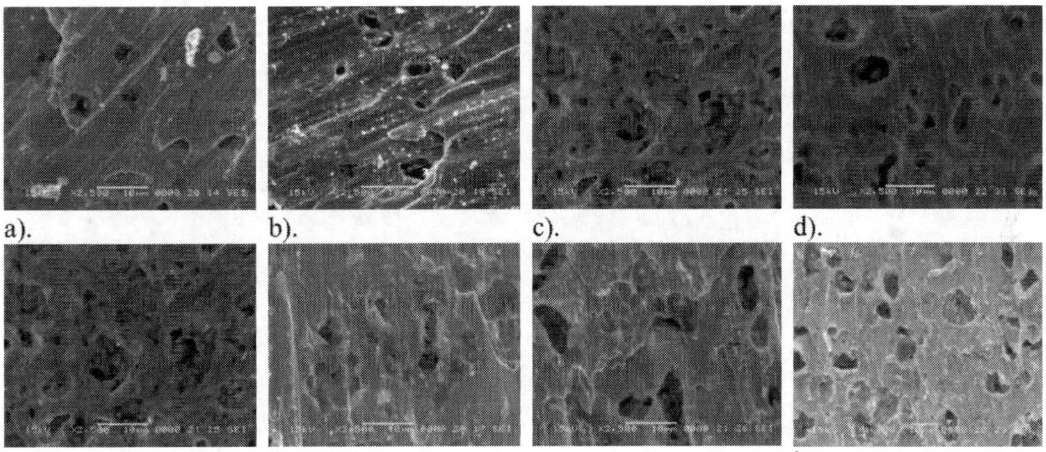
a). b). c). d).
e). f). g). h).

Figure 3. Typical SEM images for the dealloyed samples. (at: a). - 500 °C, b). - 550 °C, c). - 600 °C, d). - 650 °C/1h; holding at 550 °C for e).1h, f). - 2h, g). - 3h and h). -4h.)

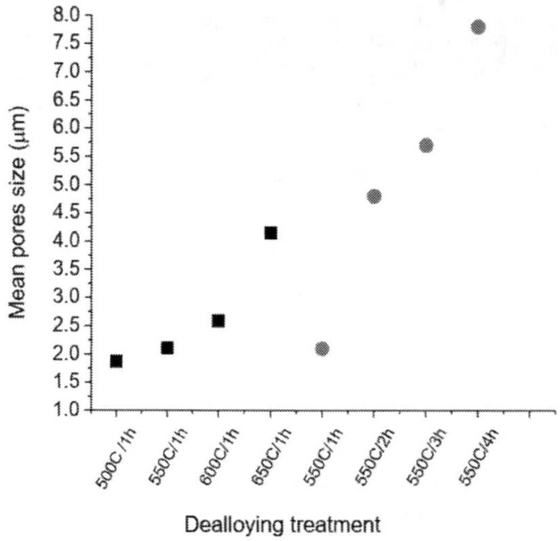

Figure 4. Mean pore size of samples dealloyed at different conditions

A linear fit between the logarithms of the mean pores size and the used time (figure 5.a.) confirms that the pores growth is a thermally controlled process. The coarsening exponent's value ~1 is an indication that the dealloying process is governed by an evaporation-condensation process fact already observed in the EDS analysis to.

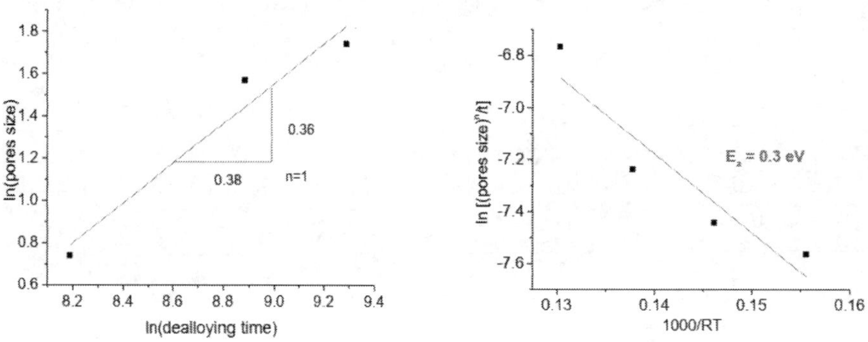

Figure 5. Coarsening exponent (a) and activation energy for the pore formation and coarsening (b)

The other important parameter of equation (1), the activation energy represents the energy barrier that needs to be overtaken in order that this coarsening of the pores to occur was estimated the data presented in figure 5.b. The activation energy for the formation and coarsening of the pores is low (~0.3 eV), so the coarsening is fast, a fact also observable in the pores size measurement results, the increased specific surface by the generated pores is also increasing the mobility of the copper atoms.

The porosity formed is directly attributed to the removal of the zinc atoms, so the overall porosity can be approximated from the sample's chemical composition variation [12], so, the samples density is calculated to be around 5.14g/cc giving a porosity of 37.4 %.

3.2 Antifungal property – Aspergillus niger

During testing the following effects were observed:

- After 24 hours from inoculation viability was proven for the mold, they developed on the nutrient substrate. No signs of fungal growth are identified on the surface of the granules of the tested material.

- After 48 hours of testing, the surface of the tested material (Cu) remains clean, without traces of fungal growth, but on the growth medium the development of mold is observed, for each system filaments that tend to get closer and closer to the sample.

- After 72 h of testing, (Fig. 6.), the surface of the tested material (Cu) remains clean, without traces of fungal growth, but on the PDA growth medium a good development of vegetation colonies, with the appearance of characteristic signs of sporulation. This mold development also takes place in the area where the tested material granules are located, the mold filaments anchoring on the PDA substrate under the material granules and developing among these granules, without however being recorded areas where the anchoring of the filaments also be realized on the surface of the Cu particles.

- After 7 days of testing, the development of mold, in both cases is intense, the surface of the nutrient substrate is covered by vegetation in proportion to more than 75%. Although in some areas adjacent to the Cu powder samples, no intense growth of Aspergillus niger mold was recorded, they could not be classified as an inhibition halo, being discontinuous, irregular in appearance and size and not being totally "clean" (free of mold filaments).

Mold filaments grow anchored on the substrate and penetrate between the Cu granules, but without being able to anchor on these granules. Further research was done to obtain more conclusive results by repeating the test using compacts of these particles.

On the tests conducted on the unsintered, high-density compacts, like in the case of loose, dealloyed chips, no clear inhibition hallows appeared, so the mode of anchoring of the mold filaments on the edges, faces and surface was traced. Only on certain faces a limited number (max 5) of filaments were observed, their growth direction was outwards from the sample.

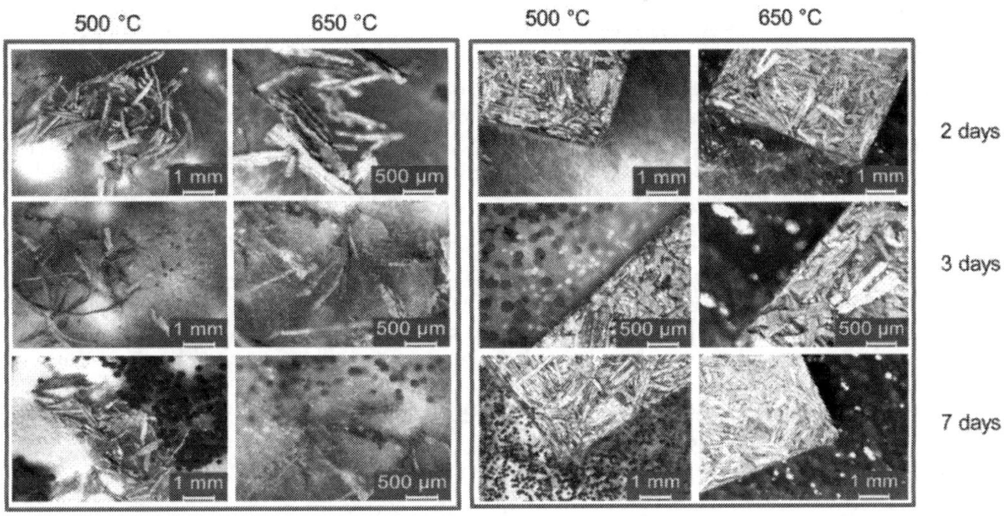

Figure 6. Mold growth with time on loose dealloyed chips and pressed samples

4. Conclusions

Porous copper powders were manufactured from reused brass chips by vacuum dealloying at different temperatures. Although low vacuum dealloying is possible the times and temperatures needed for complete zinc evaporation are considerably higher than in the case of high vacuum dealloying [14, 15], and since both time and temperature increases pore size the obtained pores are bigger than those obtained using high vacuum.

The obtainable pores size is in the micron range, generating a high surface area. The porosity obtained is dependent on the evaporated Zn quantity (P~40%) and is limited by the initial zinc content. The low activation energy for pores formation and coarsening (0.3 eV) limits the dealloying temperature and time which on their term limits obtainable structures. The low value of the coarsening exponent confirms that the formation and coarsening of the pores is controlled by the Cu diffusion at the Cu/vacuum interfaces.

The antifungal properties are not exceptional, after 7 days, the development of the mold is evident, but it cannot be said with precision if these filaments are also anchored on the analyzed material or only on the nutritional layer. Using compressed material, a limited number of filaments are bonded with the sample even after 7 days of exposure, although its viability is proven by sporulation. A good resistance in the contaminated medium was evidenced for the samples, these materials might be used as a protection layer against fungus infestation.

References

[1] Huang, M. Liu, H. Lin, Y. Xu, J. Wu, V.P. Dravid, C. Wolverton, C.A. Mirkin, Shape regulation of high-index facet nanoparticles by dealloying, Science 365 (2019) 1159–1163,

[2] Y. An, Y. Tian, et al. Heteroatom-doped 3D porous carbon architectures for highly stable aqueous zinc metal batteries and non-aqueous lithium metal batteries, Chem. Eng. J. 400 (2020) 125843,

[3] Y. An, Y. Tian, Y. Li, S. Xiong, G. Zhao, J. Feng, Y. Qian, Green and tunable fabrication of graphene-like N-doped carbon on a 3D metal substrate as a binder-free anode for high-performance potassium-ion batteries J. Mater. Chem. A 7 (2019) 21966–21975,

[4] Z. Lu, C. Li, J. Han, F. Zhang, et al., Three-dimensional bicontinuous nanoporous materials by vapor phase dealloying, Nat. Commun. 9 (2018) 276,

[5] Z. Wang, P. Liu, J. Han, C. Cheng, S. Ning, A. Hirata, T. Fujita, M. Chen, Engineering the internal surfaces of three-dimensional nanoporous catalysts by surfactant-modified dealloying, Nat. Commun. 8 (2017) 1066,

[6] Y. An, Y. Tian, C. Wei, H. Jiang, B. Xi, S. Xiong, J. Feng, Y. Qian, Scalable and Physical Synthesis of 2D Silicon from Bulk Layered Alloy for Lithium-Ion Batteries and Lithium Metal Batteries, ACS Nano 13 (2019) 13690–13701,

[7] Y. Sun, Y. Ren, K. Yang, New preparation method of micron porous copper through physical vacuum dealloying of Cu–Zn alloys, Mater. Lett. 165 (2016) 1–4,

[8] Y. An, H. Fei, G. Zeng, X. Xu, L. Ci, B. Xi, S. Xiong, J. Feng, Y. Qian, Vacuum distillation derived 3D porous current collector for stable lithium–metal batteries Nano Energy 47 (2018) 503–511,

[9] J. Han, C. Li, Z. Lu, H. Wang, Z. Wang, K. Watanabe, M. Chen, Vapor phase dealloying: A versatile approach for fabricating 3D porous materials Acta Materialia 163 (2019) 161-172,

[10] J. Li, Z.-B. Yi, N.-T. Li, N.-N. Yua, H.-R. Geng, Refinement of nanoporous copper by dealloying the Al–Cu alloy in NaOH solution containing sodium dodecyl sulfate, Phys. Chem. Chem. Phys., 25, (2023) 19492-19500,

[11] S., Zhang, H., Chen, Y. Qian, et al. Effect of dealloying temperature on microstructure and tensile properties of self-supporting nanoporous copper foil fabricated in situ. J Porous Mater 30, (2023) 267–276 ,

[12] Z. Lu, F. Zhang, D. Wei J. Han, Y. Xia, J. Jiang, M. Zhong, A. Hirata, K. Watanabe, A. Karma, J. Erlebacher, M. Chen, Vapor phase dealloying kinetics of MnZn alloys, Acta Materialia 212 (2021) 116916,

[13] Jung, D., Boan, B., Allapatt, N. et al. Nanoporous Cu created by the reduction of CuO dispersions. J Mater Sci 58, (2023).12635–12649 ,

[14] Y. Xia, Z. Lu, J. Han et al. Bulk diffusion regulated nanopore formation during vapor phase dealloying of a Zn-Cu alloy,Acta Materialia 238 (2022) 118210,

[15] Gy. Thalmaier, NA Sechel, I Vida-Simiti, M Nasui, N Cobîrzan, Micron porous copper powder through vacuum dealloying, Materials Today: Proceedings 72, (2023) 560-564.

Experimental investigations regarding the anisotropic properties of pre-impregnated woven carbon fibre-reinforced epoxy resin

Robert-Cătălin SÎRBU[1*], Alexandru-Viorel COȘA[1], Radu NEGRU[2], Lucian RUSU[2], Dan-Andrei ȘERBAN[1]

[1] Department of Mechatronics, University Politehnica Timişoara, Timisoara, Romania
[2] Department of Mechanics and Strength of Materials, University Politehnica Timişoara, Timisoara, Romania

*E-mail: catalin.sirbu@student.upt.ro

Abstract. In this work, the variation of the stiffness and strength with the weave pattern orientation of a twill carbon fibre weave pre-impregnated system was investigated. Singe-ply sheets were manufactured using vacuum-assisted curing at a temperature of 120 °C for 8 hours. Specimens were cut at 0°, 15°, 30°, 45°, 60°, 75° and 90° respectively, and subjected to tensile tests at room temperature, using a 5mm/min crosshead travel speed. The results show a symmetric variation along the 45° angle, the obtained stiffness values being between 8.53 GPa (at 45°) and around 59 GPa (at 0° and 90°), while the strength varied between 76 MPa (at 45°) and around 700 MPa (at 0° and 90°). In addition, it was observed that specimens oriented at 0° and 90° exhibited a brittle behaviour while the rest where characterized by ductile failure with significant plastic deformations.

1. Introduction

Carbon fibre-based composite materials have been developed in the late 1950's and started to gain popularity in the following decades, and have since been considered one of the top end structural materials as they offer unique combination of exceptional mechanical properties and reduced weight. Carbon fibres exhibit very good stiffness and tensile properties at relatively low densities, which gives the composite remarkable strength to weight ratios, making them ideal for weight saving structural applications [1]. Additionally, this class of materials is resistant to corrosion and fairly stable at relatively high temperatures, making them ideal for use in harsh environments [2, 3]. Due to these properties, carbon fibre-based composites are widely used in high-end applications from various domains, such as aerospace, automotive, marine, energy, sports etc., where reducing weight and improving performance are critical [4, 5].

The first applications of carbon fibres in composite materials consisted of either continuous unidirectional fibres or randomly oriented chopped fibres. In recent years, woven carbon fibre fabrics have started to become more prevalent because of the advantages they possess over the traditional variants [6]. Common types of woven fabrics include plain weave, twill weave, and satin weave, each offering different characteristics [7]. For identical tows (fibre count and dimensions), plain weaves offers greater stability when positioned in a mould (less prone to fraying, snagging and warping), but provide an initial flexibility (which determines reduced mechanical properties in the initial stages of deformation) due to a higher crimp. The twill weave provides better drapability and impact resistance at the cost of a reduced stability. Satin weaves present the greatest drapability and very low crimp at the cost of high fragility when handled and, for single plies, they determines a degree of anisotropy between the 0° and 90°.

Woven carbon fibre reinforced polymers parts are suitable for the majority of established manufacturing technologies such as hand lay-ups, resin transfer moulding (RTM) or vacuum infusion, depending on the sizes, geometries and required characteristics [8]. However, the best mechanical properties of WCFRP parts are obtained through the use of pre-impregnated composites.

Woven fibre pre-impregnated composite materials (known as prepregs) represent fabrics imbedded with a polymer matrix, which is supplied in a partially cured state (thermosetting polymers such as epoxy resins) or in a solidified state (thermoplastic polymers), which can be cured/melded and subsequently re-solidified during the thermal treatment stage [9]. This aspect determines greater fabric stability and a degree of tackiness (adhesion to the mould and between the plies), allowing for the possibility of manufacturing parts with complex geometries (i.e. sharp angles, low radius curvatures etc.) with easily obtainable higher ply count in the critical region [10]. Besides these manufacturing advantages, the main benefit of prepregs is the low polymer ratio (as low as 30%) and uniform distribution, resulting in consistent material properties such as excellent strength-to-weight ratios, high fatigue resistance, and superior surface finish [11, 12].

The goal of the study is to determine the variation of the mechanical properties with the orientation angle for a single ply of pre-impregnated woven carbon fibre reinforced epoxy composite. Even though the use of plies orientated at angle different from 0°or 90° and 45° is not common in woven fibre laminates, the goal of the study is to obtain experimental data for further use in constitutive models calibration and evaluation (as the in-service loads that structural components manufactured from woven fibre composites are seldom subjected to loads aligned with the weave/weft orientation).

2. Materials and Methods

2.1 Materials

The material investigated in this study is a pre-impregnated 2x2 twill woven carbon fibre with an epoxy matrix, having the commercial name XPREG®XC130, with its structural characteristics presented in Table 1.

Considering its mechanical properties (Table 2) and being able to withstand service temperatures up to 130°C, its recommended applications in structural components are in the fields of high-performance vehicles (unmanned aerial vehicles, motorsports, racing boats etc.), large scale components (boat hulls, wind turbine blades, light aircraft etc.), orthotics, sports equipment etc. In addition, its aesthetics, marked by superior surface finish, along with the characteristic weave texture makes this material suitable for design applications, such as vehicle interior trims, furniture, display stands etc. [13].

2.2 Manufacturing

For the sample manufacturing, a 450mm x 450mm x 10mm thermal resistant glass pale was used as mould, due to its good flatness tolerance, gloss surface and relatively high rigidity. The first step in the manufacturing process was the application of a chemical release agent (Aliphatic Hydrocarbon) in two stages (with a 15 minute curing time in between). Afterwards, prepreg laminates were positioned on the glass plate (a single ply per process), which in turn was placed on a breather cloth and enclosed in a polyamide-6 envelope bag, along with a vacuum port.

Specialized polyisobutylene-based sealant tape was use to air-tighten the bag and a vacuum was pulled using a lubricated rotary vane pump (Figure 1).

Table 1. XPREG®XC130 structural properties [1]

Nomenclature	Definition
Fibres	Pyrofil TR505 12k carbon fibres
Fibre orientation	0°/90°
Resin Matrix	Epoxy
Weave	2 x 2 Twill
Filament Count	12000
Ends	2.81/cm
Picks	2.81/cm
Consolidated Thickness	0.45 mm
Areal Weight	450 gr/m^2
Resin Weight	38%

Table 2. XPREG®XC130 mechanical properties [1]

Properties	Value
Tensile Strength	645 MPa
Tensile Modulus	55.2 GPa
Compressive Strength	615 MPa
Flexural Strength	882 MPa
Flexural Modulus	60.1 GPa
Interlaminar Shear	49.8 MPa

Figure 1. Prepreg moulding setup

After the bag was checked for leaks and a pressure of around 100 Pa was reached, the moulding system was placed in a curing oven. The curing process consisted of a controlled multi-stage temperature cycle recommended by the prepreg manufacturer, presented in Figure 2. In total, a number of four plies were manufactured.

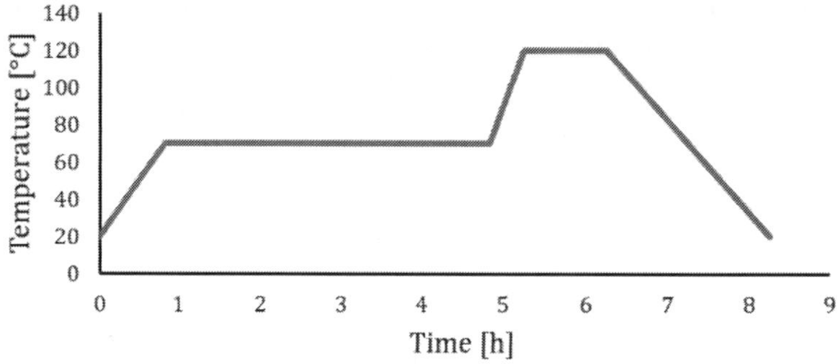

Figure 2. Curing treatment temperature-time curve

3. Experimental procedures

For this study 7 angle orientations for the specimens were chosen (0°, 15°, 30°, 45°, 60°, 75°, 90°, Figure 3), with 5 ASTM D3039M [14] specimens (250mm x 25mm) for each angle. The specimens were cut with a guillotine from the manufactured sheets and textolite tabs were fixed in the gripping area using a two-stage epoxy-based adhesive (Figure 4).

The tests were carried out on a servo-hydraulic 50 kN walter+bai testing machine at ambient temperature, using a 5mm/min crosshead travel speed, the strains being recorded with a strain-gauge extensometer of 50mm initial opening.

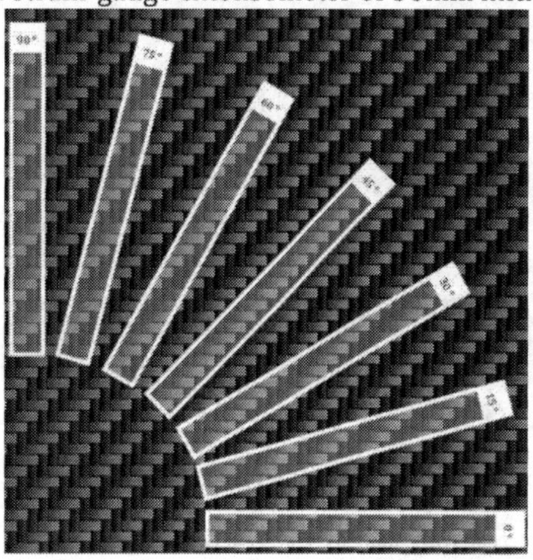

Figure 3. Schematic specimen orientation

Figure 4. Test-ready specimens

4. Results and discussion

The resulting stress-strain curves for each tested specimen are presented in Figure 5, and a comparison between representative stress-strain curves for each orientation is presented in Figure 6. From the stress-strain curves, the Young's modulus and the mechanical strength were determined, the average values for each orientation and the corresponding standard deviation being presented in Table 3.

IOP Conf. Series: Materials Science and Engineering **1319** (2024) 012031 doi:10.1088/1757-899X/1319/1/012031

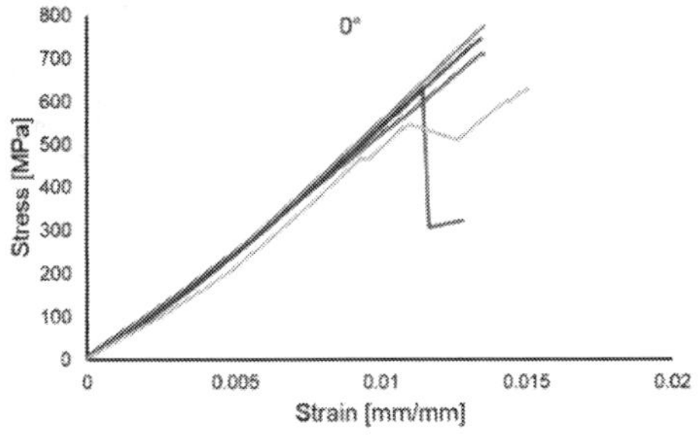

(a)

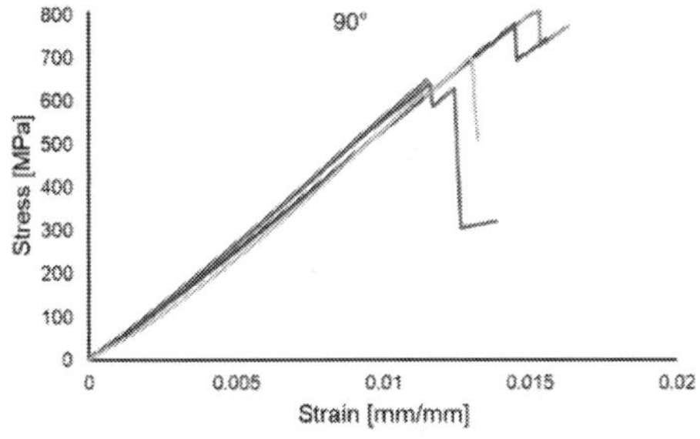

(b)

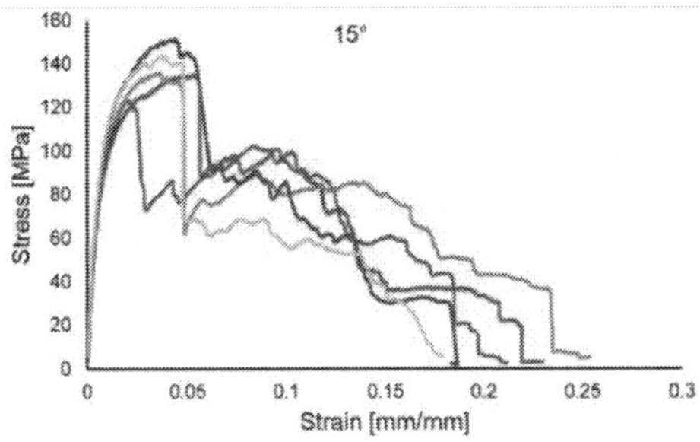

(c)

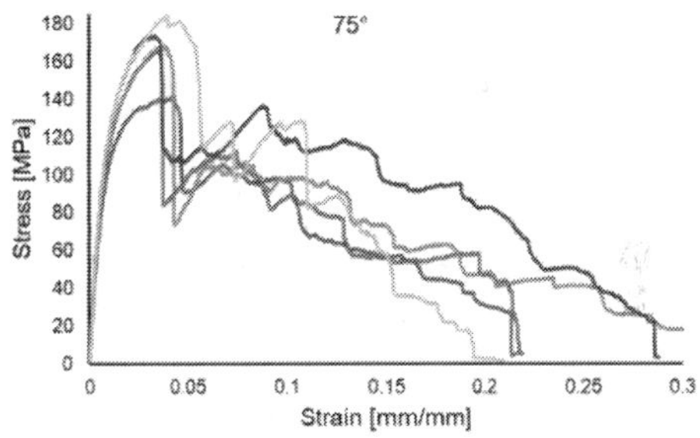

(d)

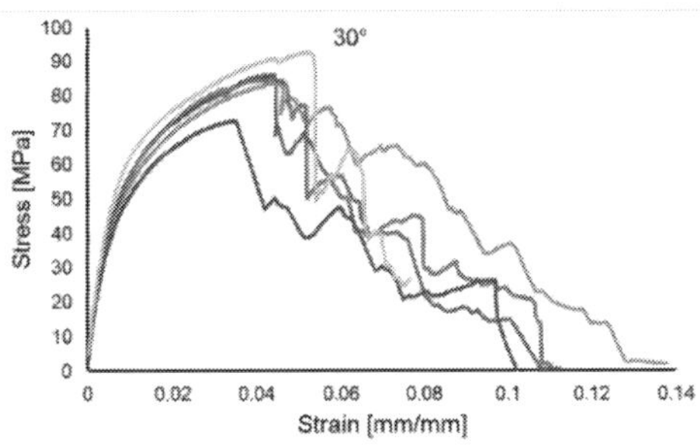

(e)

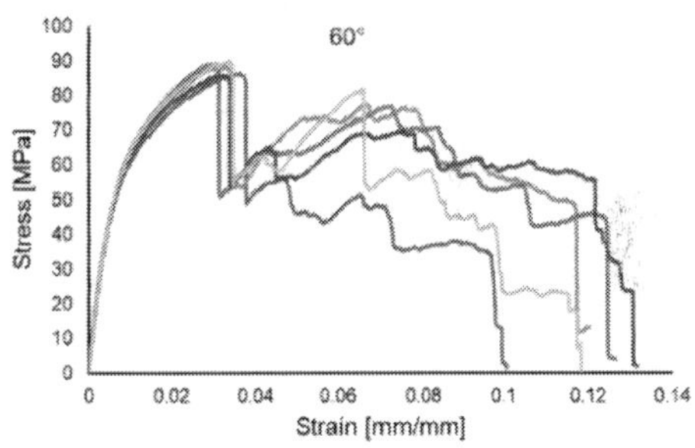

(f)

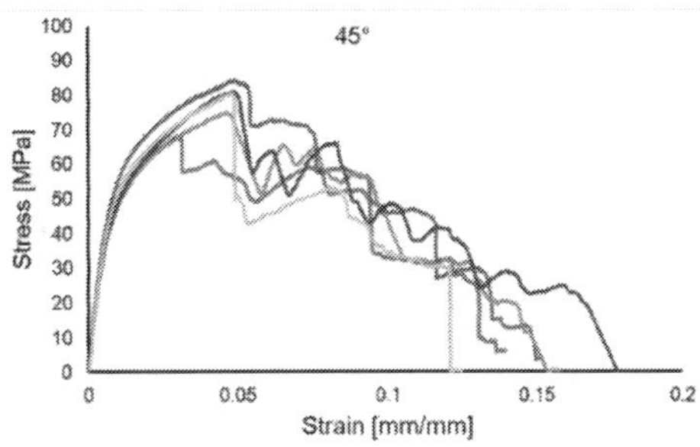

(g)

Figure 5. Stress-strain curves for the specimens tested at 0° (a), 90° (b), 15° (c), 75° (d), 30° (e), 60° (f) and 45° (g)

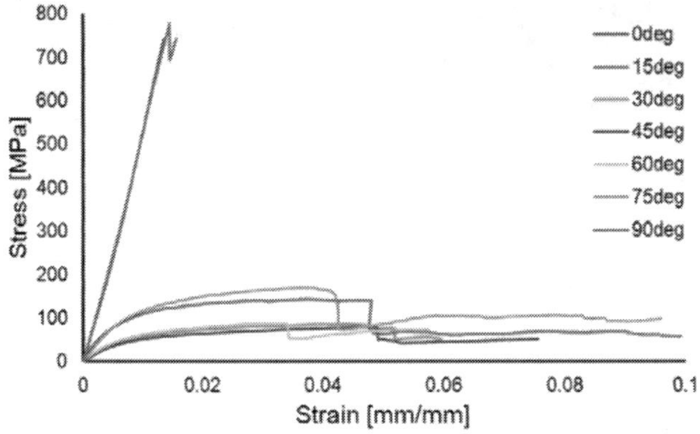

Figure 6. Comparison between representative stress-strain curves for each orientation

Table 3. Mechanical properties of XPREG®XC130 for the investigated specimen orientations

Orientation	Average Young's Modulus (GPa)	Standard deviation (%)	Average Strength (MPa)	Standard deviation (%)
0°	58.49	6.11	697.51	9.37
15°	17.35	3.16	137.8	7.61
30°	9.96	6.29	83.67	7.82
45°	8.53	3.96	76.07	6.89
60°	10.46	4.94	87.97	1.93
75°	18.25	3.2	166.68	9.52
90°	57.92	2.23	715.43	10.5

The variation of stiffness and strength with the orientation angle, plotted in polar coordinates, is presented in Figure 7.

Analysing the results, it can be observed that there is a symmetry in both the mechanical properties and the damage/fracture around the 45° axis. However, the specimens oriented at 75° and 60° respecively determined higher stiffness and strength values than the specimens oriented at 15° and 30° respectively. This phenomenon might be caused by the involuntary variations that occurred during the manufacturing process (i.e. small variations in the vacuums during curing due to better/poorer envelope sealing) than an inherent anisotropy of the material between the weave and weft directions (as no significant trend could be observed for the 0° and 90° specimens).

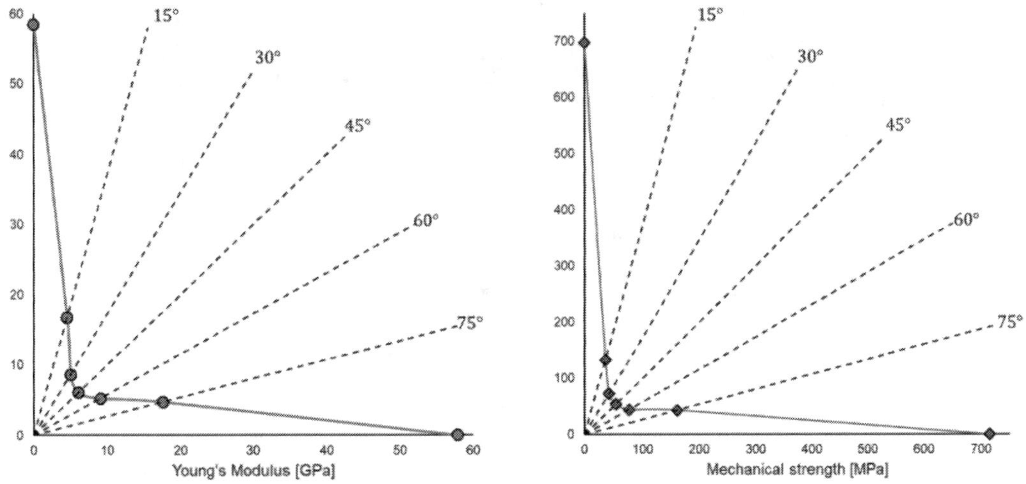

(a) (b)

Figure 7. Variation of stiffness (a) and strength (b) with the orientation angle

The mechanical properties of the 0° and 90° specimens are far superior than those of the rest of the orientaion, with only a 15° variation in angle oriantation determining a ~70% drop in stiffness and a ~75% drop in strength. The variation from 15°/75° to 45° is less significant, with only a ~54% reduction in both strenght and stiffness.

As shown in the stress-strain behaviour form Figures 6 and 7, the specimens oriented at 0° and 90° respectively exhibited brittle facture, that occurred in the linear-elastic domain. Analyzign the specimens at the end of the tests (Figure 8), it can be seen that, even though the failure occurred instantaneously, multiple fracture regions are observed on a plane normal to the loading direction. The specimens oriented at 15°, 30°, 45°, 60° and 75° showed a similar behaviour, characterized by a relatively reduced linear-elastic region, followed by a domain characterized by plastic flow. Several drops in stress are observed, trailed by subsequent strain hardening regions. This phenomenon is caused by the progressive damage of the weave pattern, where multiple failure region occur successively until the specimens lose their load bearing capabilities. As seen in Figure 8, the degradation bands are oriented at steeper angles for the 15° and 75° specimens, that gradually reduce for the 30° and 60° specimens, with the specimens oriented at 45° showing the bands with the closest orientation to the plane normal to the loading direction. It is worht mentioning that, due to the progressive failure of the weaves/wefts, the orientation of the degradation bands does not coincide with the weave orientation.

IOP Conf. Series: Materials Science and Engineering **1319** (2024) 012031 doi:10.1088/1757-899X/1319/1/012031

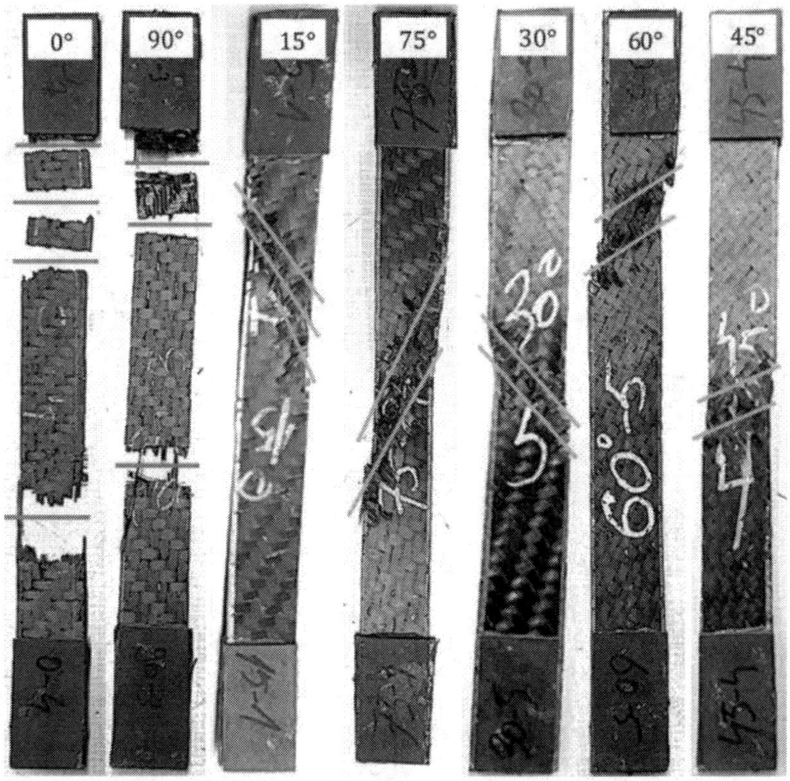

Figure 8. Representative specimens for each angle orientation photographed at the end of the tests

5. Conclusions

In this study, the influence of the orientation angle on the mechanical properties of a prepreg twill weave carbon fibre reinforced epoxy was investigated, showing a clear variation trend with a symmetry axis for the 45° orientation. The average stiffness values varied from 8.53 GPa (specimens oriented at 45°) to 58.49 GPa (specimens oriented at 0°) while the average strength values varied from 76.07 MPa (specimens oriented at 45°) to 715.43 (specimens oriented at 90°). Apart from the specimens oriented at 0° and 90°, which determined a brittle failure with multiple fracture surfaces, all other specimens exhibited ductile failure with significant plastic deformations before the onset of damage and relatively large fracture strains.

Both the recorded values for the Young's modulus and mechanical strength of the material along the 0° and 90° respectively are larger than the values specified by the producer in the technical data sheet: 5.96%/4.93% for stiffness (average values of 58.49 GPa for 0° and 57.92 GPa for 90° compared to the specified value of 55.2 GPa) and 13.42%/16.33% for strength (average values of 697.51 MPa for 0° and 715.43 MPa for 90° compared to the specified value of 615 MPa), which implies that the manufacturing process was appropriate.

6. Acknowledgements

This work was funded by the Research Grant PN-III-P2-2.1-PED-2021-1134, contract number 586PED/2022.

7. References

[1] W. Zhang, X. Ma, J. Lu, Z. Zhang, Q. Wang, X. Su, D. Zeng, M. Mirdamadi and J. Caoo, "Experimental Characterization and Numerical Modeling of the Interaction Between Carbon Fiber Composite Prepregs During a Preforming Process," *J. Manuf. Sci. Eng.,* vol. 140, no. 8, 2018.

[2] W. Sun, "Development of a testing methodology for the design and quality control of carbon fiber reinforced polymer (CFRP) anchors," *Construction and Building Materials,* vol. 164, pp. 150-163, 2018.

[3] D. Rajak, P. H. Wagh and E. Linul, "Manufacturing Technologies of Carbon/Glass Fiber-Reinforced Polymer Composites and Their Properties: A Review," *Polymers,* vol. 13, no. 3721, 2021.

[4] A. Chen, S. Baehr, A. Turner, Z. Zhang and G. Gu, "Carbon-fiber reinforced polymer composites: A comparison of manufacturing methods on mechanical properties," *International Journal of Lightweight Materials and Manufacture,* vol. 4, no. 4, pp. 468-479, 2021.

[5] J. Meredith, E. Bilson, R. Powe, E. Collings and K. Kirwan, "A performance versus cost analysis of prepreg carbon fibre epoxy energy absorption structures," *Composite Structures,* vol. 124, pp. 206-213, 2015.

[6] S. Kumar and K. Singh, "Tribological Performances of Woven Carbon Fabric/Epoxy Composites Under Dry and Oil Lubrication Condition: An Experimental Investigation," *Trends in Materials Engineering,* pp. 43-50, 2019.

[7] A. Szabo, R. Negru, A. Coșa, L. Marșavina and D. Șerban, "Multi-scale modelling of woven carbon fibre reinforced epoxy," *Materials Today: Proceedings,* vol. 45, no. 5, pp. 4298-4303, 2020.

[8] B. Suresha, G. Chandramohan, P. Samapthkumaran, S. Seetharamu and S. Vynatheya, "Friction and Wear Characteristics of Carbon-epoxy and Glass-epoxy Woven Roving Fiber Composites," *Journal of Reinforced Plastics and Composites,* vol. 25, no. 7, pp. 771-782, 2006.

[9] Y. Ma, S. Jin, M. Ueda, T. Yokozehi, Y. Yang, F. Kobayashi, H. Kobayashi, T. Sugahara and H. Hamada, "Higher performance carbon fiber reinforced thermoplastic composites from thermoplastic prepreg technique: Heat and moisture effect," *Composites Part B: Engineering,* vol. 154, pp. 90-98, 2018.

[10] Y. Yang and W. Young, "Carbon/Epoxy Composites Fabricated by Vacuum Consolidation of the Interleaved Layup of Prepregs and Dry Fibers," *Fibers Polym,* vol. 22, pp. 460-468, 2021.

IOP Conf. Series: Materials Science and Engineering **1319** (2024) 012031 doi:10.1088/1757-899X/1319/1/012031

[11] K. Kang, S. Park, A. Jo, K. Lee and H. Ju, "Development of ultralight and thin bipolar plates using epoxy-carbon fiber prepregs and graphite composites," *International Journal of Hydrogen Energy,* vol. 42, no. 3, pp. 1691-1697, 2017.

[12] S. Kim, T. Lim, N. Sottos and S. White, "Manufacture of carbon-fiber prepreg with thermoplastic/epoxy resin blends and microencapsulated solvent healing agents," *Composites Part A: Applied Science and Manufacturing,* Vols. 365-375, p. 121, 2019.

[13] Easy Composites Ltd, *XC130 Autoclave cure component prepreg technical data sheet,* Easy Composites Ltd, 2022.

[14] ASTM, *Standard Test Method for Tensile Properties of Polymer Matrix Composite Materials,* ASTM International, 2014.

Nanoporous Cu-based amorphous alloys prepared by selective dissolution in acidic media

Sebastian AMBRUŞ [1] *, Roxana MUNTEAN [1], Dragoş BUZDUGAN [1], Cosmin CODREAN [1]

[1]University Politehnica Timişoara, Victoriei square, nr. 2, Timişoara, Romania
*sebastian.ambrus@upt.ro

Keywords: dealloying, amorphous metallic alloy, nanoporous alloys

Abstract. Functional nanoporous materials are considered a significant category of nanostructured materials that exhibit distinct characteristics like high surface area, porosity, and improved mass transport properties. These qualities render them suitable for a wide range of applications including catalysis, energy storage, biomedical fields, and electrochemical sensors. Dealloying or laser-induced technologies are the primary methods employed to fabricate such nanoporous materials.
Dealloying is a dependable top-down approach used to produce hierarchical, disordered nanoporous materials with customizable pore sizes in the range of a few nanometers. The process of dealloying involves the selective elimination or dissolution of one or more elements from an alloy through a corrosion mechanism, using various dealloying techniques, such as chemical, electrochemical, liquid metal, or vapor phase dealloying.
In the present study, copper-based amorphous metallic ribbons ($Cu_{75}Ni_6Sn_5P_{10}Ga_4$) were initially manufactured using the melt-spinning method. The Cu-based amorphous ribbons were structurally investigated by X-ray diffraction and thermal analysis. Subsequently, the ribbons were subjected to a dealloying treatment, using an acidic solution to selectively dissolve the nickel from their composition and to obtain a nanoporous structure. The microstructure and chemical composition of the ribbons before and after the dealloying process were investigated using scanning electron microscopy (SEM), combined with energy-dispersive X-ray spectroscopy. The dealloying process performed in 1 M H_2SO_4 solution at 25 °C for 60 minutes leads to a large number of nanopores, uniformly distributed onto the surface of the Cu-based ribbons.

1. Introduction

After the discovery of amorphous alloys over 60 years ago, interest in these materials has significantly grown, despite facing limitations in obtaining specific geometric dimensions due to the need to preserve the amorphous structure. The development of amorphous metallic materials has expanded due to their exceptional properties, including superior hardness, mechanical strength, ductility compared to crystalline metallic materials, high magnetic permeability, low coercive field, electrical conductivity, and excellent corrosion resistance [1], [2]. Unlike traditional metals, amorphous alloys form metallic-type interatomic bonds, allowing for unique characteristics such as the ability to be processed through plastic deformation, non-fragile behavior upon breakage, thermal conductivity, and specific optical properties such as opacity and metallic luster [3], [4]. Amorphous metallic alloys, created through rapid cooling of melts, are commonly known as metallic glasses. This rapid solidification process locks the atoms in their liquid-state configuration, preventing the formation of an orderly arrangement in space, though some level of order at small interatomic distances may still exist. In recent years, these alloys have found increasing application in various industries including aerospace, automotive, electrical engineering, and electronics [3].

One technique used to rapidly solidify specific alloys is melt-spinning. This method is particularly useful for producing amorphous metallic ribbons or "metallic glasses" that cannot be achieved through traditional continuous casting approaches. By adjusting parameters such as cooling rates, amorphous structures can be attained. In this process, the alloy is melted within a quartz crucible through induction heating, then ejected via an inert gas flow through a nozzle onto a copper cooling roller where rapid solidification occurs. Critical parameters for this method include the cooling roller's rotation speed, melt outlet pressure, nozzle-to-cooling roller distance, and melting temperature. Surface characteristics like shape, uniformity, and quality can be influenced by altering additional factors like nozzle size, wheel surface roughness, ambient conditions (air or vacuum), and the type of gas used for melt discharge.

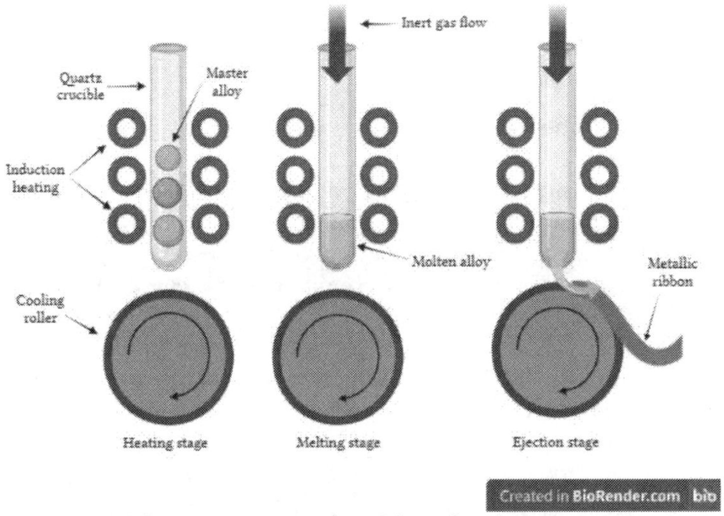

Figure 1. Schematic representation of the melt-spinning process

The dealloying process involves an electrochemical reaction where less noble chemical elements from an alloy are dissolved using an oxidizing acid solution. This selective dissolution results in the creation of a nanoporous material, which primarily contains elements with low mechanical strength. Dealloying removes one or more less noble elements from an alloy to create a porous structure. Due to this selective dissolution, the dealloying process is considered a destructive method in electrochemistry. Initial attempts to synthesize porous metallic materials through selective dissolution of amorphous structures were made in 2009. Amorphous structures offer advantages over crystalline alloy systems as they have a uniform composition and single-phase structure. Most amorphous alloys consist of multiple metals, allowing for the creation of nanoporous alloys with various elements. Unlike in crystalline structures, dealloying amorphous alloys does not rely on grain boundaries or a crystalline structure for the process to take place effectively [5]. To date, this process has been used to produce metal porous structures from Cu-based[6], Al-based [7] and Zr-based [8] systems and other alloys and solid solutions and metallic glasses. To ensure the formation of a continuous porous structure, the amount of the noble element(s) should be between 20–45 at%. Other important parameters, which play a crucial role during the dealloying process include the critical corrosion potential, the composition of the alloy, the pH and also the temperature of the oxidizing solution[5], [9]. In a recent study, successful nanoporous copper was synthesized by the dealloying route and the obtained structures were suggested as suitable for potential applications in catalysis, sensors or actuators[10].

Motivated by the remarkable properties of amorphous metallic ribbons and the unique nanoporous architecture achievable through the dealloying process, we aimed to develop a copper-based nanoporous amorphous structure. Previous research has demonstrated that a Cu-based nanoporous structure can be produced via dealloying [11], [12], [13]. Our goal was to leverage the combined advantages of amorphous metals and nanoporous structures to create a versatile copper-based nanoporous material suitable for various advanced applications, such as sensors, actuators, and catalysts. In this study, we sought to obtain a nanoporous structure starting from a Cu-based amorphous metallic ribbon, which was dealloyed using an oxidizing acidic medium.

2. Experimental and results

2.1 Materials and methods

For the purpose of this study, a $Cu_{75}Ni_6Sn_5P_{10}Ga_4$ alloy was selected. The main reason for choosing this chemical composition was that copper and tin help in reducing the melting temperature of the material, phosphorus and gallium have a high glass-forming ability, thus helping in obtaining an amorphous structure and finally nickel was chosen because it is the less-noble element of the chemical composition and can be removed or dissolved through the dealloying process. The experimental procedures are presented in Figure 2.

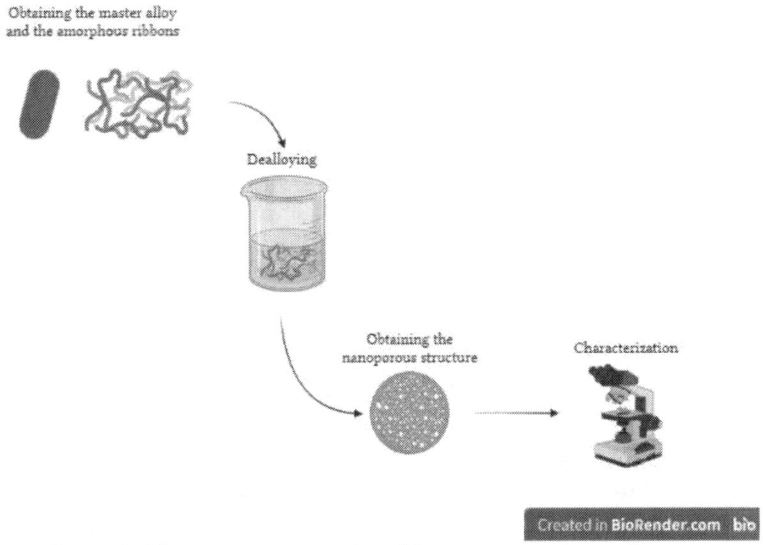

Figure 2. Schematic representation of the experimental procedures

The melt-spinning method was used to obtain the metallic ribbons, which were structurally investigated using X-ray diffraction (XRD) PANalytical X'Pert Pro Powder (Malvern Panalytical, Malvern, UK) and thermal analysis was performed by differential scanning calorimetry (DSC) using a Netzsch STA 441 Jupiter apparatus, in order to confirm the amorphous structure and to determine the critical points like glass transition temperature, crystallization temperature and melting temperature. The microstructure and chemical composition of the ribbons before and after the dealloying process were investigated using scanning electron microscopy (SEM) FEI Quanta™ FEG 250 (FEI Quanta™ Hillsboro, OR, USA), combined with energy-dispersive X-ray spectroscopy (EDX). Furthermore, the resulted Cu-based amorphous ribbons were subsequently subjected to a dealloying treatment, using an acidic solution (H_2SO_4) to selectively dissolve the nickel from their composition and to obtain a nanoporous structure.

2.2 Obtaining amorphous metallic ribbons

The elaboration of the primary alloy and the amorphous metallic ribbons was done on a melt-spinning installation developed in the laboratory. The installation includes the cooling roller, the bearing system, the mechanical transmission, an electric drive motor, a vertical inductor movement system, a vertical crucible movement system, an inductor, a quartz crucible, medium frequency generator, control panel and an argon cylinder. The frequency generator used is an electronic GSMF 10 A 8000 type with a fixed frequency of 8kH and a power of 10 kW. The elaboration technology includes the following stages: cutting the master alloy bars to lengths of up to 40 mm, inserting the master alloy into the crucible and fixing it, induction heating of the master alloy, setting the cooling roller at the desired speed, positioning the crucible above the cooling roller, application of inert gas overpressure on the melt in the crucible and ejection of the melt through the nozzle hole onto the surface of the cooling roller. The master alloy and the resulting amorphous ribbons are illustrated in Figure 3.

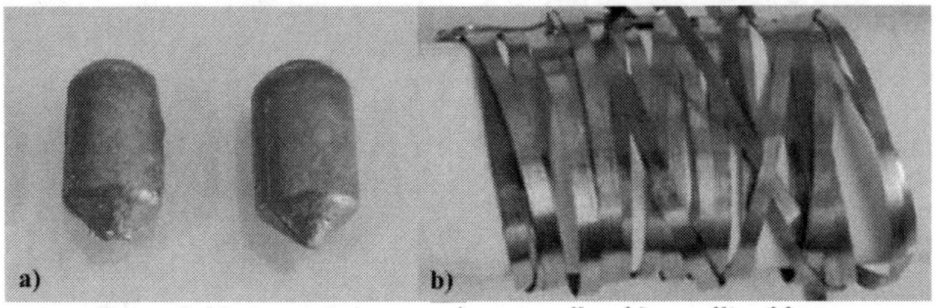

Figure 3. Macroscopic images: a) master alloy; b) metallic ribbons

2.3 Characterization of metallic ribbons

The resulting metallic ribbons were subjected to the XRD analysis to investigate if the amorphous structure was obtained.

The XRD pattern presented in Figure 4 shows no crystallinity, the amorphous structure being indicated by the diffuse peak found between 40 and 50 (2θ) degrees. The increased cooling rate provided by the copper cooling roller during the melt-spinning process contributes to the formation of the amorphous structure.

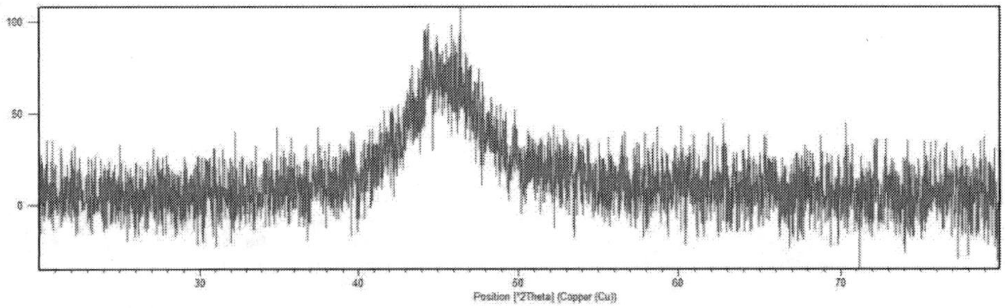

Figure 4. XRD pattern of the metallic ribbons

2.4 Thermal analysis

The thermal analysis was performed on the amorphous ribbons for both crystallisation and melting processes. The glass transition temperature Tg, the crystallization temperature Tx, and the melting temperature Tm, were determined as the onset temperatures of the peaks, during heating of a sample that weighed approximately 10mg with a rate of 20 K · min-1 under a flow of purified nitrogen at a flow rate of 40 ml · min-1.

Figure 5 shows the DSC curves of the ribbons for the crystallization (a) and for melting (b). The first event is the glass transition for which Tg = 169.5 °C followed by a two-stage crystallization of the amorphous phase, first one begins at Tx1 = 194.9 °C and the other at Tx2 = 271 °C. The supercooled region ΔTx defined as the difference between the crystallization and glass transition temperatures was found to be 25.4 °C, which indicates a good glass forming ability. The DSC curves show two endothermic processes that mark the sample's melting (Fig. 5b). The first process starts at Tm1 = 513 °C and the second at Tm2 = 603 °C. It is noted that this alloy possesses a low melting temperature.

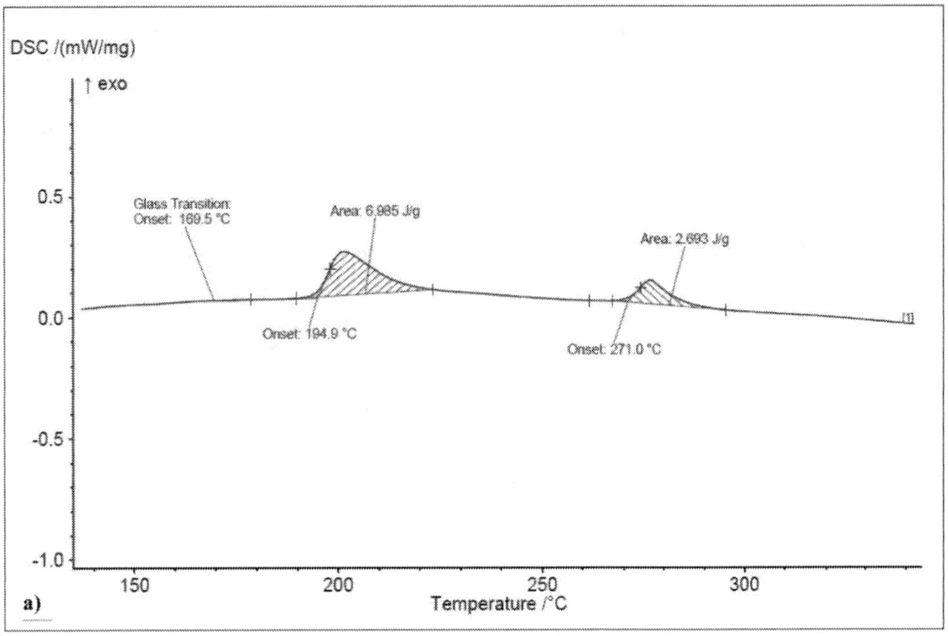

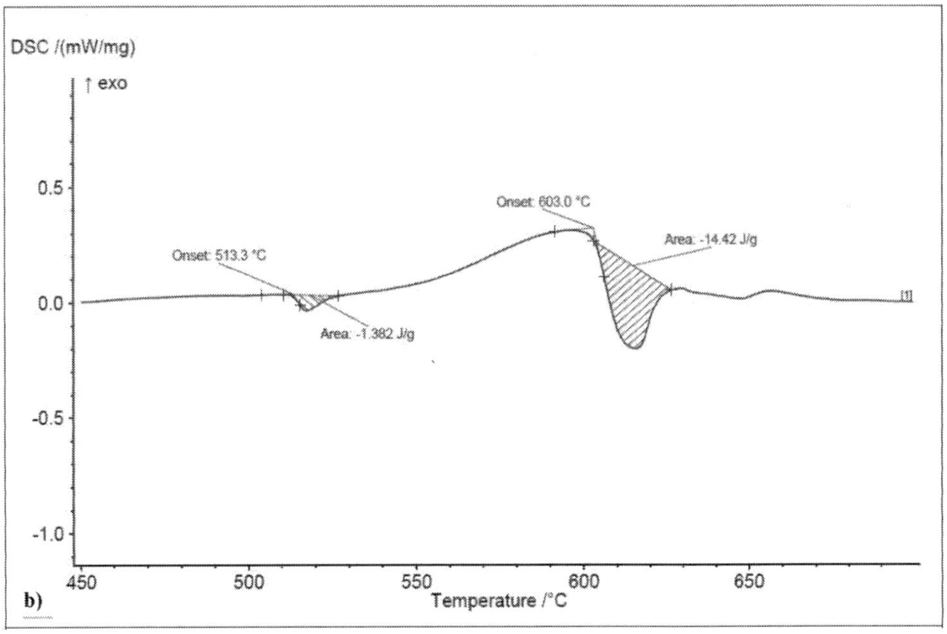

Figure 5. DSC curves: a) crystallization curves; b) melting curves

2.5 Dealloying process

The obtained metallic ribbons were subsequently subjected to dealloying treatment in an oxidizing acidic solution in order to selectively remove the nickel from the composition and to obtain a nanoporous structure. For this stage, a sulfuric acid solution (1M H_2SO_4) was selected due to the increased sensitivity of certain chemical elements in this specific oxidizing environment. The solution was prepared by mixing H_2SO_4 (98% purity, Sigma Aldrich Solutions) with distilled water. Also, different exposure conditions were used, varying the exposure time, the solution temperature and solution concentrations.

Figure 6. Dealloying process

The dealloying process performed in 1M H2SO4 solution at 25 °C for 60 minutes leads to the optimum results in terms of porosity. A large number of nanopores, uniformly distributed onto the surface of the Cu-based ribbons were obtained when compared to other treatment solutions.

2.6 Characterization of the dealloyed metallic ribbons

The microstructure and chemical composition of the ribbons before and after the dealloying process were investigated using SEM and EDX analysis.

As can be observed from the EDX spectra presented in Figure 7, after performing the dealloying process, the nickel was mostly removed from the initial composition.

The SEM micrographs presented in Figure 8 reveals a highly porous structure, with the pores ranging in the nanometric scale, uniformly distributed in the structure of the amorphous ribbon.

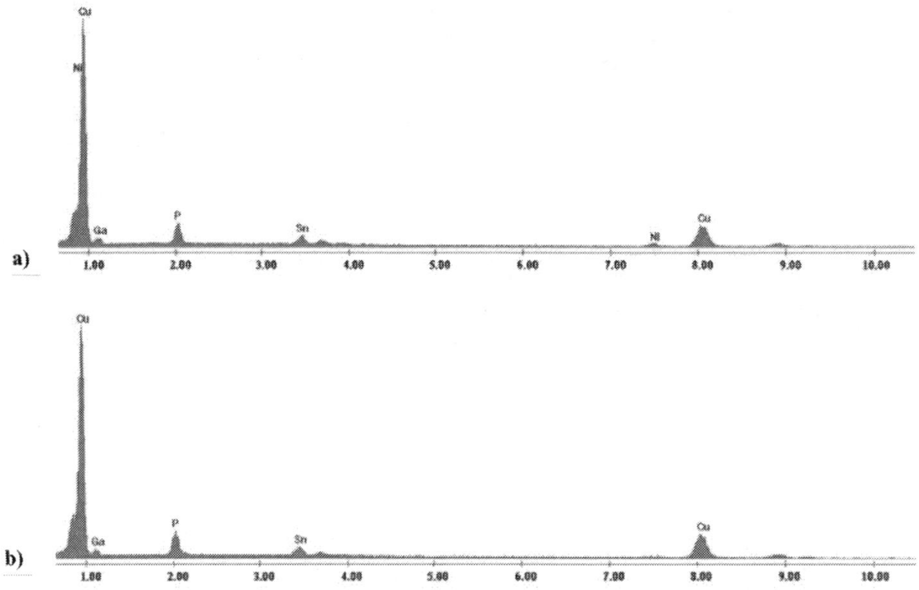

Figure 7. EDX spectra: a) before dealloying; b) after dealloying

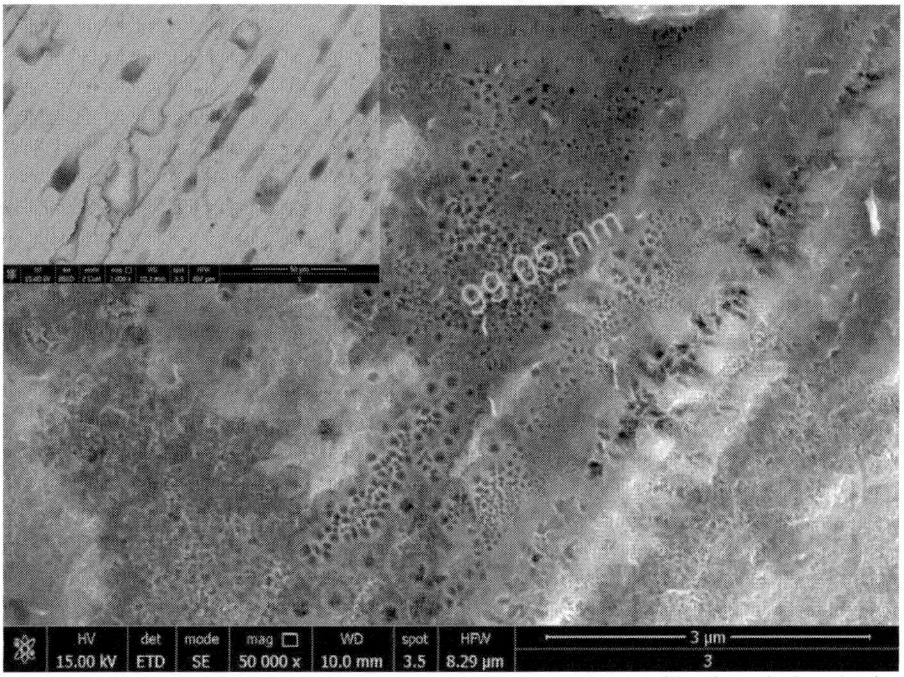

Figure 8. SEM image of the obtained nanoporous structure (in inset: the metallic ribbon before the dealloying process)

3. Conclusions

The present study focused on obtaining nanoporous structures starting from an amorphous metallic ribbon which was dealloyed using an oxidizing acidic media. The X-ray diffraction and DSC analysis certified the amorphous structure and helped to determine the glass transition temperature, the crystallization temperature and the melting temperature of the as-cast ribbons. The SEM images and EDX spectra confirmed the fact that a nanoporous structure was obtained using a H_2SO_4 solution by dissolving the nickel from the chemical composition of the alloy.

References

[1] H. Jiang *et al.*, "Structures and functional properties of amorphous alloys," *Small Struct.*, vol. 2, no. 2, p. 2000057, 2021.
[2] L. Guo, *Amorphous nanomaterials: preparation, characterization and Applications*. John Wiley & Sons, 2021.
[3] M. Hadden, D. Martinez-Martin, K.-T. Yong, Y. Ramaswamy, and G. Singh, "Recent Advancements in the Fabrication of Functional Nanoporous Materials and Their Biomedical Applications," *Materials*, vol. 15, no. 6, 2022, doi: 10.3390/ma15062111.
[4] F. C. Li *et al.*, "Amorphous–nanocrystalline alloys: fabrication, properties, and applications," *Mater. Today Adv.*, vol. 4, p. 100027, Dec. 2019, doi: 10.1016/j.mtadv.2019.100027.
[5] Z. Zhang, Y. Wang, Z. Qi, W. Zhang, J. Qin, and J. Frenzel, "Generalized Fabrication of Nanoporous Metals (Au, Pd, Pt, Ag, and Cu) through Chemical Dealloying," *J. Phys. Chem. C*, vol. 113, no. 29, pp. 12629–12636, Jul. 2009, doi: 10.1021/jp811445a.
[6] F. Yue *et al.*, "Dealloyed nano-porous TiCu coatings with controlled copper release for cardiovascular devices," *Biomater. Adv.*, vol. 157, p. 213728, Feb. 2024, doi: 10.1016/j.bioadv.2023.213728.
[7] Q. Zhang and Z. Zhang, "On the electrochemical dealloying of Al-based alloys in a NaCl aqueous solution," *Phys. Chem. Chem. Phys.*, vol. 12, no. 7, pp. 1453–1472, 2010.
[8] M. Al-Harbi, A. Abdel-Daiem, and E. Shalaan, "Self-supporting electrodes obtained by electrochemical dealloying of Zr-based metallic glass alloys for energy storage applications," *J. Mater. Sci. Mater. Electron.*, vol. 34, no. 17, p. 1358, 2023.
[9] Y. Ding, "Dealloyed Nanoporous Metals for Catalysis," in *Reference Module in Materials Science and Materials Engineering*, Elsevier, 2016. doi: 10.1016/B978-0-12-803581-8.03801-7.

IOP Conf. Series: Materials Science and Engineering **1319** (2024) 012032 doi:10.1088/1757-899X/1319/1/012032

[10] M. Nicolaescu *et al.*, "Synthesis of nanoporous copper by dealloying CuZrAl and CuZrAlAg amorphous ribbons in acidic solution," *12th Int. Conf. Mater. Sci. Eng.*, vol. 72, pp. 565–571, Jan. 2023, doi: 10.1016/j.matpr.2022.10.050.
[11] S. Yamada, "Nanoporous Cu Prepared through Dealloying by Selectively Etching an Alkaline Metal with Saline," *ACS Appl. Nano Mater.*, vol. 6, no. 9, pp. 7229–7233, May 2023, doi: 10.1021/acsanm.3c00320.
[12] Y. Gao *et al.*, "Nanoporous Cu-based metamaterial for fenton-like catalysis," *Chem. Eng. J.*, vol. 477, p. 146902, 2023.
[13] E. Hengge *et al.*, "Porosity evolution and oxide formation in bulk nanoporous copper dealloyed from a copper–manganese alloy studied by in situ resistometry," *Nanoscale Adv.*, vol. 5, no. 2, pp. 393–404, 2023.

Evaluation of Artichoke Extract as a Corrosion Inhibitor for Copper and OL45 in Aggressive Environments

George-Daniel DIMA [1,a], Mircea Laurențiu DAN[1,b*], Nataliia RUDENKO[2,a], Alin FAUR[1,c], Cristian George VASZILCSIN[3,a]

[1]University Politehnica Timisoara, Faculty of Industrial Chemistry and Environmental Engineering, Laboratory of Electrochemistry, Corrosion and Electrochemical Engineering, 6 Pârvan, 300223 Timisoara, Romania

[2]University Politehnica Timisoara, Innovation and Technology Transfer Center, 2 V. Pârvan, 300223 Timisoara, Romania

[3]National Institute of Research and Development for Electrochemistry and Condensed Matter, Dr. A. P. Podeanu 144, 300569, Timişoara, Romania

*E-mail: mircea.dan@upt.ro

Abstract. Metal corrosion is a significant issue in areas such as transport, construction, or production, with effects on equipment productivity, user safety, and finally, the costs related to the replacement of affected components. Research in recent years has focused on the discovery of substances capable of significantly reducing the corrosion rate of metals and alloys with a minimal impact on the environment. A promising category of green corrosion inhibitors represents active components from natural extracts. Corrosion experiments have been focused on the use of artichoke (Cynara scolymus) stem extract as a corrosion inhibitor for copper and mild steels such as OL45 in H_2SO_4 0.5 mol L^{-1} and NaCl 3.5% environments. The electrochemical behaviour of the active components in the artichoke extract was studied using cyclic voltammetry on Pt electrode, and their inhibitory effect has been demonstrated by using electrochemical techniques such as linear voltammetry at a speed of 1 mV s^{-1} and chronoamperometry method in the absence, respectively, presence of 1, 2, 4, and 8 g L^{-1} extract. The variation of parameters such as corrosion potential E_{corr}, corrosion current i_{corr}, and impedance values with increasing concentrations shows that the artichoke extract greatly reduces the corrosion rate. To complement the corrosion studies carried out, quantum chemical calculations were performed by using the B3LYP/6-31G* basis set, and the descriptors E_{HOMO}, E_{LUMO}, dipole moment, ionization energy, and molecular hardness substantiated the inhibitory effect. Considering artichoke extract components, this study presents a potential prospect for using natural extracts in anti-corrosion protection as well as an ecological alternative to standard corrosion inhibitors.

Keywords: green corrosion inhibitor, molecular modelling, cyclic voltammetry, linear sweep voltammetry.

1. Introduction

Corrosion is the degradation of a metal or alloy as a result of its contact with a corrosive environment. This can have a negative impact on the costs of operating and managing infrastructure and industry, as well as user safety. Annually, this has led to the destruction of a large amount of metal and the decommissioning of machinery used industrially or in road infrastructure [1]. One of the solutions to the corrosion problem is the use of corrosion inhibitors. Over time, inorganic inhibitors, such as chromates, have been used, which have been shown to be harmful to health, having a carcinogenic effect [2]. At the same time, in the context of the requirements of the Paris Commission and the Organization for the Registration, Evaluation and Restriction of Chemicals, for which a substance is accepted as an inhibitor, it must simultaneously meet three conditions, namely biodegradability, biocumulativity, and minimum toxicity to the aquatic environment. These conditions are usually met by natural compounds such as flavonoids, tannins, alkaloids, or polyphenols [3]. From this point on, most of the research in the field of corrosion protection focuses on green corrosion inhibitors in natural products [4]. Their mode of action at the level of the metal surface consists of the formation of a protective film or by making an adsorption layer between the metal and the aggressive solution [5]. The second category mentioned maintains its trend in scientific research through the diversity of organic compounds, metals, and last but not least, as is the case of adsorption inhibitors from natural extracts, the multitude of natural compounds that can simultaneously contribute to the formation of the adsorption layer therefore to the inhibition of corrosion [6]. As examples of green inhibitors from natural extracts, the literature mentions the use of black pepper extracts (Piper nigrum) and black cumin (Nigella sativa) as inhibitors for corrosion of A335 steels in a simulated seawater environment (with a high NaCl content) [7], papaya leaf extract (Carica papaya) for AISI 4140 steel in HCl medium [8], white-seed watermelon bark extract (Cucumeropsis mannii) for carbon steels in NaCl environment [9], okra stem (Okra stem), and plum seed extract (Prunus domestica) proved inhibitory effects for corrosion of steels in an HCl environment [10]. Leaf extract of the pink trumpet tree (tabebuia heterophylla) has been reported as a corrosion inhibitor for steels with a low carbon content in HCl environment. Here, the electrochemical studies were completed by theoretical calculations, respectively, surface morphology studies [11]. Nettle leaves have active compounds with the role of corrosion inhibitor, a fact demonstrated by Maizia et. al, which used the extract of Urtica dioica L. to inhibit the corrosion of mild steels of type X38 in sulfuric acid medium [12]. ASTM A36 steel experienced an corrosion inhibition effect in environments such as H_2SO_4 and NaOH after the addition of different concentrations of pineapple leaf extract [13]. High performance as a corrosion inhibitor of mild steels in environments of 0.5 mol L^{-1} also manifests extracts of Amoracia rusticana [14], Alkanna tinctoria [15], Salvia officinalis [16], Cassia bark [17], Magnolia kobus [18], and Myristica fragans, respectivelly [19]. Artichoke, or taxonomic Cynara scolymus, is a plant that grows in southern Europe and the Mediterranean and has anti-carcinogenic medical applications, in gastrointestinal disorders, anti-inflammatory action, and the treatment of liver diseases [20]. In the literature, the inhibitory effect of artichoke extract has been studied in an acidic environment for weak, low-carbon steels by Salmasifar et al. [21]. Next, we aim to focus the research on the inhibitory effect of the active compounds in artichoke extract on OL45 steel and copper in aggressive sulfuric acid environments, respectively sodium chloride 3.5%.

2. Materials and methods

2.1 Materials

The corrosive environment consisted of electrolyte support solutions 3.5% NaCl, respectively, 0.5 mol L^{-1} H_2SO_4 prepared from analytical purity reagents, Merck 99% sodium chloride, and 96% sulfuric acid produced by Sigma Aldrich. To investigate the influence of concentration on corrosion rate, solutions were prepared using comercially available artichoke alcoholic extract produced by Favisan, at concentrations of 1, 2, 4 and 8 g L^{-1}. The experimental determinations were carried out using a BioLogic SP150 potentiostat connected to a 3-electrode assembly cell, in which the working electrode was the metal under study (OL45 and Cu pure), with an active surface of 1 cm^2, two graphite electrodes for circuit closure, and an Ag/AgCl reference electrode. For cyclic voltammetry measurements, a Pt electrode with an active surface area of 0.5 cm^2 was used. The percentage composition of the OL45 steel used in the experimental determinations is shown in Table 1.

Table 1. The composition of OL45 electrode [22].

Fe (%)	C (%)	Si (%)	Mn (%)	P (%)	S (%)	Cr (%)	Ni (%)	Cu (%)
97.68	0.45	0.27	0.65	0.04	0.045	0.3	0.3	0.3

2.2 Methods

Cyclic voltammetry experiments were performed using Pt as a working electrode, at polarization rates of 500, 100, 50, and 5 mV s^{-1} in a potential window between -1.5 to 2.5 V/Ag/AgCl. The experiments of chronopotentiometry, potentiodynamic polarization, and chronoamperometry were conducted to stabilize the working electrodes (OL45 and Cu), as well as to track the evolution of the Tafel slope allure and the oxidation current with the increase of the inhibitor concentration. Prior to conducting these techniques, the electrode was sanded, using Si-C abrasive sheets of grits 120, 600, 1200, and 2400, followed by ultrasonication, acetone cleaning, and drying. In order to eliminate any dissolved oxygen from the electrolyte, inert gas (N_2) was bubbled before the start of each determination for 10 minutes. The reproductibility of each measurement was confirmed by successsively tracing at least three voltammograms under identical working conditions. The theoretical molecular modeling calculations were performed using the Hyperchem 8.0 release software (Hypercube Inc.). The structures of the major organic compounds in the artichoke extract were optimized by applying the DFT method to a B3LYP function, on a 6-31G* base set. Based on the optimized structure, the Frontier molecular orbitals (H.O.M.O. and respectively L.U.M.O) were plotted, and molecular descriptors such as energy difference ΔE, μ dipole moment, χ electronegativity, η hardness, and σ sensitivity were calculated, using equations 1-5 [23].

$$I = -E_{HOMO} \quad (1)$$

$$A = -E_{LUMO} \quad (2)$$

$$\chi = \frac{I+A}{2} \quad (3)$$

$$\eta = \frac{I-A}{2} \quad (4)$$

$$\sigma = \frac{1}{\eta} \tag{5}$$

Properties such as molecular volume, molecular surface area and volume-to-surface ratio were obtained using the QSAR Properties module within the same program.

3. Results and discussions

3.1 Cyclic voltammetry

To determine how the active compounds of the studied artichoke extract (Cynara scolymus) behave in the potential range between the oxygen and hydrogen release potentials, cyclic voltammograms were performed on Pt in the electrolyte support (BS) solutions of H_2SO_4 0.5 mol L^{-1} and NaCl 3.5%, respectively, in the presence and absence of different concentrations of extract.

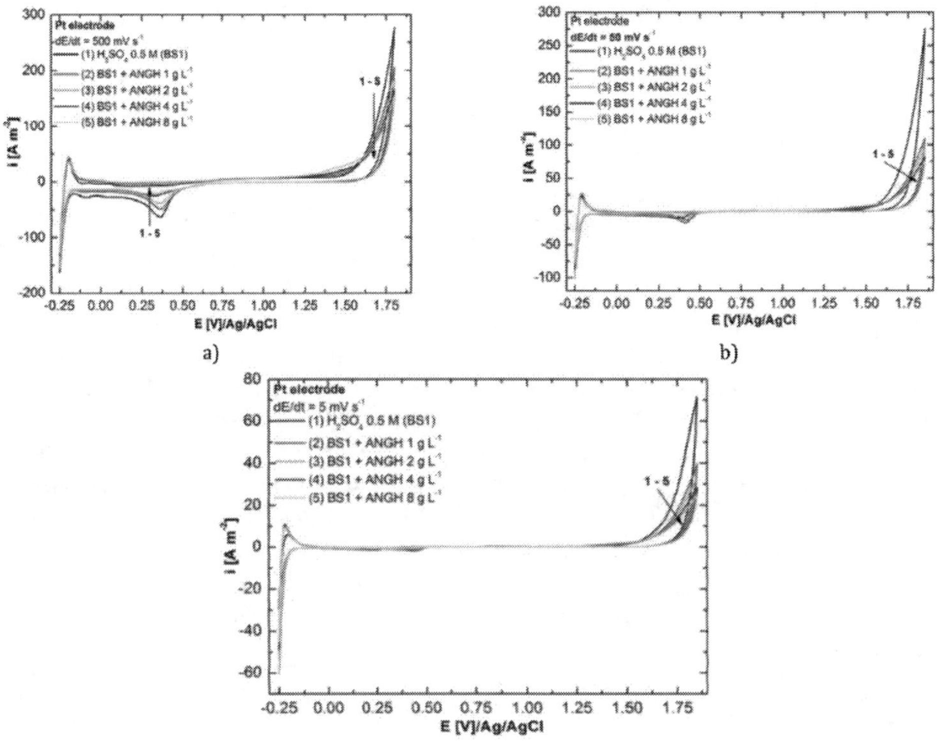

Figure 1. Cyclic voltammograms recorded on the Pt electrode, in the medium of H_2SO_4 0.5 mol L^{-1}, in the presence and absence of different concentrations of Artichoke extract, at the polarization rate of (a) 500 mV s^{-1}, (b) 50 mV s^{-1}, (c) 5 mV s^{-1}

Figure 1 (a-c) depicts the cyclic voltammograms recorded in the H_2SO_4 medium at the scanning rates of 500 mV s^{-1}(a), 50 mV s^{-1} and 5 mV s^{-1}. They cover both the anodic and cathodic potential ranges. In this figure, on the curve corresponding to the absence of the addition of artichoke extract, at values below 0 V, the process of oxidation of adsorbed hydrogen can be noticed, as well as the reduction of positive hydrogen ions to molecular hydrogen. Between the potentials +0.75 and 1.5 V, vs. the Ag/AgCl electrode, the process of formation of the surface layer of platinum oxides is observed, as is their decrease on the cathodic side, in symmetry with it. Moving towards values more positive than the open circuit potential (OCP), the peak associated with oxygen release is recorded. At the same time, increasing the concentration of artichoke extract causes the specific potentials for oxygen release to shift to the right, emphasizing the extract's inhibitory effect on the anodic process. On the cathodic branch, the allure of the curves remains almost unchanged as the extract addition rises and they are almost overlaid.

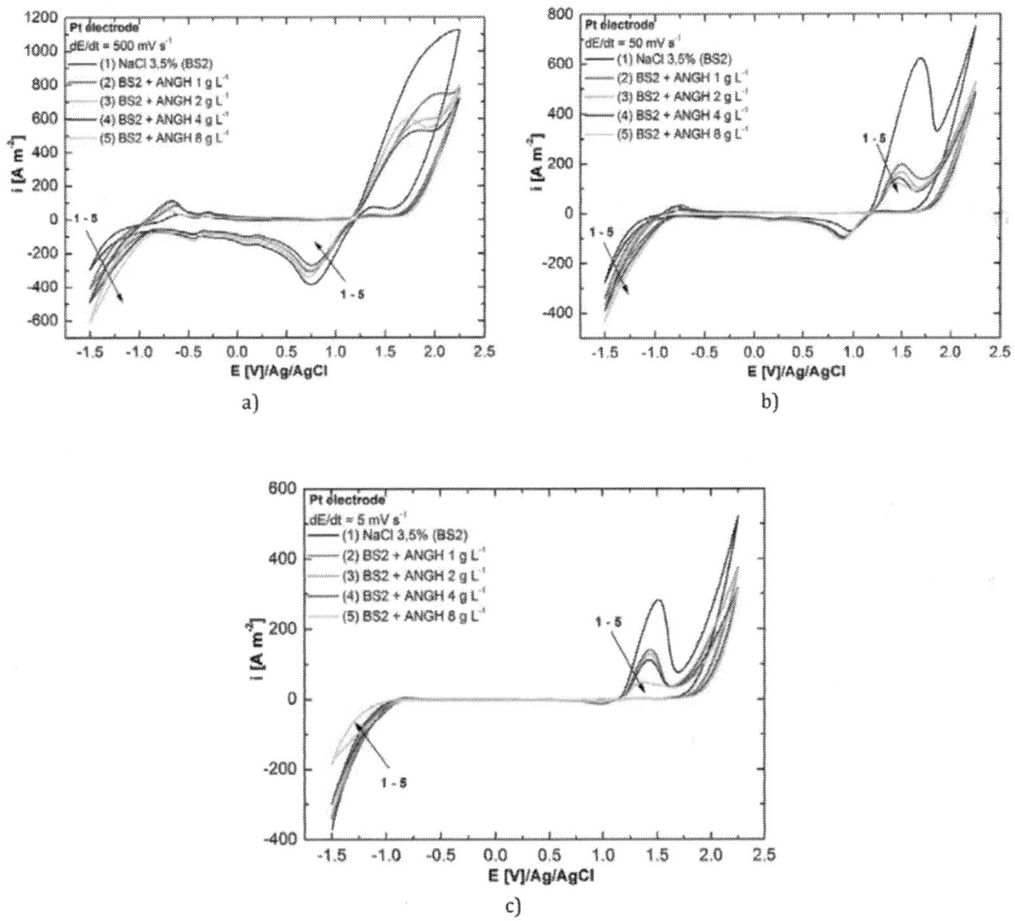

Figure 2. Cyclic voltammograms recorded on Pt electrode, in 3.5% NaCl medium, in the presence and absence of different concentrations of Artichoke extract, at polarization rate of (a) 500 mV s^{-1}, (b) 50 mV s^{-1}, (c) 5 mV s^{-1}

For the other corrosion media of 3.5% sodium chloride, the cyclic voltammograms plotted on the Pt electrode are represented by Figure 2 (a-c), in electrolyte support solution, respectively at the addition of different concentrations of artichoke extract, for the polarization rates of 500 mV s^{-1} (a), 50 mV s^{-1}, and (b), 5 mV s^{-1} (c). Figure 2 analysis shows that in the anodic domain the

IOP Conf. Series: Materials Science and Engineering **1319** (2024) 012033 doi:10.1088/1757-899X/1319/1/012033

peak at 1.5 V/Ag/AgCl is caused by the oxidation process of the negative chloride ion to molecular chlorine, which then dimerizes to the diatomic chlorine molecule. As in the case of voltammograms on sulfuric acid medium, in the anodic domain, with the increase in the concentration of extract, the oxygen release potentials sweep to the right, i.e. towards more positive values, whereas in the cathodic domain the hydrogen release potentials move towards more negative values, indicated that it has an inhibitory effect on both the cathodic and anodic processes.

To highlight additional processes, the anodic and cathodic domains were displayed separately at polarization rates of 100 mV s^{-1}. For the H_2SO_4 medium, in Figure 3, the recording of voltammograms started from the open circuit potential, towards positive values. The anodic range, between 1.1 and 1.6 V/Ag/AgCl, shows the oxidation peak of the active compounds in the artichoke extract, while the cathodic range, at 0.3 V, shows the peak associated to the reduction of the compounds from the above-mentioned process.

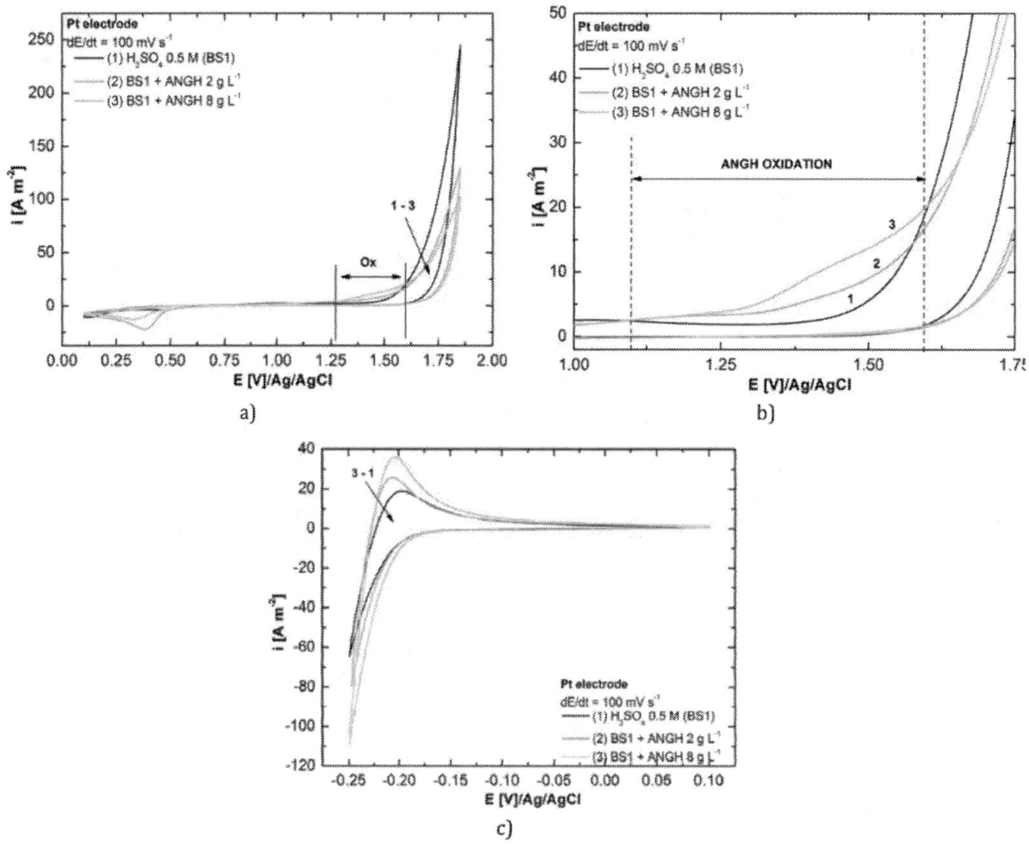

Figure 3. The anodic (a, b) and the cathodic domain (c) traced on the Pt electrode, in H_2SO_4 electrolyte solution, at a polarization rate of 100 mV s^{-1}.

The cathodic peak in Figure 3 (c), at the -0.4 V potential, demonstrates that the oxidation described above is almost quasi-reversible, and its electrochemical oxidation mechanism is practically unknown, being driven by the presence of several natural compounds in the system.

The natural extract tested exhibits stability in the potential regions between the open-circuit potential and the initiation of hydrogen release.

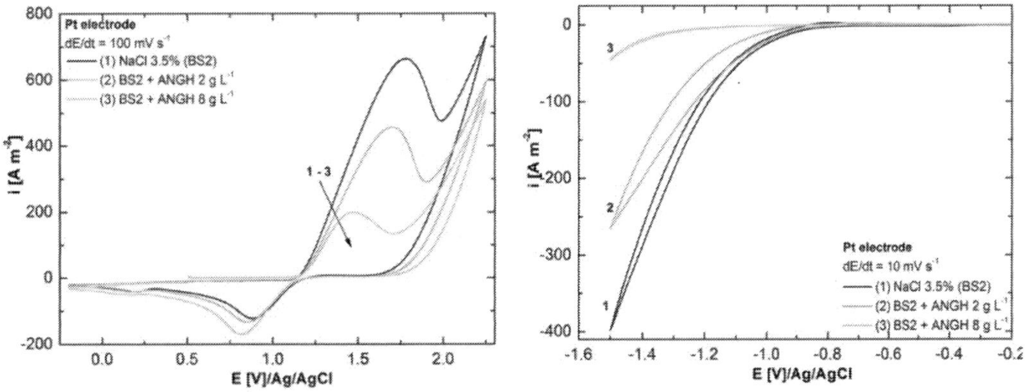

a) b)

Figure 4. The anodic domain (a, b) and the cathode range (c) drawn on the Pt electrode, in 3.5 % NaCl electrolyte solution, at a polarization rate of 100 mV s^{-1} and 10 mV s^{-1} respectively.

Cyclic voltammograms for the anodic and cathodic ranges (Figure 4) in the saline environment reveal that the artichoke extract undergoes no transformations (oxidation or reduction), indicating its stability. The present differences only have an inhibiting influence on the hydrogen and oxygen release mechanisms.

3.2 Potentiodynamic polarization method

To determine how the inhibitor concentration influences the corrosion rate, logarithmic linear voltammograms were drawn, at a speed of 1 mV s^{-1}, at 25°C on the OL45 electrode (Fig. 5), respectively on the Cu electrode (Fig. 6), in H_2SO_4 (a) and NaCl (b) environments, respectively.

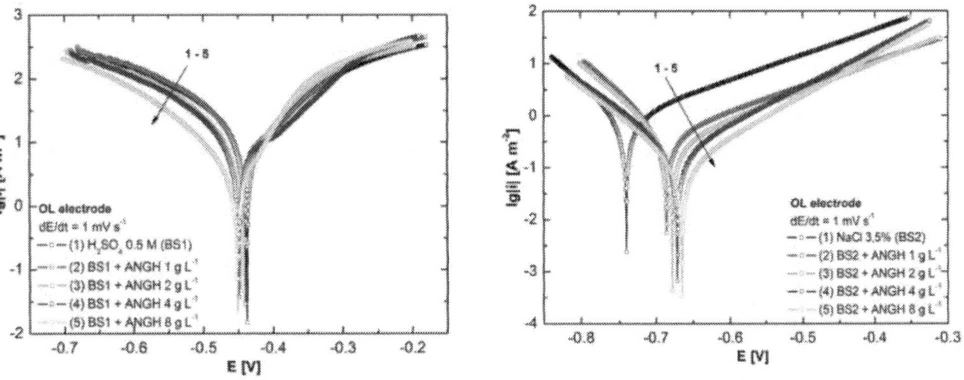

(a) (b)

Figure 5. Tafel diagrams plotted on the OL45 electrode in the absence and presence of different concentrations of artichoke extract, in H_2SO_4 0.5 mol L^{-1} (a), NaCl 3.5% (b) media.

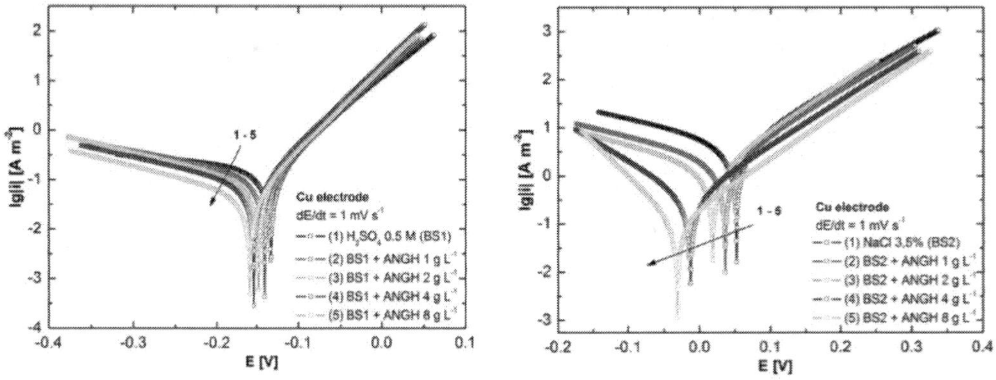

(a) (b)

Figure 6. Tafel diagrams plotted on the Cu electrode in the absence and presence of different concentrations of artichoke extract, in H_2SO_4 0.5 mol L^{-1} (a), NaCl 3.5% (b) media.

By fitting the logarithmic forms of the polarization curves with the help of the BioLogic 150 potentiostat software, the parameters such as: i_{corr} corrosion current, E_{corr} corrosion potential, values of anodic slopes b_c, respectively cathodic b_a, polarization resistance R_p, and based on them, the v_{corr} corrosion rate was calculated, the values obtained being presented in Table 2 (OL45) and 3 (Cu).

Table 2. Parameters of Tafel slopes on OL45 electrode in H_2SO_4 0.5 mol L^{-1} and 3.5% NaCl media

Electrolite solution	i_{corr}	E_{corr}	$-b_c$	b_a	R_p	v_{cor}	IE	θ
	[mA cm^{-2}]	[mV]	[mV dec^{-1}]	[mV dec^{-1}]	[Ω]	[mm y^{-1}]	[%]	
SB(0.5M H_2SO_4)	1.95	-437	178	196	22.2	22.83	-	-
SB + 1 g L^{-1} ANGH	1.54	-441	153	160	24.8	17.98	21.23	0.21
SB + 2 g L^{-1} ANGH	1.09	-445	134	124	28.6	12.75	44.09	0.44
SB + 4 g L^{-1} ANGH	0.71	-448	128	112	32.2	8.32	63.53	0.64
SB + 8 g L^{-1} ANGH	0.35	-452	86.9	108	38.2	4.19	81.94	0.82
SB (3.5% NaCl)	101	-740	208	87	210	1.18	-	-
SB + 1 g L^{-1} ANGH	47.6	-685	210	81	382	0.57	52.87	0.53
SB + 2 g L^{-1} ANGH	32.1	-678	185	82	552	0.37	68.22	0.68
SB + 4 g L^{-1} ANGH	21.4	-671	140	97	949	0.25	78.81	0.79
SB + 8 g L^{-1} ANGH	11.6	-664	129	91	1512	0.14	88.51	0.89

Table 3. Parameters of Tafel slopes on Cu electrode in H_2SO_4 0.5 mol L^{-1} and 3.5% NaCl media.

Electrolite solution	i_{corr}	E_{corr}	$-b_c$	b_a	R_p	v_{cor}	IE	θ
	[mA cm^{-2}]	[mV]	[mV dec^{-1}]	[mV dec^{-1}]	[Ω]	[mm y^{-1}]	[%]	
SB(0.5M H_2SO_4)	501	52.1	127	310	72.3	5.82	-	-
SB + 1 g L^{-1} ANGH	218	36.2	113	283	167	2.52	56.49	0.56
SB + 2 g L^{-1} ANGH	155	18.3	106	263	283	1.81	69.06	0.69
SB + 4 g L^{-1} ANGH	37.8	-13.1	102	230	561	0.44	92.46	0.92
SB + 8 g L^{-1} ANGH	12.8	-34.6	99	202	907	0.13	97.45	0.97
SB (3.5% NaCl)	34.4	-132	73.3	335	822	0.41	-	-
SB + 1 g L^{-1} ANGH	14.6	-140	67.6	311	998	0.18	57.56	0.58
SB + 2 g L^{-1} ANGH	9.03	-147	63.8	277	1112	0.10	73.75	0.74
SB + 4 g L^{-1} ANGH	3.85	-153	62.3	264	1291	0.06	88.81	0.89
SB + 8 g L^{-1} ANGH	1.28	-158	61.8	248	1473	0.02	96.28	0.96

The analysis of the Tafel slopes presented above shows that as the artichoke concentration increases, the corrosion rate decreases, increasing the inhibitory efficiency, reaching for the maximum concentration of 8 g L^{-1} for OL45 to 81.94% (H_2SO_4 media), 88.51% (NaCl media), and for Cu to 97.45% (H_2SO_4 media), respectively 96.28% (NaCl media). In comparison to the corrosion rate accepted for industrial applications, at the maximum inhibitor concentration, the corrosion rate of OL45 is close to the maximum accepted value for industrial applications (0.1 mm y^{-1}), while the corrosion rate of Cu is below this threshold.

3.3 Chronoamperometry studies

Chronoamperometry consists of recording the oxidation current over time, while maintaining a constant potential. The working oxidation potentials for OL45 samples, were set to +25 mV/OCP and +250 mV/OCP for 900 seconds, respectively. The chronoamperograms recorded on the OL45 electrode are presented in Figure 7.

After analyzing the chronoamperograms presented in Figure 7, it is observed that increasing the concentration of artichoke extract results in a decrease in the values of oxidation current densities, which is interpreted as a result of the adsorption process of the artichoke extract's active components at the electrode surface.

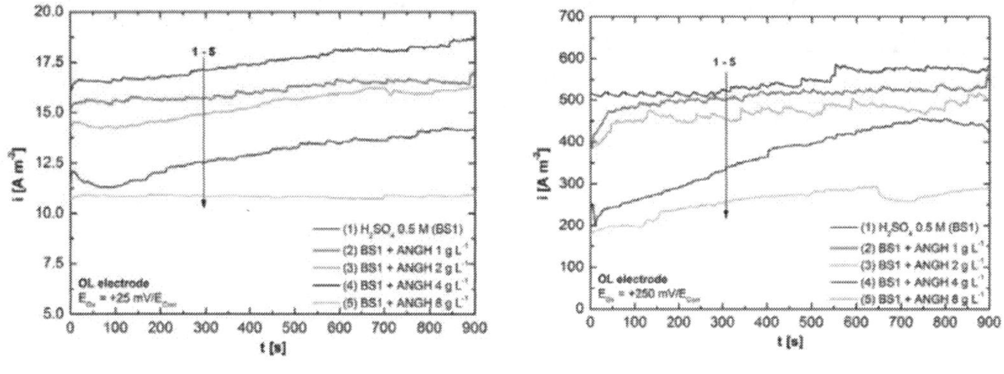

a) b)

Figure 7. Chronoamperograms plotted on the OL45 electrode in the presence and absence of different concentrations of artichoke extract at the potential of (a) 25 mV/E_{OCP}, (b) 250 mV/E_{OCP}

3.4. Molecular modelling

In order to complete the voltametric studies, and to understand the way in which organic compounds are adsorbed at the level of the metal surface, achieving the inhibitory effect, theoretical molecular modelling calculations were performed for the major compounds in the structure of artichoke extract, namely cynarine, respectively chlorogenic acid.

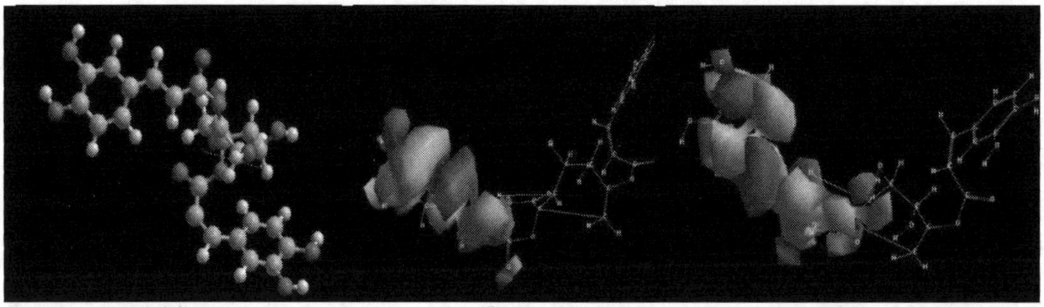

(a) (b) (c)

Figure 8. Optimized structure (a) and H.O.M.O orbital (b) and L.U.M.O orbital (c) respectively for the Cynarine molecule

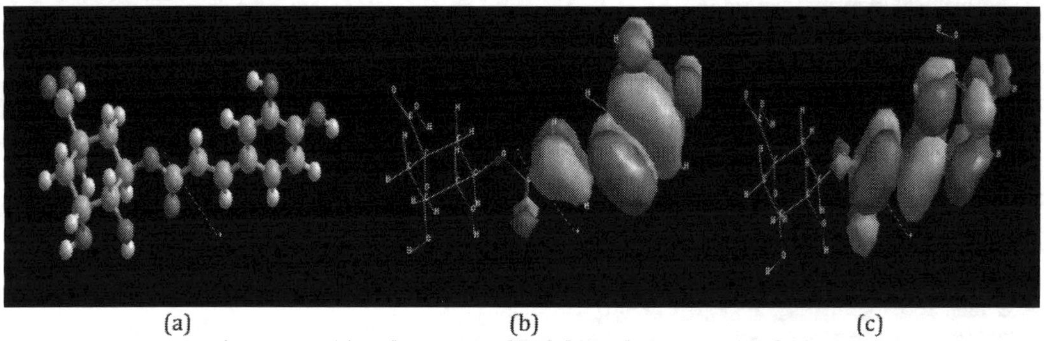

(a) (b) (c)

Figure 9. Optimized structure (a) and H.O.M.O orbital (b) and L.U.M.O orbital (c) respectively for the chlorogenic acid molecule

Figures 8 and 9 respectively show the optimized structures (a), the structure of the highest energy occupied molecular orbital H.O.M.O (b), respectively the lowest energy unoccupied molecular orbital (L.U.M.O) (c) obtained using the Hyperchem software. Carbon atoms are represented in light blue, hydrogen atoms in white, and oxygen atoms in red.

Based on the level values of the Frontier molecular orbitals and the formulas discussed in Chapter 2.2, the values of the molecular descriptors in Table 4 and 5 respectively were obtained.

Table 4. Molecular Descriptors for Cynarine molecule.

E_{HOMO}(eV)	E_{LUMO}(eV)	ΔE (eV)	μ (Debye)	χ (eV)	η (eV)	σ (eV^{-1})	V [Å3]	S [Å2]	V/S [Å]
-9.20	-0.96	8.24	3.27	4.12	5.08	0.19	1307.79	754.28	1.73

Table 5. Molecular Descriptors for Chlorogenic Acid molecule.

E_{HOMO}(eV)	E_{LUMO}(eV)	ΔE (eV)	μ (Debye)	χ (eV)	η (eV)	σ (eV^{-1})	V [Å3]	S [Å2]	V/S [Å]
-8.94	-0.95	7.99	3.61	3.99	4.95	0.20	933.69	563.68	1.66

Following the analysis of the molecular descriptors for Cynarine, the high value in the E_{HOMO} mode corresponds to a high ionization potential of the -OH functional groups, whilst the E_{LUMO} value indicates the possibility of chemical interactions between metal and substrate. The values of the QSAR properties demonstrate that the super unit value of V/S ratio reflects the concentration-dependent rise in inhibitory efficiency. The values for the chlorogenic acid molecule are close to those of Cynarine since the two structures are similar and vary only by one unit of caffeic acid. Molecular model studies explain the inhibitory process by forming a coordinative bond between the hydroxyl functional groups of the two active compounds and the vacant d-orbital of iron and copper, respectively, by creating a protective adsorption layer, which prevents further attacks by the corrosive environment.

4. Conclusions

The analysis of the cyclic voltammograms, in the acidic environment underlines that the artichoke extract undergoes transformations in the anodic range, influences oxygen release, but its cathodic influence is diminished. In the NaCl environment it influences both cathode and anode processes, the artichoke extract being stable on the potential range between the hydrogen and the oxygen release.

Tafel slopes and chronoamperometry revealed that artichoke extract functions as a corrosion inhibitor for carbon steel and copper in harsh sulfuric acid and sodium chloride environments. The concentration of the inhibitor is critical to its efficacy. For OL45 inhibitor efficiency of 82% (H_2SO_4) and 89% for NaCl are obtained; for Cu, in H_2SO_4 IE = 97% and in NaCl 96%.

The molecular modelling studies emphasizes that, the values of E_{HOMO} and E_{LUMO} as well as the other descriptors indicate a high ionization potential and the presence of concurrent physical and chemical interactions of the active compounds in the extract on the metal surface, interactions that play an important role in the formation of an adsorption layer which helps to prevent further attacks of the aggressive environment on the metals studied.

References

[1] Verma C., Alfantazi A., Quraishi M.A.,Rhee K.Y., 2023, Sustainable Chemistry and Pharmacy, 31, 100943.
[2] Zakeri A., Bahmani E., Sabour Rouh Aghdam A., 2022, Corrosion Communications, 5 ,25-38.
[3] Umoren Saviour A., M. Moses Solomon, Obot I. B, Suleiman R. K., 2019, Journal of Industrial and Engineering Chemistry, 76, 91-115.
[4] Izadi M., Shahrabi T., Mohammadi I., Ramezanzadeh B., 2019, Progress in Organic Coatings, 135, 135e147.
[5] Salleh S. Z., Yusoff A. H., Zakaria S. K., Ali Azhar Taib M., Abu Seman A., Najmi Masri M., Mohamad M., Mamat S., Ahmad Sobri S., Ali A., Ter Teo P., 2021, Journal of Cleaner Production, 304, 127030.
[6] Chauhan D.S., Quraishi M.A., Qurashi A., 2020, Journal of. Molecular Liquids 326, 15117.
[7] Nawaz A., Deen K. M., Ishraq S., Haider W., 2024, International Journal of Electrochemical Science, 19(8), 100680.
[8] Oghenerukevwe P.O., Ajuwa C.I.,. Samuel O.D., Adepoju T.F., 2023, Case Studies in Chemical and Environmental Engineering, 8, 100479.
[9] Taofeek Popoola L., Sikiru Yusuff A., Ikumapayi O. M., Chima O. M., Ogunyemi A. T., Obende B. A., 2023, Scientific African, 21, e01860.
[10] Abd-El-Nabey B. A., Mohamed M. E., Helmy A. M., Elnagar H., Abdel-Gaber A. M., 2024, International Journal of Electrochemical Science, 19 (8), 100695.
[11] Pai G. D., Rathod M. R., Sk R., Kittur A. A., 2024, Results in Surfaces and Interfaces, 15, 100203.
[12] Maizia R. et al., 2023, Arabian Journal of Chemistry, 16 (8), 104988.
[13] Utami S. P. et al., 2023, Materials Today: Proceedings, 87, 355–359.
[14] Haldhar R., Prasad D., Saxena A., 2018, Journal of Environmental Chemical Engineering 6 ,5230–5238.
[15] Haldhar R., Prasad D., Saxena A., Kaur A., 2018, The European Physical Journal Plus. 133, 1–18.
[16] Rodríguez-Torres A., Valladares-Cisneros M.G., Gonzalez-Rodríguez J.G., 2014, International Journal of Electrochemical Science 10, 4053–4067.
[17] Abdallah M., Altass H.M., Jahdaly B.A.AL,. Salem M.M, 2018, Green Chemistry Letters and Reviews, 11, 189–196.
[18] Chung I.M., Malathy R., Priyadharshini R., Hemapriya V., Kim S.H, Prabakaran M., 2020, Materials Today Communications 25, 101687.
[19] Haldhar R., Prasad D., Saxena A., 2018, Journal of Environmental Chemical Engineering, 6, 2290–2301.
[20] Feiden T, Valduga E, Zeni J, Steffens J., 2023, Food Technology and Biotechnology, 61(3), 312-327.
[21] Salmasifar A., Edraki M.,. Alibakhshi E, Ramezanzadeh B., Bahlakeh G., 2021, Journal of Molecular Liquids, 327, 114856.
[22] "OLC45 Datasheet," Steel Grades, [Online document], Available: https://www.steel-grades.com. [Accessed Jun. 15, 2024].
[23] Alibakhshi E., Ramezanzadeh M., Bahlakeh G., Ramezanzadeh B., Mahdavian M., Motamedi M., 2018, Journal of Molecular Liquids, 255, 185–198.

IOP Conf. Series: Materials Science and Engineering **1319** (2024) 012034 doi:10.1088/1757-899X/1319/1/012034

Tensile properties of Romanian "Țurcana" sheep wool farm-waste fibers

Corina Sosdean[1] and Sergiu-Valentin Galatanu[2*]

[1] Department of Mechatronics, Politehnica University Timisoara, M. Viteazu Blvd., No. 1, Timisoara 300 222, Romania
[2] Department of Mechanics and Strength of Materials, Politehnica University Timisoara, M. Viteazu Blvd., No. 1, Timisoara 300 222, Romania

*E-mail: sergiu.galatanu@upt.ro

Abstract. With a long tradition in sheep farming, Romania is still one of the major sheep breeding countries in Easter Europe with the "Țurcana" sheep breed playing a central role. In recent decades, wool has been increasingly considered waste from the shearing process, yet its potential as a natural, renewable, biodegradable fiber offers opportunities for its use as raw construction material due to its beneficial characteristics and eco-friendly nature. With a focus on reducing environmental impact and optimizing its functionality, this study conducts tensile testing on "Țurcana" wool to evaluate its strength and explores the potential of utilizing wool fibres as reinforcement in newly developed composites.

1. Introduction

Due to their good mechanical characteristics (high strength-to-weight ratio) and green attributes (recyclable, biodegradable and renewable), natural fibers (NFs) are of growing interest in various industries [1, 2]. In general, NFs are extracted from three categories subdivided into animal fibers (wool, silk, camel, alpaca etc.), plant fibers (cotton, jute, hemp, flax, etc.) and mineral sources fibers (asbestos, graphite, glass) [3-5]. Despite being one of the NFs known to humanity since ancient times, and having been used for centuries to produce carpets, blankets, clothing, and interior textiles [6], over the past few decades, sheep wool (SW) has been increasingly seen as waste material, often discarded or left unused as a by-product of the shearing process [7]. Currently, SW has attracted interest as a construction material owing to its multitude of advantageous characteristics and eco-friendliness [8].

SW contains several proteins including: silk, collagen, gelatin, fibrinogen, elastin, keratin, actin, and myosin. Wool keratin has a high abundance of cysteine residues (7–20% of the total amino acid residues), making the fibers strong and stiff [9]. Generally, keratin fibers consist of two major morphological parts: a cortex (the inner part of the fiber) and the cuticle outer layer, which comprises 10% of the total weight and is composed of overlapping, nearly rectangular scales, or sheets that protect the cortex [10]. The interlocking structure of keratin fibers provides excellent tensile strength, enabling wool to endure stretching and bending without losing its shape.

SW is a versatile material with numerous potential applications in the field of building materials. Nowadays its main use is as thermal and acoustical insulation material and as reinforcement fiber for bio composites to improve their mechanical behaviors, as ductility and shrinkage rate [11]. Earlier studies were carried out to develop a sandwich panel made from SW with various testes conducted in order to assess its acoustic, thermal, and mechanical properties and determine its suitability as an insulation material [12]. In the studies conducted by

Szatkowski et al. [13] was concluded that SW, as a fiber with unique physicochemical properties, can be used as an element of composite materials, in line with the strategies of bio economy and sustainable development. Alyousef et al. [14] studied the mechanical and microstructural properties of concrete containing SW fibers. Dénes et al. [15] analyzed the properties of SW through the lens of a thermal insulation material, concrete reinforcement, and carbon fiber precursor by the means of comparative analysis against a few materials from the considered domains.

Sheep farming in Romania has deep-rooted cultural, economic, and environmental significance, with the "Ţurcana" sheep breed serving as a central and emblematic component of this heritage [16]. According to Statista Research Department [17], the sheep population in Romania in 2022 was over 10.2 million, which increased by approximately 127 thousand sheep since 2021. The 'Ţurcana' breed, the most common and oldest local sheep in Romania, is a hardy species well-suited to the hill and mountain regions [18]. It belongs to the group of long coarse-SW breed [19]. With a focus on reducing environmental impact and optimizing its functionality, this study conducts tensile testing on "Ţurcana" wool to evaluate its strength and explores the potential of utilizing SW fibers as reinforcement in newly developed composites.

2. Test specimens

In the present study, "Turcana" breed single wool fibers were collected from a batch of sheared wool that was previously washed with plain water and let to dry. A total of 20 samples having different lengths, were considered. For each sample the mean width was measured using the optical microscope (Table 1).

Table 1. Average width of the SW specimens

Sample	Mean width (μm)	Sample	Mean width (μm)	Sample	Mean width (μm)	Sample	Mean width (μm)
1	50.67	6	47.33	11	44.22	16	51.66
2	53.00	7	50.55	12	46.44	17	51.66
3	54.22	8	47.33	13	53.22	18	45.33
4	48.44	9	50.55	14	48.00	19	53.00
5	48.55	10	45.44	15	50.00	20	49.45

Figure 1 shows the width variation for each investigated specimen. As it can be seen, the smallest width value is 44.22 μm and the highest value is 54.22 μm. The average of the width of the specimens is 49.45 μm.

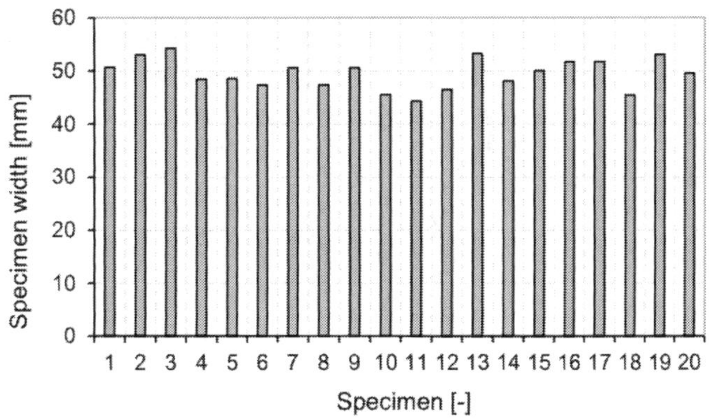

Figure 1. Mean width of the SW samples.

3. Experimental procedure

Tensile tests were conducted on the METROTECH testing machine, at a speed of 2mm/min, the gauge's length being set at 40 mm as shown in Fig. 2. An AEP transducer type TCA with a maximum load capacity of 5 kg was used for the experimental tests.

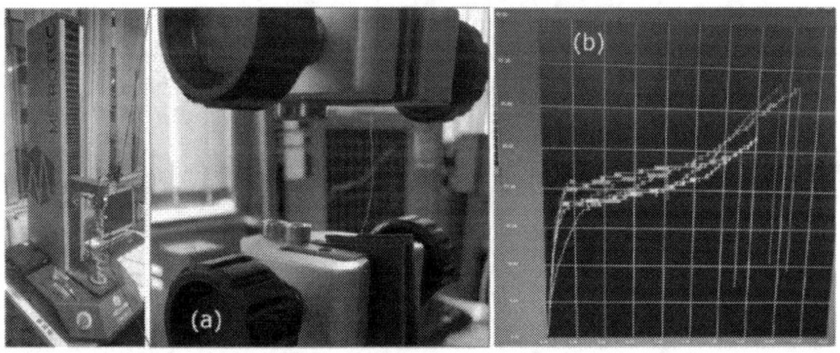

Figure 2. Experimental setup (a) and obtained data (b).

After conducting the tensile tests, the variation of force, time and displacement were obtained. With this data, the stress-strain curves were plotted. The stress, measured in MPa, indicates the load that the fiber can withstand, while the strain represents the ratio of the fiber's elongation to the gauge length. The stress was determined by dividing the force variation by the initial area, which was calculated based on the diameter of the fibers measured with the microscope. Likewise, the strain was obtained by dividing the variation of the standard travel to the initial distance between the clamping dies.

4. Results and discussions

In Figure 3, the stress-strain curves for the tested wool fibers are presented. As can be seen, the tensile behaviors of the investigated SW can be grouped into 4 distinct classes based on their tensile strength. The plotted curves are the most representative of their class. Regardless of the class they are in, the SW fibers show a linear-elastic behavior up to a stress of about 100 MPa, after which they show a yield zone. After the yield point, the samples plastically deform forming a long plateau. The length of the plateau is mainly due to the delamination of the fibers that occurs at the microstructural level. For all four classes, the plateau shows a slight hardening and ends with the maximum stress, defined and tensile strength. Once the maximum tension is reached, the samples suddenly break.

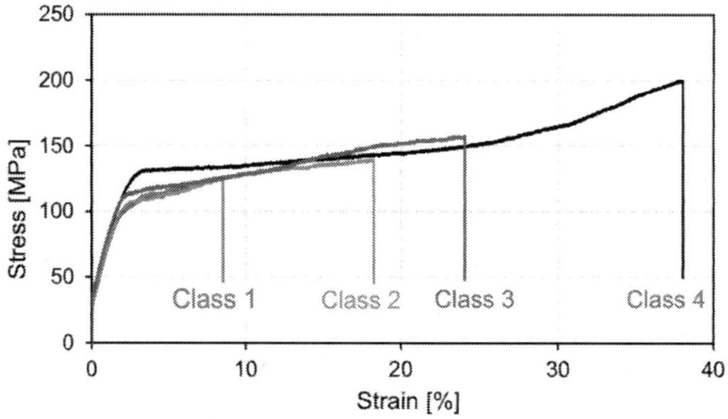

Figure 3. Stress-strain curves of the representative WS samples.

The variability in fiber strength is due to the inherent distribution of flaws along each fiber [24] and the inconsistency in strength between different fibers within the same batch. Based on the tensile behavior (see Fig. 3), the tensile strength was determined. Figure 4 shows the average results of each class.

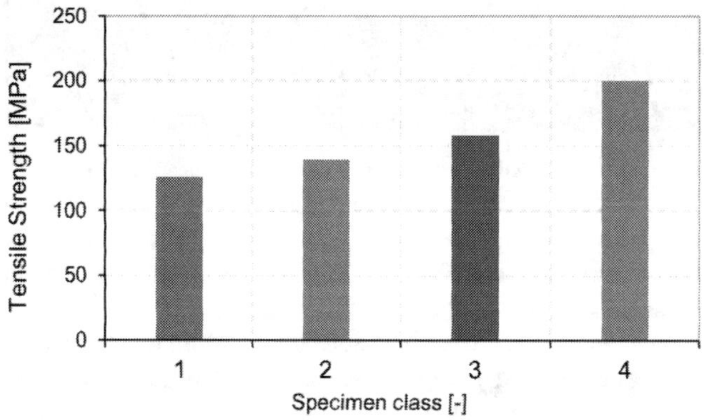

Figure 4. Tensile strength depending on the SW class.

An increase in strength was obtained with the chosen class. More exactly, the tensile strength increases from 125.4 MPa for Class 1 (red), to 139.5 MPa for Class 2 (green), 157.8 MPa for Class 3 (blue) reaching the maximum value of 200.1 MPa for Class 4 (orange). Class 4 stands out for its superior tensile strength, which could indicate better material properties or processing methods.

Thus, between the extreme values of the strength (Classes 1-4, a percentage difference of 59.5% is obtained. The smallest difference, of 11.2%, is identified between classes 1-2.

At the same time, in Fig. 5, a clear increase in deformation can be observed with the change of class. Elongation increases from 8.5% (red, Class 1), to 183% (green, Class 2), 24.1% (blue, Class 3) reaching the maximum value of 38.0% (orange, Class 4). Class 4 stands out for its superior ductility, which could indicate a material that is more flexible and capable of undergoing more deformation before breaking. It was obtained that the strain at strength of Class 4 is up to 4.5 times higher than that of Class 1.

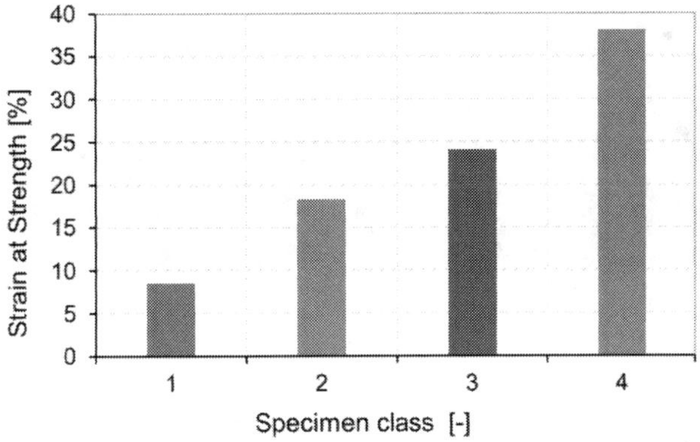

Figure 5. Strain at strength depending on the SW class.

Figure 6 shows a column graph regarding the tensile modulus of the SW samples tested by class. Modulus values between 4581.2 MPa (Class 1) and 5043.3 MPa (Class 4) were obtained, with a maximum difference of 10% between them. It seems that the type of fiber (called class) does not significantly influence the stiffness of the samples. In fact, this aspect can also be observed in Fig. 3, where the linear-elastic zone almost overlaps for all four classes.

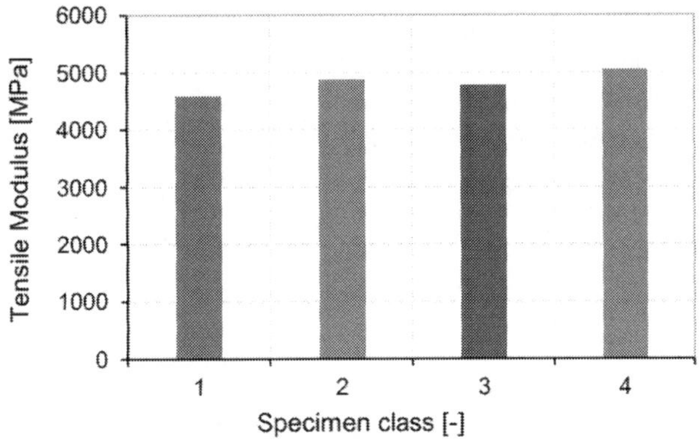

Figure 6. Tensile modulus depending on the SW class.

For comparison, plastics like high-density polyethylene (HDPE) and polypropylene (PP) have tensile moduli between 1000 and 4000 MPa, while Glass Fiber reinforced composites show a Young's modulus ranging from 3000 to 5000 MPa.

After fracture, the tested SW fibers were measured again and it was found that their thicknesses are in the range of 17-25 μm. It was observed that some fibers presented a transverse fracture (Fig. 7a), and others a combination of delamination + transverse fracture (Fig. 7b). In the same figure, the damage of the fiber's keratin cortex can be easily seen.

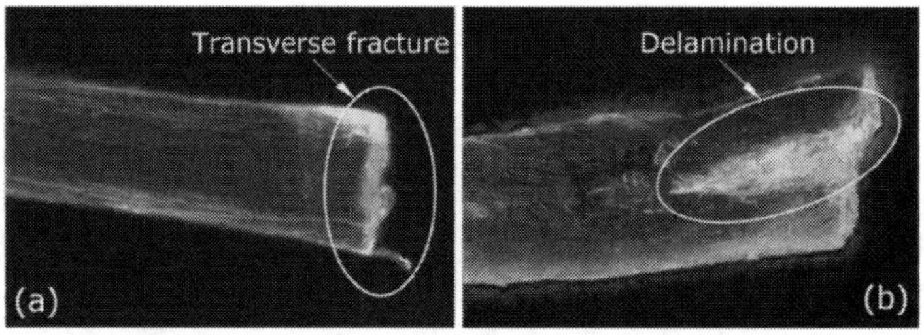

Figure 7. Microstructural analysis of fractured specimens.

5. Conclusions

In this work the tensile strength of "Turcana" single wool fibers were tested.

- We do not have information on whether the wool fibers we examined came from a single sheep or multiple individuals, nor do we have details about the specific body parts of the sheep from which the samples were taken. According to [20], over an entire fleece, or even within a representative wool sample, fiber diameter is not homogenous, but ranges from 10-70 μm.
- Based on their tensile behavior, the samples analyzed in this study were divided in four main classes: Class 1: 125 MPa, 8.5% elongation; Class 2: 139 MPa, 18.25% elongation; Class 3: 158 MPa, 24% elongation and Class 4: 200 MPa and 38% elongation. In the studies conducted by Rottenbury et al. [21] and Ross et al. [22] it was concluded that there were quite large differences in staple strength across the body of a sheep. This could explain the four patterns of tensile strength of the tested wool fibers and the necessity to further research the importance of the wool sampling location.
- This research has tried to highlight the tensile properties of "Turcana" single wool fibers. We emphasize that despite extensive research into the use of natural fibers for composite reinforcement [23], there is a notable lack of information regarding the use of wool as a reinforcement material in the construction industry. Thus, the innovative application of wool as a composite element presents a valuable opportunity.

References

[1] F. Khan, N. Hossain, F. Hasan, et al., 2024, Applications in Engineering Science, 18, 100184.
[2] D.K. Rajak, D.D. Pagar, P.L. Menezes, et al., 2019, Polymers 11 (10), 1667.
[3] S. Palanisamy, K. Vijayananth, T.M. Murugesan, et al., 2024, International Journal of Lightweight Materials and Manufacture, 7, 496-506.
[4] D.K. Rajak, P.H. Wagh, E. Linul, 2022, Materials 15 (14), 4790.
[5] DK Rajak, PH Wagh, E Linul, 2021, Polymers 13 (21), 3721
[6] J. Broda, A. Gawlowski, M. Rom, et al., 2023, Journal of Natural Fibers. 18, 20(2), 2200047.

[7] B.D. Chereji, F.D. Munteanu, 2022, Earth Observation & Surveying, Environmental Engineering. XI.
[8] M.C.M. Parlato, S.M.C. Porto, F. Valenti, 2022, Building and Environment, 225, 109596.
[9] S.G. Giteru, D.H. Ramsey, Y. Hou, et al., 2023, Comprehensive Reviews in Food Science and Food Safety, 22(1), 643-87.
[10] R. Silva, B. Fabry, A.R. Boccaccini, 2014, Biomaterials, 35(25), 6727-38.
[11] M.C. Parlato, M. Cuomo, S.M. Porto, 2022, Construction and Building Materials, 326, 126855.
[12] D.R. Tămaş-Gavrea, T.O. Dénes, R. Iştoan, et al., 2020, Coatings, 10, 148.
[13] P. Szatkowski, A. Tadla, Z. Flis, et al., 2021, Roczniki Naukowe Polskiego Towarzystwa Zootechnicznego, 17(4).
[14] R. Alyousef, H. Alabduljabbar, H. Mohammadhosseini, et al., 2020, Journal of Building Engineering, 30, 101216.
[15] O. Dénes, I. Florea, D.L. Manea, 2019, Procedia Manufacturing, 32, 236-41.
[16] L. Udrea, G. Teodorescu, S.V. Morărita, et al., 2021, Agrobiodiversity & Agroecology, 1, 79-95.
[17] https://www.statista.com/
[18] D.I. Neaţă, T. Vintilă, 2023, Scientific Papers Animal Science and Biotechnologies, 56(1), 203.
[19] E. Ilişiu, S. Dărăban, R. Radu, et al., 2013, Landbauforsch–Appl. Agric Forestry Res., 2, 161-70.
[20] E. Wood, 2003, Textile properties of wool and other fibres, 272-290.
[21] R.A. Rottenbury, M.R. Bow, W.J. Kavanagh, et al., 1981. Wool Technol. Sheep Breed. 29, 143–148.
[22] D.A. Ross, H.E. Meikle, 1985. Proceedings of the 7th International Wool Textile Research Conference, Tokyo, pp. 45–54.
[23] I. Elfaleh, F. Abbassi, H. Habibi, et al., 2023,. Results in Engineering, 1, 101271.
[24] F. Wang, J. Shao, 2014, Polymers, 6(12), 3005-18.

304L AUSTENITIC STAINLESS STEEL CORROSION STUDIES IN NEUTRAL ENVIRONMENT IN PRESENCE OF SULFITE IONS

Mihaela-Alexandra LĂBOȘEL[1,a], Nataliia RUDENKO[2,b], George-Daniel DIMA[1,c], Mircea Laurențiu DAN[1,d*], Nicolae VASZILCSIN[1,e]

[1]University Politehnica Timisoara, Faculty of Industrial Chemistry and Environmental Engineering, 6 V. Pârvan, 300223 Timisoara, Romania
[2]University Politehnica Timisoara, Innovation and Technology Transfer Center, 2 V. Pârvan, 300223 Timisoara, Romania
[a]mihaela.labosel@gmail.com, [b]nataliia.rudenko@upt.ro, [c]george.dima@student.upt.ro, [d]mircea.dan@upt.ro, [e]nicolae.vaszilcsin@upt.ro

Abstract

This paper examined the influence of the sulfite concentration introduced in the electrolyte added to the neutral solution (1 mol L^{-1} Na_2SO_4) on the corrosion process of 304L stainless steel material. 304L. The accelerated effect of sulfite ions under corrosion rate of 304L austenitic stainless steel was demonstrated using electrochemical techniques such as linear sweep voltammetry at a low scan rate - 1 mV s^{-1}, the Tafel slope method, chronoamperometry, and electrochemical impedance spectroscopy. Electrochemical studies were conducted using a BioLogic SP150 potentiostat/galvanostat.

Stainless steels are the most varied and intricate group among all steels. Among them, 304L austenitic stainless steel is one of the most commonly utilized alloys, prized for its excellent corrosion resistance and strength, hygienic and aesthetic properties, affordability, good formability, and effective weldability. [1-3]. In modern industry, the amount of austenitic stainless steel represents only about 2% of the total steel production manufactured globally. This type of steel is commonly employed in chemical engineering, metal extraction, desalination and wastewater treatment plants, the oil and gas sector, transportation and aerospace industries, the food and beverage sector, as well as in architecture and construction engineering. [1].

Keywords: sulfite ions, stainless steel corrosion, Tafel method, chronoamperometry

1. Introduction

Sulfur is generally recovered in refineries in the form of volatile or liquid gaseous compounds, such as hydrogen sulfide and sulfur oxides, which makes it feasible to use them in fuel cells. Anodic oxidation of SO_2 is a component of the hybrid sulfur cycle for hydrogen production and is also used for fuel gas desulfurization [1-3].

Regarding trends in the study of corrosion in environments containing sulfur atoms, Zhang et al investigated the effect of the corrosion behavior of sulfite ions on stainless steels 316SS and C276 in supercritical environments, where the corrosion behavior is due to the composition of the studied material [4]. Cabrini et al studied the effects of thiosulfates and sulfite ions on various steels [5]. Another approach, this time in the field of construction, was carried out by Klein by studying the effect of sulfate in the corrosion of reinforcing steel in concrete [6]. Other materials studied in the literature subjected to electrochemical behavior tests in sulfite environments were: smooth nickel [7], copper – where a synergistic study was carried out with a sulfate-chloride solution [8], cobalt layer perovskite type $Y_{0.5}Ca_{0.5}BaCo_4O_7$ [9], skeletal nickel with 3 layers of Pt nanoparticles [10], respectively graphite [11] and lead [12]. At this time, the field of anodic materials for use in air sulfite fuel cells represents another niche that needs to be investigated.

The theme addressed in our research has practical importance in the current context of environmental protection, taking into account that, the large amounts of sulfur dioxide resulting from waste in combustion processes are absorbed and neutralized in the form of sulfites, which are then passed into an amorphous state in the form of sulfate. Current research has been oriented towards the oxidation of residual sulfites in fuel cells, so that in addition to protecting the environment, electricity is also obtained [13,14]. In order for electrochemical reactors to be profitable from an economic point of view, it is necessary for the very expensive anode materials based on precious metals to be replaced with common metals, which, along with catalytic activity, also show corrosion resistance under the conditions of use [15].

Our research was oriented towards the use of stainless steel as anode material in fuel cells used for sulfite oxidation. In this work, we present the results obtained in the study of the corrosion resistance of 304L stainless steel in electrolyte solutions containing sulfite.

2. Materials and methods

The experimental setup used for the electrochemical measurements included an SP150 Biologic potentiostat/galvanostat connected to an undivided cell containing three electrodes: a 304L steel working electrode with an active surface of 1 cm^2, two graphite counterelectrodes with a cumulative surface submerged in solution of 16.1 cm^2 and an Ag/AgCl reference electrode. The solutions tested were prepared from Na_2SO_4 analytically pure reagents (Merck p.a. c = 99%), sodium sulfite Na_2SO_3 (Merck p.a. c = 98%). Electrochemical experiments were carried out in 1 mol L^{-1} Na_2SO_4 solution in the absence and presence of different sulfite concentrations, namely 10^{-3}, 10^{-2}, 10^{-1}, $5 \cdot 10^{-1}$, and 1 mol L^{-1}. 304L SS samples used in the electrochemical tests had the following chemical composition (wt.%): 18.04% Cr, 8.30% Ni, 0.02% Mo, 1.41% Mn, Fe balance [16].

Before each determination, the working electrode was sanded with 120, 400, 600, 1200 grit Si-C paper, rinsed with distilled water, ultrasonicated, cleaned with acetone and dried. For reproducibility, each curve was traced at least three times. The electrochemical impedance spectroscopy was carried out by tracing 60 impedance points in the frequency range between 10^6 and 10^{-2} Hz and the results were analyzed using the Levenberg-Marquardt technique and the Zview 2 software (Scribner Associates).

3. Results and discussions

3.1. Chronopotentiometric studies

In the corrosion studies, the first technique used was to determine of the evolution over time of the electrode potential at zero current. Thus, based on some potentiometric recordings, the evolution of the open circuit potential (OCP) of the 304L SS samples in the test solutions at the four working temperatures for 60 minutes was determined, which was considered sufficient for quasi-equilibrium installation at the metal-electrolyte interface.

Figure 1 (a-c) shows graphs of the evolution of the potential of the stainless steel electrode in the support electrolite solution without and with the addition of 0.1 mol L^{-1} and 1 mol L^{-1} sulfite.

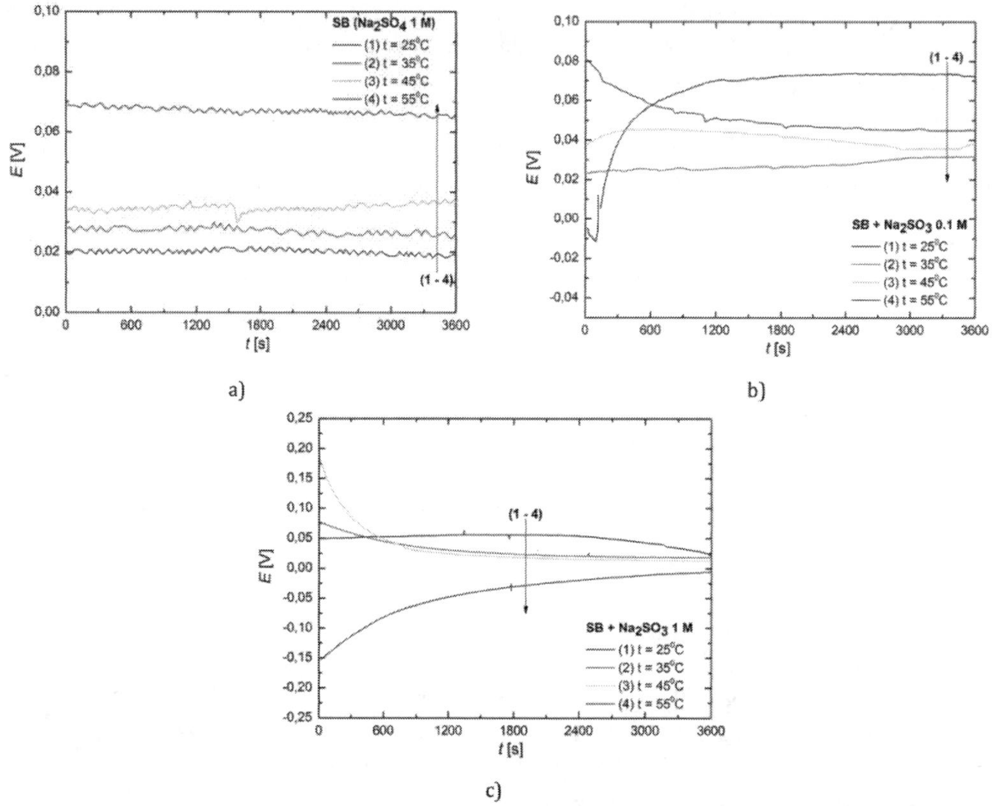

Figure 1. Evolution of the 304L stainless steel electrode potential in the support electrolite solution (a), support electrolite solution + 0.1 M sulfite (b), and support electrolite solution + 1 M sulfite at various temperatures

The study of the OCP curves reveals a distinct behavior in the presence, specifically the absence of the sulfite addition. Thus, when the temperature rises in the support electrolite solution, the potential shifts to positive values, indicating that the corrosion rate in this environment increases in tandem with the temperature. In the presence of sulfite addition, the evolution reverses, with increasing temperature causing a shift towards negative electric potential values, suggesting that the corrosion process of the investigated alloy is inhibited in these environments. The inhibition is primarily caused by a reduction in the concentration of molecular oxygen dissolved in the electrolyte due to its consumption during the chemical process of of sulfite ions oxidation.

3.2. Linear sweep voltammetry

To gather information on the effect of sulfite additions on type 304L steels, linear sweep voltammetry measurements were performed at a polarization speed of 1 mV s^{-1}. The determination of the corrosion rate was carried out by the Tafel extrapolation method, and the results that were the basis for obtaining the parameters i.e. the corrosion potential *E*, the i_{corr} corrosion current, the anodic and cathodic slopes, respectively, were obtained by extrapolating the logarithmic form of the Tafel diagrams (Figure 2), with the help of the EC Lab Express software, with which the BioLogic SP150 potentiostat is equipped, are shown in Table 1 for the working temperatures 298, 308, 318, 328 K

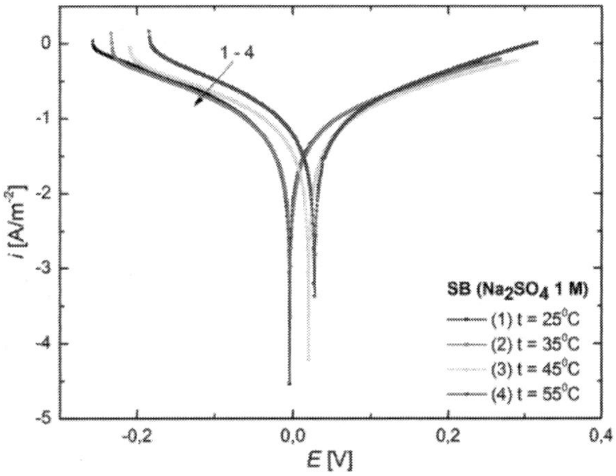

Figure 2. Tafel plot for anodic oxidation of sulfite on a 304L electrode in support electrolite solution at the testing temperatures

Table 1. Tafel slope parameters of the corrosion process of 304L in support electrolite solution at various temperatures

Sample	T [K]		E [mV/Ref]	i_{corr} [μA]	β_a [mV dec^{-1}]	βc [mV dec^{-1}]	Corr rate [mm y^{-1}]
304L SB	298	1	-4.885	5.743	238.6	202.7	0.059
		2	-0.662	6.379	245.1	220.1	0.066
		3	-0.336	5.772	229.7	217.3	0.060
	308		-3.243	6.895	264.2	237.1	0.071
	318		21.16	7.373	267.0	256.2	0.076
	328		27.93	8.339	238.0	215.0	0.086

The analysis of Figure 2 and the parameters presented in Table 1 shows a shift towards more positive values of the corrosion potential. The corrosion current increases somewhat, as does the corrosion rate, which rises in tandem with the temperature.

When 0.1 mol L^{-1} sulfite is added, the polarization curves move slightly to the negative side, and the anodic, cathodic, and anodic slopes decrease with temperature (Figure 3 and Table 2). When sulfite is added to the electrolyte support solution, the corrosion rate decreases as compared to the Tafel curves

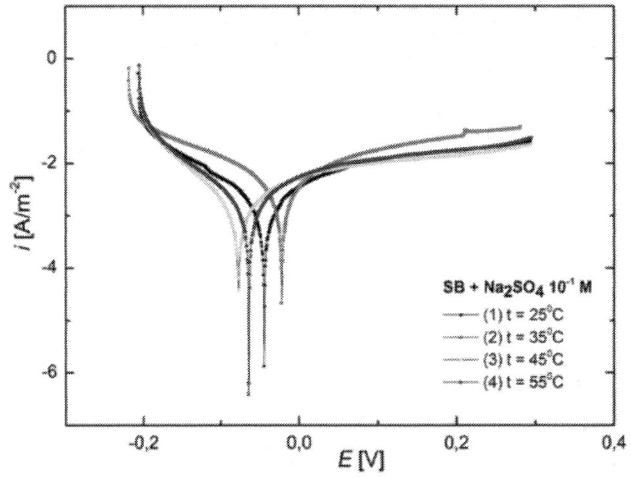

Figure 3. Tafel plot for anodic oxidation of sulfite on 304L electrode in support electrolite solution with 0.1 M sulfite at the testing temperatures

Table 2. Tafel slope parameters of the corrosion process of 304L in support electrolite solution + 0.1 M sulfite at various temperatures

Sample	T [K]		E [mV/Ref]	i_{corr} [μA]	β_a [mV dec^{-1}]	βc [mV dec^{-1}]	Corr rate [mm y^{-1}]
304L SB + 10^{-1}M SO_3^{2-}	298	1	-42	0.212	165	126	$2.21 \cdot 10^{-3}$
		2	-32	0.220	172	142	$2.29 \cdot 10^{-3}$
		3	-25	0.230	183	140	$2.4 \cdot 10^{-3}$
	308		-22	0.466	200	170	$4.85 \cdot 10^{-3}$
	318		-76	0.122	105	76	$1.27 \cdot 10^{-3}$
	328		-66	0.163	127	93	$1.7 \cdot 10^{-3}$

Finally, at a concentration of 1 mol L^{-1} (Figure 4, respectively Table 3), the downward trend of the corrosion rate and the corrosion current is maintained. Increasing the concentration of sulfite, demonstrates that 304L steel is resistant to corrosion in the sulfate/sulfite environment, showing a decrease in the corrosion rate even at the maximum temperature studied (328 K). This decrease occurs with with the increase of the sulfite concentration, in the order of 0.086> $1.7 \cdot 10^{-3}$ >$0.8 \cdot 10^{-3}$ mm y^{-1}.

IOP Conf. Series: Materials Science and Engineering **1319** (2024) 012035 doi:10.1088/1757-899X/1319/1/012035

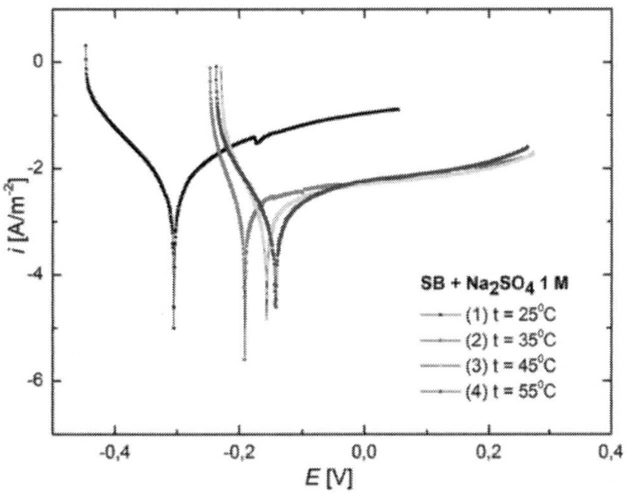

Figure 4. Tafel plot for anodic oxidation of sulfite on 304L electrode in support electrolite solution with 1 M sulfite at the testing temperatures

Table 3. Tafel slope parameters of the corrosion process of 304L in support electrolite solution + 1 M sulfite at various temperatures

Sample	T [K]		E [mV/Ref]	i_{corr} [μA]	β_a [mV dec^{-1}]	βc [mV dec^{-1}]	Corr rate [mm y^{-1}]
304L SB + 1 M SO_3^{2-}	298	1	-306	0.405	105.6	80.2	$4.19 \cdot 10^{-3}$
		2	-268	0.371	137	121	$3.86 \cdot 10^{-3}$
		3	-245	0.179	98.7	119	$1.86 \cdot 10^{-3}$
	308		-191	0.087	71.3	29.5	$0.96 \cdot 10^{-3}$
	318		-157	0.081	87.0	39.6	$0.84 \cdot 10^{-3}$
	328		-143	0.080	99.1	13.4	$0.8 \cdot 10^{-3}$

The decrease in corrosion rate with the increase in the concentration of sulfite in the system, makes austenitic steel type 304L usable as an anode in sulfite-air fuel cells.

3.3. Electrochemical impedance spectroscopy

In order to confirm the trend of decreasing corrosion, with the increase of the sulfite concentration, which was obtained from the previous investigations (chronopotentiometry and linear voltammetry), electrochemical impedance spectra were drawn for the working concentrations. Figure 5 shows the impedance spectra obtained for sulfite oxidation on 304L in 1 M Na_2SO_4 solution, at different sulfite concentrations are represented as complex plane (Nyquist) plots, and in the same figure it is represented as absolute impedance and phase angle versus frequency (Bode) plots.

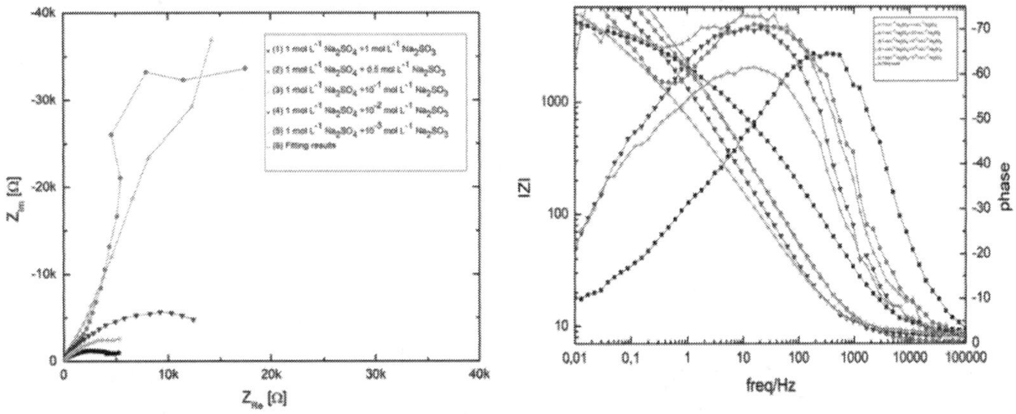

Figure 5. Impedance spectra of sulfite oxidation on SS 304L in 1 mol L^{-1} Na_2SO_4 at 25°C presented as Nyquist and Bode plots for sulfite concentrations of: 1 mol L^{-1} , 0.5 mol L^{-1}, 10^{-1} mol L^{-1}, 10^{-1} mol L^{-1} and 10^{-3} mol L^{-1} .

In order to study the influence of temperature on the strength of the studied steel, the impedance spectra were drawn at temperatures 298, 308, 318 and 328 K, for the three concentrations of sulfite studied. Figure 6. represents the impedance spectra of sulfite oxidation on SS 304L in 1 mol L^{-1} Na_2SO_4 with 10^{-3} mol L^{-1} sulfite at the mentioned temperatures, presented as Nyquist and Bode plots.

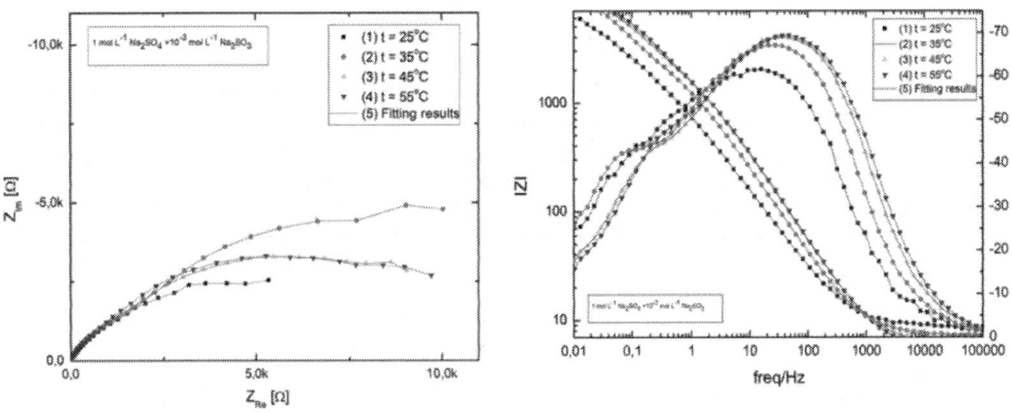

Figure 6. Nyquist and Bode plots of sulfite oxidation on SS 304L in 1 mol L^{-1} Na_2SO_4 with 10^{-3} mol L^{-1} sulfite concentration added at testing temperatures

Figure 7 depicts the impedance spectra of sulfite oxidation on SS 304L in 1 mol L^{-1} Na_2SO_4 with a 10^{-1} mol L^{-1} sulfite at the working temperatures. The same picture shows a comparison with the Nyquist diagram for the electrolyte support solution at 25°C. In comparison, the diameter of the semicircle corresponding to the load transfer process grows, resulting in an increase in resistance as both the working temperature and sulfite concentration increase.

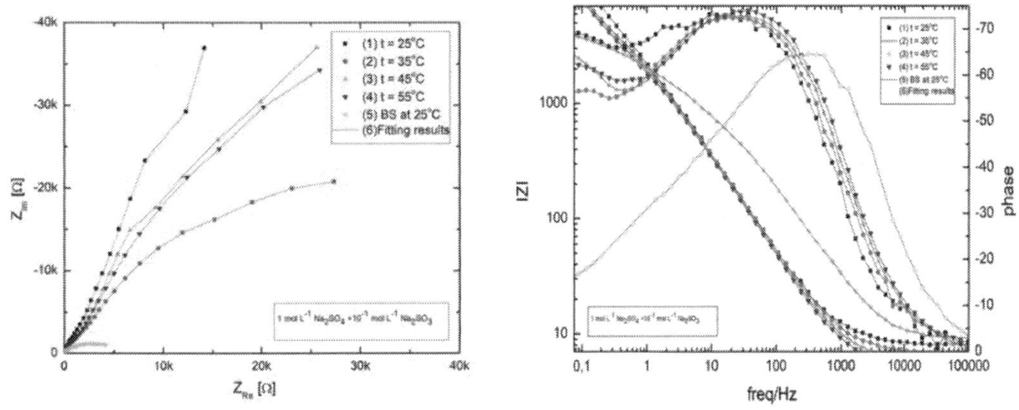

Figure 7. Nyquist and Bode plots of sulfite oxidation on SS 304L in 1 mol L^{-1} Na_2SO_4 with 10^{-1} mol L^{-1} sulfite concentration added at testing temperatures and BS at 25°C

Temperature has no major impact on the maximum sulfite concentration, which is 1 mol L^{-1} Na_2SO_4 with 1 mol L^{-1} sulfite (Figure 8), although their order is obvious in the Bode graphs. The difference in impedance between the electrolyte support solution at 25°C and the last added sulfite concentration makes this electrolyte concentration ideal for preventing corrosion of the anode constructed of 304L steel.

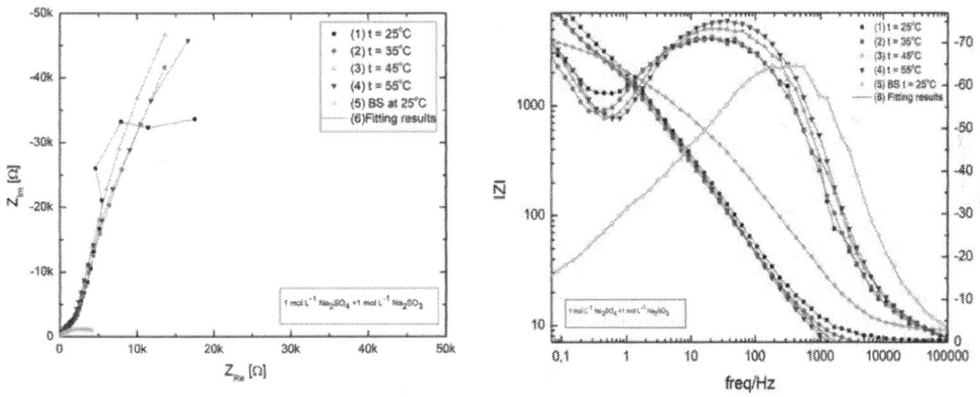

Figure 8. Nyquist and Bode plots of sulfite oxidation on SS 304L in 1 mol L^{-1} Na_2SO_4 with 1 mol L^{-1} sulfite at testing temperatures and BS at 25°C

To represent the impedance data, a simple electric equivalent circuit was used, consisting of the solution resistance R_s in series with a parallel connection of the double layer capacitance C_{dl} and the polarization resistance R_p. Because actual systems sometimes depart from ideal capacitive behavior, a constant phase element (CPE) replaces the double-layer capacitance to account for surface inhomogeneity. In the impedance of a CPE, we note that T is connected to C_{dl}, and n is an exponent between 0 and 1 that describes the CPE's constant phase angle.

Table 4 shows the results of fitting for 304L steel samples, as well as the impedance parameter errors and the quality of fit, as measured by the Chi-squared values.

Table 4. Calculated data of the circuit elements and fitting errors for a 304L steel sample in 1 M Na_2SO_4 with various sulfite concentrations added at 4 different temperatures

t [°C]	R_s [Ω cm²]	R_{ct} [Ω cm²]	CPE1-T [F s^{n-1} cm^{-2}]	CPE1-n	χ^2
			c = 1 mol L^{-1}		
25	7.165	7283	$9.36 \cdot 10^{-4}$	0.82	$1.9 \cdot 10^{-3}$
35	5.931	5912	$10.95 \cdot 10^{-3}$	0.82	$7.0 \cdot 10^{-3}$
45	4.866	4805	$9.28 \cdot 10^{-4}$	0.84	$1.4 \cdot 10^{-3}$
55	4.157	4042	$8.60 \cdot 10^{-4}$	0.86	$2.3 \cdot 10^{-3}$
			c = 0.5 mol L^{-1}		
25	2.568	2616	$1.58 \cdot 10^{-3}$	0.54	$1.9 \cdot 10^{-3}$
			c = 10^{-1} mol L^{-1}		
25	8.122	13988	$8.78 \cdot 10^{-5}$	0.83	$1.99 \cdot 10^{-3}$
35	6.734	8800	$9.36 \cdot 10^{-5}$	0.82	$3.94 \cdot 10^{-3}$
45	5.623	7915	$9.92 \cdot 10^{-4}$	0.82	$5.83 \cdot 10^{-3}$
55	4.829	8102	$10.1 \cdot 10^{-4}$	0.83	$7.90 \cdot 10^{-3}$
			c = 10^{-2} mol L^{-1}		
25	8.036	14070	$1.82 \cdot 10^{-4}$	0.78	$1.30 \cdot 10^{-2}$
			c = 10^{-3} mol L^{-1}		
25	8.46	4848	$3.33 \cdot 10^{-4}$	0.72	$4.91 \cdot 10^{-3}$
35	7.18	3674	$1.52 \cdot 10^{-4}$	0.79	$2.16 \cdot 10^{-3}$
45	6.0	3787	$1.07 \cdot 10^{-4}$	0.80	$2.71 \cdot 10^{-3}$
55	6.02	3588	$1.05 \cdot 10^{-4}$	0.80	$1.87 \cdot 10^{-3}$

Based on the analysis of the equivalent circuit whose parameters are shown in Table 4, it can be concluded that increasing the temperature accelerates electrochemical processes, resulting in a decrease in the observed resistances, confirming that 304L steel becomes more resistant to corrosion in the sulfate/sulfite environment as the sulfite concentration increases. The general pattern of lowering R_{ct} and R_s values with rising temperature and sulfite concentration indicates that 304L steel has improved corrosion resistance under these conditions, which supports the results of the linear voltammetry and OCP techniques. These results indicate that 304L steel may be utilized well as an anode in sulfite-oxygen fuel cells.

Conclusions

The study's findings from diverse approaches contribute to a thorough knowledge of 304L steel corrosion behavior in sulfate/sulfite environments. OCP graphs show that, without sulfite addition, rising temperatures in the electrolyte solution cause a positive shift in potential, suggesting higher corrosion rate. However, with sulfite addition, rising temperature reduces the potential, implying corrosion inhibition due to lower molecular oxygen concentration, which is consumed during sulfite ion oxidation. The Tafel slope method supports this, showing that at a concentration of 1 mol L^{-1}, 304L steel exhibits a maintained downward trend in corrosion rate and current, even at high temperatures (328 K), with corrosion rates decreasing as sulfite concentration increases ($0.086 > 1.7\cdot10^{-3} > 0.8\cdot10^{-3}$ mm y^{-1}).

EIS data analysis supports these findings, demonstrating that higher temperatures accelerate electrochemical processes, resulting in lower resistances and improved corrosion resistance of 304L steel in higher sulfite concentrations. Furthermore, experimental evidence support the use of 304L stainless steel as an anodic material in sulfite-oxygen fuel cells, providing a low-cost, chemically stable, and readily available alternative with considerable electrocatalytic effects. The use of stainless steel anodes can significantly lower fuel cell manufacturing costs by using the material's electrocatalytic characteristics for oxidation processes in neutral conditions. Overall, our findings confirm that 304L steel is very resistant to corrosion in sulfate/sulfite environments and may be used efficiently as an anode in sulfite-oxygen fuel cells.

Acknowledgements

This work was partially supported by University Politehnica Timisoara in the frame of PhD studies.

References

[1] A.F. Enache, M.L. Dan, N. Vaszilcsin, 2018, Int. J. Electrochem. Sci., 13, 4466 – 4478.
[2] A. Enache, M. Dan, A. Kellenberger, N. Vaszilcsin, 2017, Bulg. Chem. Commun., 49, 139 – 146.
[3] M. Talebian, K. Raeissi, M. Atapour, B.M. Fernández-Pérez, A. Betancor-Abreu, I. Llorente, S. Fajardo, Z. Salarvand, S. Meghdadi, M. Amirnasr, R.M. Souto, 2019, Corros. Sci., 160, 108130.
[4] J. Zhang, B. Li, Y. Li, J. Lu, K. Wang, H. Zhang, 2021, International Journal of Hydrogen Energy, 46(42), 22222-22233.
[5] Cabrini, M., Lorenzi, S., & Pastore, T. 2018, Corr. Sci., 135, 158–166.
[6] N. Klein, E. D. Gómez, G. S. Duffó, S. B. Farina, 2022, Constr Build Mater, 354, 129214.
[7] M. Lăboşel, A-F. Enache, G-D. Dima, M. L. Dan, N. Vaszilcsin, 2023, Glob. Nest J, 26(1), 05374.
[8] G.S., Sajadi, V. Saheb, M. Shahidi-Zandi, et al., 2022, Sci Rep 12, 14384.
[9] A. Enache, M. Dan, M. Lăboşel, N. Vaszilcsin, 2018, IOP Conf. Ser.: Mater, 416, 012064.
[10] Enache, A., Dan, M., Vaszilcsin, N., 2017, Analele Universităţii din Oradea, XXVII.
[11] A.G. Zelinsky, B.Ya. Pirogov, 2017 Electrochimica Acta, 231, 371-378.
[12] M.-A. Lăboşel, G.-D. Dima, D.-A. Duca, N. Vaszilcsin, M. L. Dan, 2023, Mater. Today: Proc, 78, Part 2, 302-307.
[13] Y. Yan, B. Lin, L. Zhang, Y. Wang, H. Zhang, H. Zheng, T. Zhou, Y. Zhan, Z. Yu, Y. Kuang, J. Tang, 2024, Chem. Eng. J, 487,150447.
[14] R.L. Plaut, C. Herrera, D.M. Escriba, P.R. Rios, A.F. Padilha, 2017, Mater. Res., 10, 453-460
[15] M.L. Dan, A. Kellenberger, D.A. Duca, N. Vaszilcsin, C.M. Craciunescu, I. Mitelea, A. Ercuta, S. Lædre, T. Khoza, 2023, Materials. 16(4), 1501.
[16] "304L Datasheet," Steel Grades, [Online document], Available: https://www.steel-grades.com. [Accessed Jun. 22, 2024].

IOP Conf. Series: Materials Science and Engineering **1319** (2024) 012036 doi:10.1088/1757-899X/1319/1/012036

Evaluation of the layering adhesion between PLA and TPE materials

Gabriel MURARIU[1], Matei MARIN-CORCIU[1] and Sergiu-Valentin GALAȚANU[1*]

[1]Department of Mechanics and Strength of Materials, Polytechnic University, Timisoara, Romania

E-mail: sergiu.galatanu@upt.ro

Abstract.
The additive manufacturing industry has witnessed rapid growth and has applications across various sectors such as aerospace, automotive, healthcare, and consumer goods. Both Thermoplastic Elastomer (TPE) and Polylactic Acid (PLA) hold significant relevance in these industries because of their unique properties and environmental sustainability, thus driving the demand for research aimed at improving their compatibility and performance in multi-material printing processes. This study delves into the interlayer adhesion aspects of multi-material filament fabrication, particularly focusing on the incorporation of TPE and PLA. TPEs offer elasticity and flexibility, while PLA provides rigidity and biodegradability, making them an intriguing combination for a wide range of applications. Experimental investigations of mechanical properties, specifically tensile, compression, and Charpy impact tests, are conducted to evaluate the interlayer adhesion between TPE and PLA. This study aims to contribute to the understanding of interlayer adhesion in multi-material filament fabrication, focusing on the integration of TPE and PLA. Through experimental analysis and exploration of industrial applications, we seek to shed light on the potential of these materials in additive manufacturing and outline future directions for research and development in this field.

1. Introduction

Additive manufacturing (AM) is a technology that is considered new and produces three-dimensional parts by extruding molten polymeric materials and depositing them into sequential layers. AM is becoming the most desirable choice in the field of prototyping because of its resulting dimensional accuracy in relation to its manufacturing time.

Initially adopted and used by home users as a hobby, AM was quickly implemented and used by researchers for prototyping one-of-a-kind designs, building test benches for their experiments. Because of this, it is starting to be implemented in various industries because of its cost and production time advantages over other fabrication technologies [1].

In the 1980s, the United States of America went through a manufacturing crisis, and due to this additive manufacturing was developed to satisfy the need to produce parts with complex geometries.

The main constraint of additive manufacturing is the build volume, as it is directly related by the equipment and the type of additive manufacturing in question. Due to the rising popularity and recognisable versatility of additive manufacturing, steady advancements are being made to the machines that are used, materials and, subsequently, the technology itself.

Designing and prototyping of parts or 3D models is the act of conceptualising a part that serves and fulfils the desired function depending on the application at hand. This process can be classified in three distinct stages: rapid prototyping, virtual prototyping (conceptualising), and physical prototyping. Rapid prototyping (RP) is the actual fabrication of a three-dimensional model using additive manufacturing. This method requires the least amount of user input and is arguably the most versatile and straight forward method of manufacturing. [2]

In this paper, the additive manufacturing technology used is fused deposition modelling (FDM). The fundamental principle of FDM is the extrusion of polymeric material consisting of one continuous plastic thread wound around a reel, which has gained popularity due to its accessibility and reduced cost. The FDM process entails in the extrusion of polymeric material that is the form of filament through a melting zone located on the print head, which can move independently or synchronously on three axes within the boundaries of the build volume of the machine at hand. [3]

In this paper, experimental investigations of the mechanical properties, specifically tensile, compression, and Charpy impact tests, are conducted to evaluate the interlayer adhesion between TPE and PLA. Understanding the interplay between these materials at the interface is crucial for optimizing the printing process and enhancing the structural integrity of fabricated parts. Looking ahead, the prospects for multi-material filament fabrication are promising. As advances in material science and AM technology continue, we anticipate further enhancements in the mechanical properties and printability of composite filaments. This paves the way for the creation of more intricate and functional parts, expanding the horizons of additive manufacturing in diverse industrial applications.

2. Methods and Materials

In this paper, we analysed the behaviour of mixed material test specimens produced through additive manufacturing. The materials chosen for this analysis are polylactic acid (PLA) and thermoplastic elastomer (TPE). The chosen materials differ significantly in their mechanical and chemical properties, which was also a challenge and a point of interest for us. PLA is a biodegradable bioplastic derived from plant starches produced through the lactic acid condensation, which is derived from bacterial fermentation of sugars found in natural products such as corn, potatoes, sugar cane, milk products, and even agricultural waste.

Taking into consideration the global state and the effort towards reducing the level of pollution, PLA is an advantageous choice of material not only because of its biodegradability but also because it is superior to petroleum-based polymers such as polyethylene (PE) and polypropylene (PP), which are pervasive presently. For these reasons PLA is steadily being used in various fields and industries for various applications, especially regarding the food industry. In the medical field, PLA biocompatibility makes nit ideal for applications such as degradable sutures, porous scaffolds for cellular applications, drug-releasing microparticles and biodegradable polymer implants, eliminating the need for a second surgical intervention [4].

The PLA filament used in this study is manufactured by CREALITY, a popular hobbyist 3D printer and 3D printing filament manufacturer. PLA's main mechanical and 3D printing properties include high surface quality, strong layer adhesion and negligible impact strength [5].

Polyether-polyurethane thermoplastic elastomers (TPE) are a category of materials that combine thermoplastic and elastomeric mechanical properties. TPE is the desired choice by multiple industries due to its more cost-effective manufacturing process compared to vulcanized elastomers. Its unique chemical structure, consisting of alternating flexible and rigid polymer segments, providing a mixture of properties: flexibility and elasticity from the flexible segments, and a variable hardness and modulus of elasticity from the rigid segments. [6-7]

The main industries of application are the textile industry to create fibers with various properties, but also in the biomedical field. The TPE used in this article is a TPE filament manufactured by Filaflex with a shore hardness of 70A, making it a high temperature printable thermoplastic elastomer, comparable to the hardness of a tyre or the heel of a shoe [8-9].

A comparison between these two materials reveals a clear difference in their mechanical properties. The aim of this paper is to combine the rigidity of PLA with the flexibility and impact resistance of TPE.

2.1 Tensile Investigation

The materials used in this investigation are polylactic acid (PLA) from CREALITY and a thermoplastic elastomer (TPE) from FilaFlex. Tensile analysis was performed with five test specimens for each material according to the standard D638-14, International Standard Test Method for Tensile Properties of Plastics [10]. The test specimens (Type IV specimens, Table2) were modelled in Fusion 360 and the software for generating the machine code for the printing process was Prusa slicer. The machine used to print the test specimens was the Prusa i3 MK3 3D and the printing parameters used can be found in Table1.

Tabel 1. Properties used for printing

Printing properties PLA/TPE	PLA	TPE
Layer height [mm]	0.2	0.2
Number of vertical shells [-]	2	2
Filling density [%]	100	100
Speed [mm/s]	20	5
Fan speed [%]	100	40
Retraction of material [-]	YES	NO
Filament diameter [mm]	1.75	1.75
Printing bed temperature [°C]	55	60
Extruder temperature [°C]	200	230

Table 2. Type IV tensile specimen dimensions

Width of narrow section [mm]	6 ± 0.5
Length of narrow section [mm]	33 ± 0.5
Width overall [mm]	19
Length overall [mm]	115
Distance between grips [mm]	65 ± 5
Radius of fillet [mm]	14 ± 1
Outer radius [mm]	25 ± 1

Tensile tests were carried out on the universal tensile/compression machine, Zwick/Roell Z005. To determine the breaking strength of the material and the modulus of elasticity of the tested material, the machine was equipped with two clamps, one fixed and one movable.

IOP Conf. Series: Materials Science and Engineering **1319** (2024) 012036 doi:10.1088/1757-899X/1319/1/012036

Fig.1 Universal tensile/compression machine Zwick/Roell Z005

The test conditions for the tensile tests with PLA and TPE materials were performed according to the standards. For PLA specimens, the specimen was preloaded with a force of 0.1 N and a speed of 10 mm/min. The machine position control was performed at a speed of 5mm/min. For the TPE samples, the parameters used in the tensile test were different due to the elastomeric behaviour of the samples. To calculate the displacement of the specimens for both materials under the action of force, a clip-on extensometer with an active surface of 10 mm and an initial opening of 30 mm was used for the test. All tests were performed under the same standard conditions to eliminate possible errors [10-11].

2.2 Charpy Impact Investigation

Impact test specimens were manufactured with the identical parameters presented in the tensile investigation section. In the case of the mix material test specimens manufacturing, because of alternating materials, each layer the adhesion of these may be imperfect. To mitigate possible adhesion issues between the two layers, the printing speed was lowered, equal to the normal printing speed of TPE of 5 mm/s and after a microscopic analysis of the adhesion between the layers of the mixt material test specimens was performed.

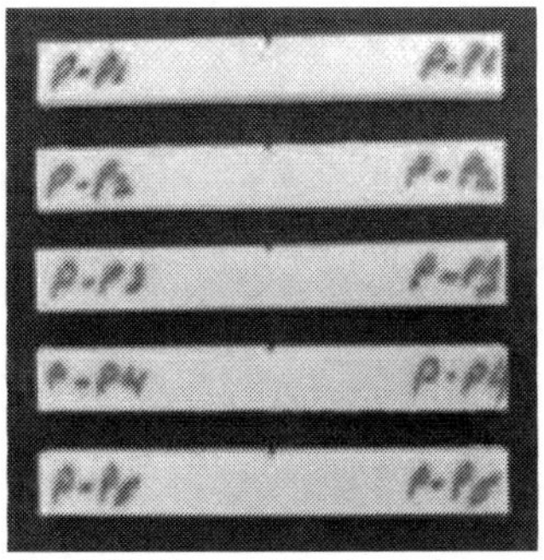

Fig.2 PLA Impact test specimens

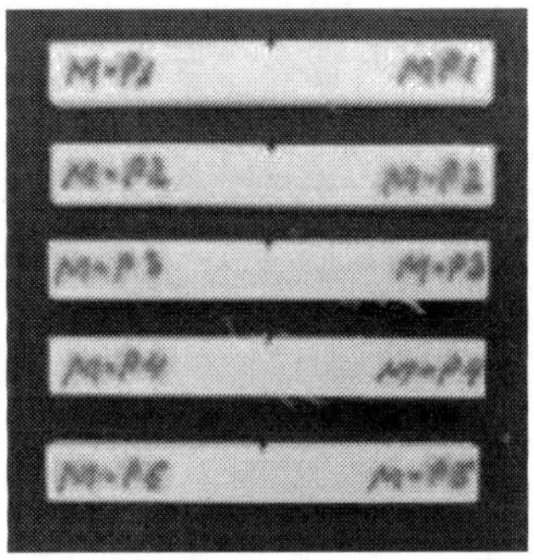

Fig.3 PLA/TPE Impact test specimens

Impact tests were performed on the CEAST 9050 Pendulum Impact System. The test conditions for the impact tests with PLA and PLA/TPE materials were carried out according to the ISO 1779-1 standards.

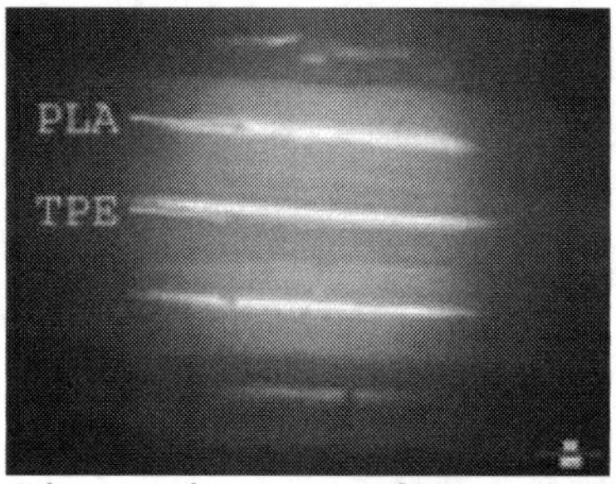

Fig.4 Side view of mixt material impact specimens

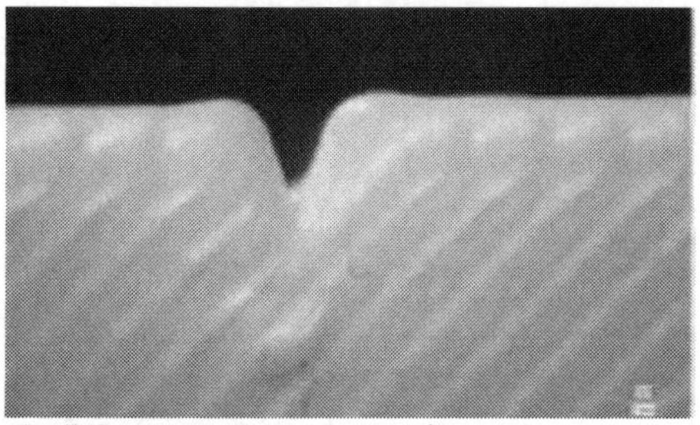

Fig.5 Top view of mixt material impact specimens

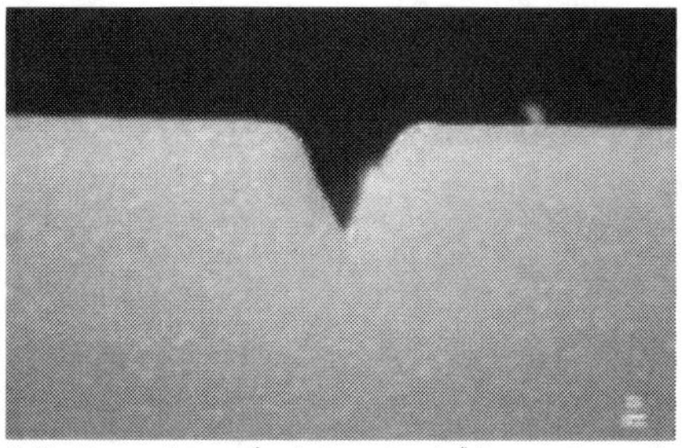

Fig.6 Bottom view of mixt material impact specimen

3. Results and Discussion

To analyse the behaviour of the three types of specimens, PLA and PLA/TPE Mix, tensile, compression, and Charpy impact samples were printed, and due to the hyper-elastic behaviour of the elastomer it was decided to print samples only to test compression. All of which are subjected, according to the standards to analyse the individual performance of the two main materials, but also the mix between the two.

After the printing process, all the samples were marked, and the dimensions were measured with a digital calliper according to precision standards (two consecutive measurements made by the same operator with the same measuring object). All of these data were collected and centralised using the professional Windows Excel software.

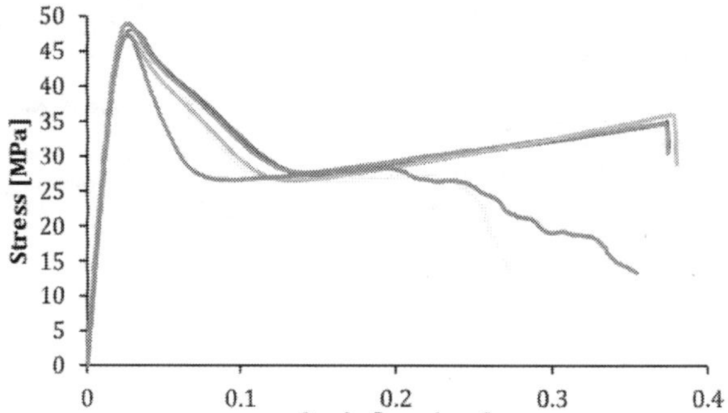

Fig.7 Engineering Stress-Strain curves for PLA tensile specimens

IOP Conf. Series: Materials Science and Engineering **1319** (2024) 012036 doi:10.1088/1757-899X/1319/1/012036

The results of the tensile tests for the PLA material are visible in Figure 7 as stress-strain curves engineering. From what can be seen, there are discrepancies between the pairs formed by specimens P1, P3, P4, and P2, P5. This is due to the premature breakdown of the PLA layers for samples P2, P5. For the rest, the fracture was symmetrical, the material having a strain value in the calibrated area of approximately 38% at a stress of 35MPa. On the linear-elastic domain, it can be observed that the samples have the same behaviour, reaching a maximum stress of approximately 48MPa at a 2% strain.

The results of the tensile tests of the mixed material samples are presented in Figure 8. It can be observed that the same linear-elastic slope is present in all of the samples, the difference coming from the PLA layers. The difference in the behaviour of the 5 samples is due to the integrity of the PLA layers. In combination with TPE, they fractured prematurely, leaving only the TPE layers intact during the test, Figure 9.

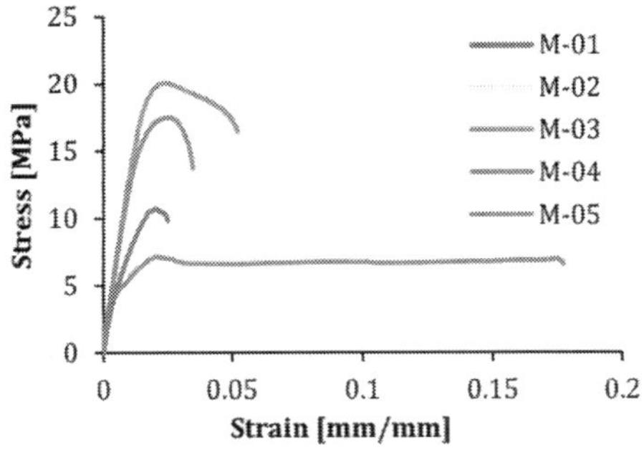

Fig.8 Engineering stress-strain curves, tensile mix

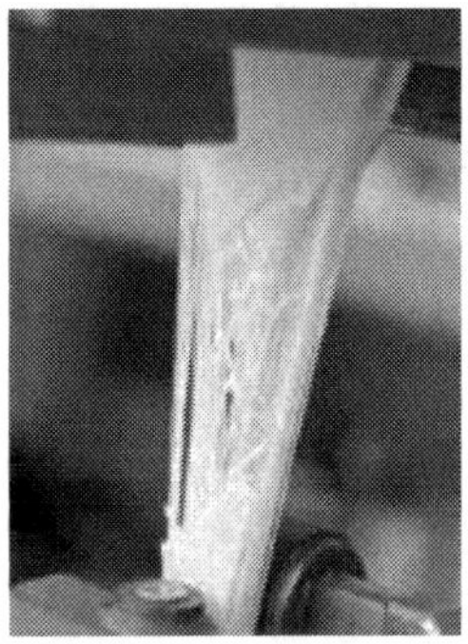

Fig.9 During test snapshot of mix tensile specimens

The results of the Charpy impact test for the PLA base material are presented in Figures 10 and 11.

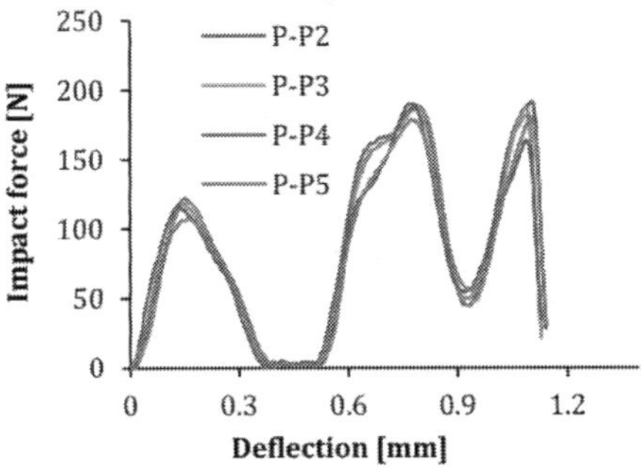

Fig. 10 Impact force-deflection for PLA

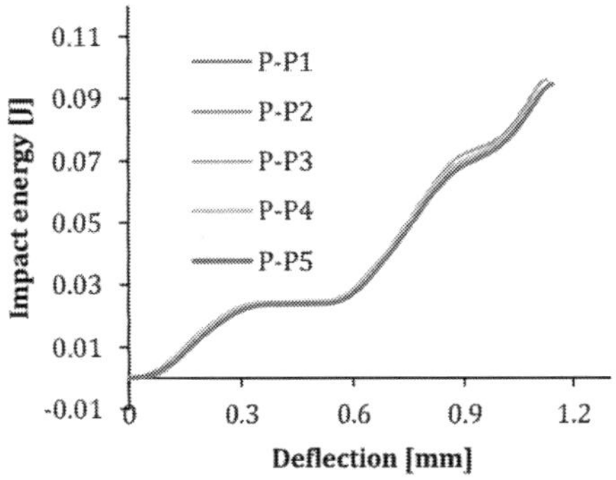

Fig.11 Impact toughness for PLA

In Figure 12 we can see the impact force-deflection for PLA Charpy specimens and impact toughness in Figure 13. Similar behaviour was observed for all five impact specimens, with an impact force average of 125 N and a total average deflection of 1,1mm. The impact energy has an average value of 0.09 J. A. Milovanović et al. (2023) showed in their research a similar behaviour for impact force and impact energy [13].

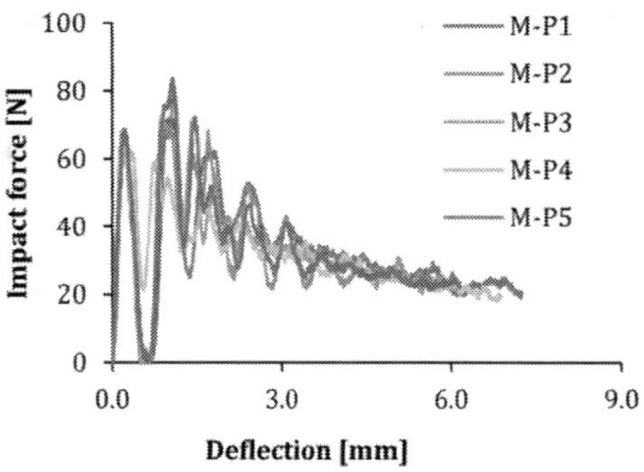

Fig. 12 Impact force-deflection mixt

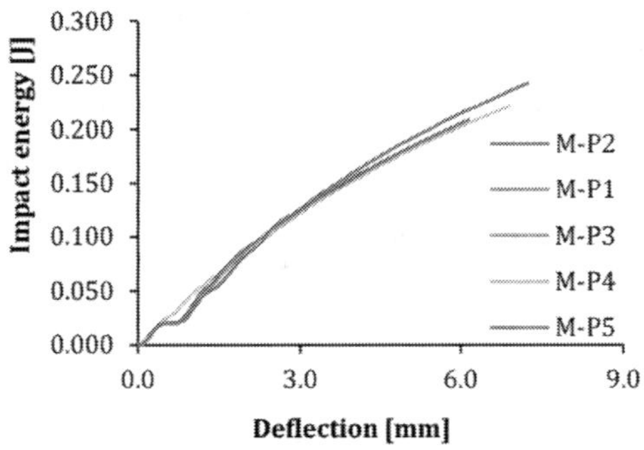

Fig. 13 Impact toughness mixt

The impact force-deflection for the Mix material shows a different behaviour in comparison with that of PLA. We can see similar force peaks, but with a significantly lower required force, and a much greater deflection because of the hyper-elastic behaviour for TPE after entire PLA layers were fractured. The impact energy recorded shows an almost 0.25 J with an average of 6.5 mm deflection of the specimens.

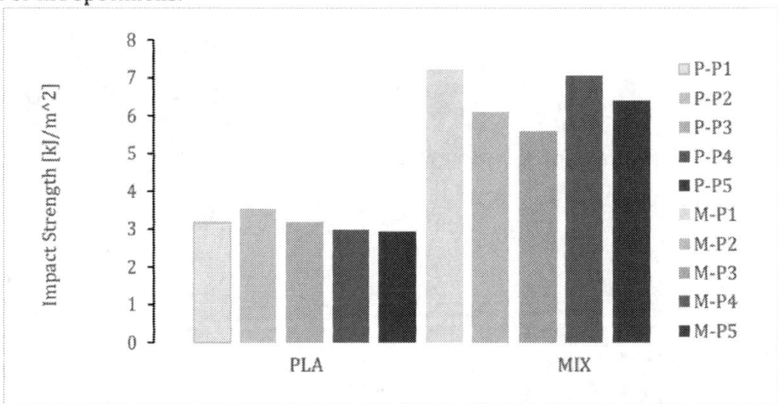

Fig.14 Impact strength results for PLA and Mixt specimens

The impact strength results for Charpy specimens are presented in Figure 14. An average value for Impact strength around 3.7 kJ/m^2 and almost double the value for the mix specimens, gives us information about the energy absorption for the Mix material.

4 Conclusions

Tensile and Charpy impact tests were conducted for PLA, TPE, and Mix between PLA and TPE. All specimens were prepared according to international standards for experimental testing. The results are interpreted on engineering stress-strain charts for tensile tests, and impact force/impact energy-deflection charts along with impact strength chart. The following conclusions could be drawn:

- Good repeatability for the PLA tensile test for 3 specimens, giving a consistent prediction of the PLA tensile behavior for the PLA material;

- Regarding the mix tensile specimens, because of the different base material behavior for the two materials, a ductile PLA and a hyper-elastic TPE, we see a completely different behavior for each specimen, but a good linear-elastic slope;
- High repeatability of results is present concerning PLA and Mix Charpy specimens;
- Different behavior between Mix and PLA in the Charpy impact test, with a consistent difference between impact strengths
- Overall higher impact force for PLA, and a higher deflection in Mix specimens.

References

[1] Abdulhameed, O., Al-Ahmari, A., Ameen, W., & Mian, S. H., 2019. Additive manufacturing: Challenges, trends, and applications. Advances in Mechanical Engineering, 11(2), 1-27. doi: 10.1177/1687814018822880
[2] Rapid Prototyping Association of the Society of Manufacturing Engineers, 1997. International Technology Research Institute - European and Japanese Development, Rapid Prototyping in Europe and Japan
[3] GIBSON, Ian, et al. Design for additive manufacturing. *Additive manufacturing technologies*, 2021, 555-607.
[4] REZVANI GHOMI, Erfan, et al. The life cycle assessment for polylactic acid (PLA) to make it a low-carbon material. *Polymers*, 2021, 13.11: 1854.
[5] INTAN, NI Nik, et al. Preliminary study of new thermoplastic elastomer (TPE) for fused deposition modelling (FDM) 3D printing. *International Journal of Advanced Research in Technology and Innovation*, 2021, 3.1: 35-42.
[6] AWASTHI, Pratiksha; BANERJEE, Shib Shankar. Fused deposition modeling of thermoplastic elastomeric materials: Challenges and opportunities. *Additive Manufacturing*, 2021, 46: 102177.
[7] Tessa, Jane; Gordelier, Philipp Rudolf, Thies; Louis, Turner; Lars, Johanning, 2019. Optimising the FDM additive manufacturing process to achieve maximum tensile strength: a state-of-the-art review, Rapid Prototyping Journal, doi: 10.1108/RPJ-07-2018-0183
[8] MUSA, Luqman, et al. A review on the potential of polylactic acid based thermoplastic elastomer as filament material for fused deposition modelling. *journal of materials research and technology*, 2022, 20: 2841-2858.
[9] Andrzej Puszka, 2018. Thermal and Mechanical Behavior of New Transparent Thermoplastic Polyurethane Elastomers Derived from Cycloaliphatic Diisocyanate, doi:10.3390/polym10050537
[10] D638-14; ASTM International Standard Test Method for Tensile Properties of Plastics. ASTM: West Conshohocken, PA, USA, 2014
[11] Information on https://www.zwickroell.com/products/static-materials-testing-machines/universal-testing-machines-for-static-applications/proline/
[12] information on CEAST 9000 Series | Pendulum Impact Systems (instron.com)
[13] MILOVANOVIĆ, Aleksa, et al. Layer thickness influence on impact properties of FDM printed PLA material. *Procedia Structural Integrity*, 2024, 56: 190-197.

Finite element modeling and analysis of tire creep test

Sadjiep Tchuigwa Baurice Sylvain[1], **Jan Krmela**[1,2], **Jan Pokorný**[1], **Vladimíra Krmelová**[2]

[1] Department of Transport means and Diagnostics, Faculty of Transport Engineering, University of Pardubice, Studentska 95, 532 10 Pardubice, Czech Republic
[2] Faculty of Industrial Technologies in Púchov, Alexander Dubček University of Trenčín, Ivana Krasku 491/30,02001 Púchov, Slovak Republic

E-mail: bauricesylvain.sadjieptchuigwa@student.upce.cz,sylvainsadjiep@gmail.com, jan.krmela@tnuni.sk, jan.pokorny@upce.cz, vladimira.krmelova@tnuni.sk

Abstract. The tire is one of the most complex components of the vehicle, which justifies strict standards and regulations set by governments worldwide to ensure the performance, durability, and reliability of tires approved for use in the transport sector. Among these requirements, tire durability testing involves the assessment of short- and long-term tire response to determine optimum conditions of use and service life. Therefore, this paper focuses on the creep performance of tires using the finite element method in ABAQUS. As a case study, a selected tire with visco-hyperelastic rubber compounds and linear elastic reinforcements is subjected to a constant prescribed force for a period of time, and the resulting strain evolution is recorded till the release of the applied load. Several parameters are accounted for, such as the inflation pressure, material characteristic times, and load amplitude. The post-processing of results enabled the prediction of the overall creep response of the tire over time with respect to the prescribed load. The knowledge gained in this study can be exploited to predict tire failure caused by excessive strain and pinpoint the areas of the tire that are the most exposed to creep failure.

Keywords— creep test, tire modeling, static loading, FEM, ABAQUS

1 Introduction

In today's global context, the transport industry is constantly seeking innovative solutions to tackle challenges in ecology, energy, and access to production resources. The quest for innovation encompasses all aspects of vehicle systems, including transmission, braking, steering, emissions, cooling, and suspension. Tires play a crucial role in the suspension system and heavily influence the ride quality and stability of a vehicle[1]. The tire industry is continuously working to enhance the strength and durability of its products[2] due to the demanding operational conditions that tires are subjected to. Ongoing research [3], [4] focuses on improving tire structure to ensure better distribution of stresses and deformations during normal and exceptional driving conditions, such as encounters with obstacles or diverse climatic conditions. Additionally, durability research aims to comprehend and predict tire behavior in prolonged situations, also known as creep. By definition, creep is the increase in deformation of a material under the effect of a long-term imposed stress that is less than the material's ultimate strength[5]. While structural analysis focuses primarily on stress-induced failure[4], durability analysis addresses strain-induced failure, which occurs when the deformation state exceeds the material's allowable strain. Consequently, tire creep testing is a relevant step in the tire design validation process, as it can significantly affect the durability and performance of tires, leading to premature failure and suboptimal performance. This requires thorough testing and analysis to predict and counteract these effects. However, tire experimental creep testing can be very fastidious and costly, often requiring a significant amount of time ranging from hours to days or even months to collect the data. Thanks to advances in finite element analysis, it is possible to minimize the experimental costs while generating a reliable prediction model. From a methodological point of view, the creep test is preceded by preconditioning of the structure under static load, analogous to that carried out on material samples [5]. In this way, the structure is statically loaded up to the set load level and then held for the duration required to observe the creep response.

As investigated by authors such as Chen [6] and Sagar, Zheng, Suwannachit, *et al.* [7], material viscoelastic properties are responsible for the creep behavior. This phenomenon is predominant in rubber compounds used in tire making and plays an important role in providing the tire with its damping, which is essential for withstanding and absorbing shocks and vibrations while traveling in complete safety and comfort. For this sake, the time range of the test is not known in advance since it should be larger than at least the shortest characteristic creep time of the material. Among others, Johnson, Tanner, and Mason [8] have proposed a tire model with a linear viscoelastic mixte shell formulation. In this model, the laminated theory was adopted for layers reinforced with rebars, and linear elastic constitutive law was chosen for the rubber part. Wang, Dong, Xu, *et al.* [9] proposed an uncoupled 3D tire model with a hyperelastic material model to simulate contact stress distribution on a viscoelastic multilayer pavement. He, Li, Yang, *et al.* [10] dove into the finite element analysis of the effect of the driver's maneuver on the distribution of contact pressure at static, free rolling and braking conditions as per the vertical load and the inflation pressure. Before the development of tire models with hyperelastic rubber compounds [11]–[13], incompressible linear elastic rubber material models [14], [15] were the most popular. However, they are inadequate when dealing with large deformations, as is the case for tire deformation. Furthermore, their linearity makes them less accurate for dynamic simulation.

In this paper, a finite element model is proposed to simulate and predict the creep response of the tire with the designation 175/75R14. The proposed model takes into account the real physical and mechanical properties of tire components (rubber compounds and reinforcements). Therefore, the structure of this paper is as follows: first, the modeling process is outlined along with assumptions, material and mesh properties. Secondly, the setup of the creep test is presented. Lastly, the results of the creep test are described, and a prediction model is derived.

2 Material and method

Before proceeding further, it is worth outlining the assumptions that sustain the current study:

- The homogenized rubber approach is adopted. That is to say, tread, inner liner, sidewall, undertread, and apex, are assumed to have the same mechanical properties;
- The rim is treated as a rigid body and is in coupling contact with the center of the wheel.;
- The simplified bead approach with a homogenized property is adopted [16];
- A rigid road model is utilized;
- Since preconditioning and the creep phase are carried out in a static configuration, frictional effects are omitted.

2.1 Geometry of the tire:

175/75R14 tire is a passenger car radial (PCR) tire with half cross-section sketched in Fig. 1. It has a 618mm diameter, 175mm tread width, and is designed to be mounted on a 355.6mm rim diameter.

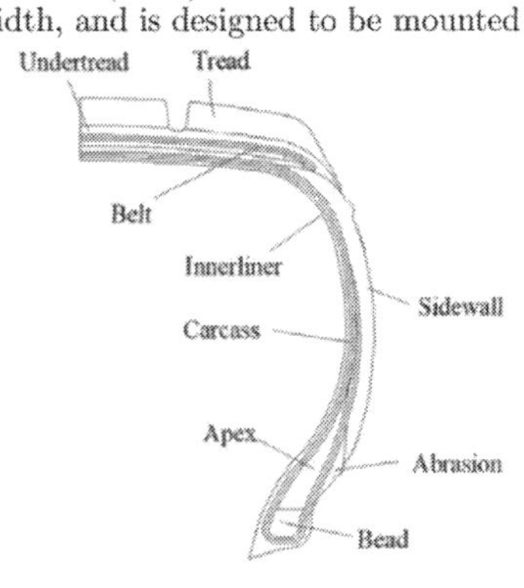

Figure 1: Half cross-section of 175/75R14 tire[4]

In order to generate tire cross-section in ABAQUS[17], the scanned cross-section was imported in AUTOCAD and utilized as layout to sketch the cross-section. Afterward, the dxf file was brought into ABAQUS and utilized for creating the 3D tire (see Fig. 3) by employing various commands like revolution, symmetry, and mirror.

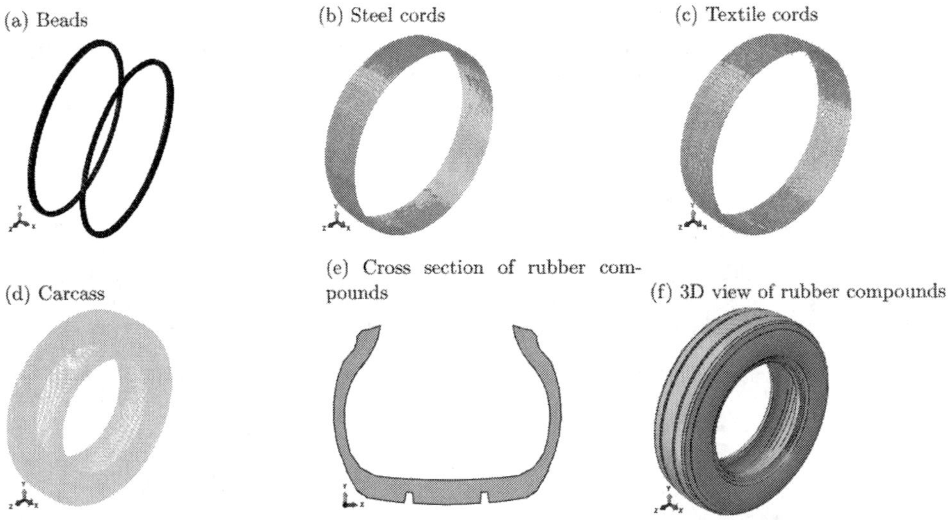

Figure 2: Parts of the tire

To simplify the model, we have assumed that the mechanical properties of rubber compounds, such as tread, inner liner, sidewall, undertread, and apex, are the same, allowing the homogenize rubber matrix approach to apply.

2.2 Material properties:

It is important to have the mechanical properties of the parts or structural components when building the FE model. Typically, a series of uniaxial tensile tests are conducted in the laboratory on steel cords, textile cords, and beads in line with relevant test standards. Additionally, a cyclic tensile test is necessary to capture the hyper-viscoelastic response of the rubber part (see, for instance in [18]). As a result, the mechanical properties of the reinforcements used in this study are listed in Table 1.

Table 1: Reinforcement material propeties

Tire parts	E [MPa]	ν	$\varnothing$ [mm]	ρ [Kg/m^3]
Steel cords	15.724	0.3	0.60	7 850
Textile cords	1900	0.35	0.80	1 250
Carcass	1900	0.40	1.00	1 250
Bead	164499	0.29		7 850

As for the rubber part, we consider the hyper-viscoelastic material model defined as a combination of the Money Rivlin model (see Eq. (1)) and linear viscoelasticity (see Eq. (2)).

$$W\left(\overline{I}_1, \overline{I}_2, J\right) = C_{10}\left(\overline{I}_1 - 3\right) + C_{01}\left(\overline{I}_2 - 3\right) + \frac{1}{D_1}(J-1)^2 \qquad (1)$$

Where C_{10}, C_{01} and D_1 are material constants. $J = det\,(J)$, $\overline{I}_1$ and $\overline{I}_2$ are respectively the determinant jacobean, the first invariant and the second invariant of the left Cauchy tensor $\overline{B} = \overline{F}\overline{F}^T$. The linear viscoelasticity of the material is commonly defined using the Prony series of the shear and bulk components, respectively g_r and k_r.

$$g_r\,(t) = 1 - \sum_{n=1}^{n} g_i\left(1 - e^{-t/\tau_i}\right); \quad k_r\,(t) = 1 - \sum_{n=1}^{n} k_i\left(1 - e^{-t/\tau_i}\right); \qquad (2)$$

The parameters of each of these models are summarized in Table 2

Table 2: Material properties of rubber compounds [3], [4]

Hyperelasticity		viscoelasticity	
Density ρ [Kg/m^3]	1200	g_1	0.50
C_{01} [MPa]	0.417	g_2	0.49
C_{10} [MPa]	0.032	τ_1 [s]	7.6125
D_1 [MPa^{-1}]	0.0018	τ_2 [s]	235.62
		k_1	0.50
		k_2	0.49

2.3 Mesh properties:
The meshing of parts highlighted in Fig. 3 was done separately so to avoid unnecessary mesh densification in rubber compounds near the rebars. Furthermore, we partitioned the tire into two portions: the first meshed with elements of bigger size and the second with smaller elements since the latter may come into contact with the road. Table 3 reports mesh properties of the tire components generated above.

Table 3: Mesh properties

Parts	**Elements**	**Element type**	**Number of elements**
Rubber compounds	3D solid	C3D8R: 8-node linear brick, reduced integration with hourglass control	20 090
Bead	3D solid	C3D8R: 8-node linear brick, reduced integration with hourglass control	3 248
Textile cords	1D beam	B32H: 3-node quadratic hybrid elements	3 082
Carcass	1D beam	B32H: 3-node quadratic hybrid elements	19 800
Steel cords	1D beam	B32H: 3-node quadratic hybrid elements	5 768
Road	2D Shell	S4R: 4-node linear shell with hourglass control	16 950

It should be noted that for the stability of the simulation, we activated enhanced hourglass control and distortion to prevent mesh distortion. Besides, we opted for reduced integration to prevent volumetric locking, particularly in the rubber component. A view of the created mesh is given in Fig. 3.

(a) Mesh view on the cross-section (b) Longitudinal view on the tire

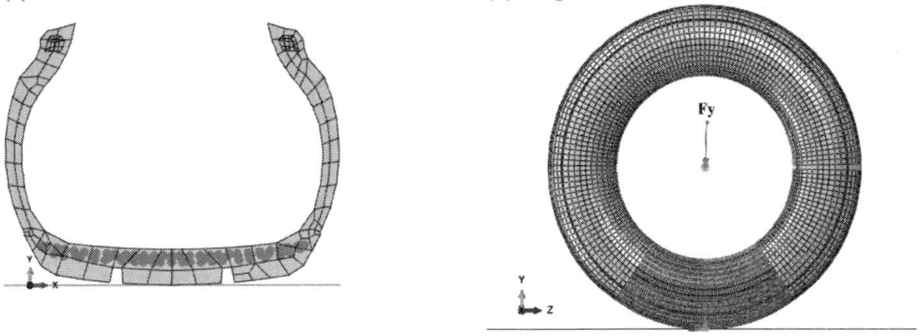

Figure 3: Views on the tire model

2.4 Boundary conditions:
Three types of boundary conditions were enforced: Prior to this, the so-called hard contact type was created.

IOP Conf. Series: Materials Science and Engineering **1319** (2024) 012037 doi:10.1088/1757-899X/1319/1/012037

- *Rim-tire interaction*: Rim-tire interaction was established through a coupling constraint of type kinematic using the wheel center as the reference point. Since the analysis assumed a rigid rim, there was no need to model the rim;
- *Tire-road interaction*: This contact constraint was enforced using both the general and surface-to-surface contact features. While the former enables the restriction of the set of surfaces to check specifically for possible contact, the latter helps check self-contact.
- *Embedded constraints*: This constraint enables the incorporation of rebars(carcass, steel cords, textile cords and bead) into rubber;
- Road's degrees of freedom were all blocked;
- For stability reasons, U_x at the wheel's center was fixed;
- *Vertical applied load* F_y was applied linearly at the reference point located at the wheel's center in Fig. 4(c) as per the preconditioning process outlined in the introduction;
- *Inflation pressure*: a uniformly distributed pressure of 0.35 MPa was applied on the inner liner surface;
- *Tire self-weight*(see Fig. 4(b)) was taken into account thanks to gravity effects defined on all model's elements except the road.

The abovementioned loads were applied successively in four different steps such that the total step time is given by

$$T = [0, t_1] \cup [t_1, t_2] \cup [t_2, t_3] \cup [t_3, t_4] \quad (3)$$

Where $t_1 = 0.1\ s$, $t_2 = 0.6\ s$, $t_3 = 1.6\ s$ and $t_4 = 2.6\ s$. It is worth pointing out that the choice of such a small step size requires special treatment of the load application since, in reality, the creep test is conducted in static (i.e., F_y is applied progressively so that kinetic effects are negligible). In order to circumvent this problem, at the beginning, each load is applied using the smooth step function, as illustrated by blue curves in Fig. 4. The creep test is conducted during the step in cyan color in Fig. 4(c).

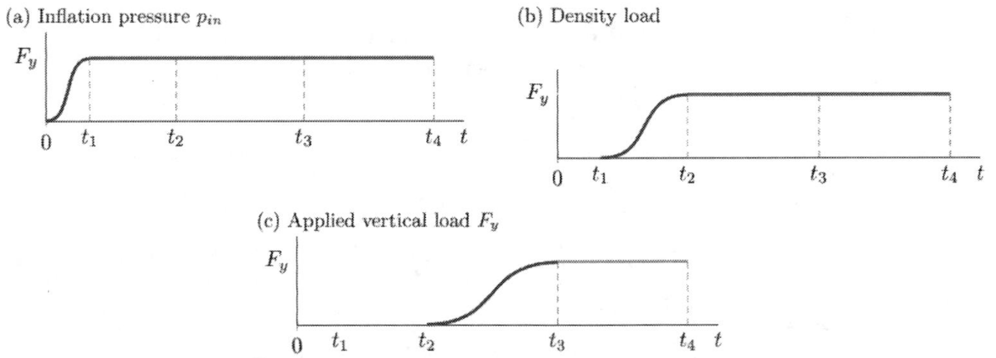

Figure 4: Definition of loads

Five levels of preconditioning force F_y were selected for recording the creep response: 1 kN, 2.5 kN, 3.5 kN, 4.5 kN, and 6.5 kN. Thereby, five separate jobs of the model were created with each of these values.

2.5 Solver setting:

The dynamic implicit solver of type "Moderate" was selected with default convergence parameters. Since dynamic implicit schemes are unconditionally stable, a starting time increment of 0.1 s was prescribed.

2.6 Location of observation point:

Throughout the creep test procedure, two directions of deformations (ϵ_{xx} and ϵ_{yy}) prevail, respectively, in the vertical and transverse directions. Therefore, strain history outputs at all times $t \in [t_3, t_4]$ were recorded.

IOP Conf. Series: Materials Science and Engineering **1319** (2024) 012037 doi:10.1088/1757-899X/1319/1/012037

3 Results and discussion

Fig. 5 presents the distribution of the Von Mixes stress [MPa] for each level of preconditioning.

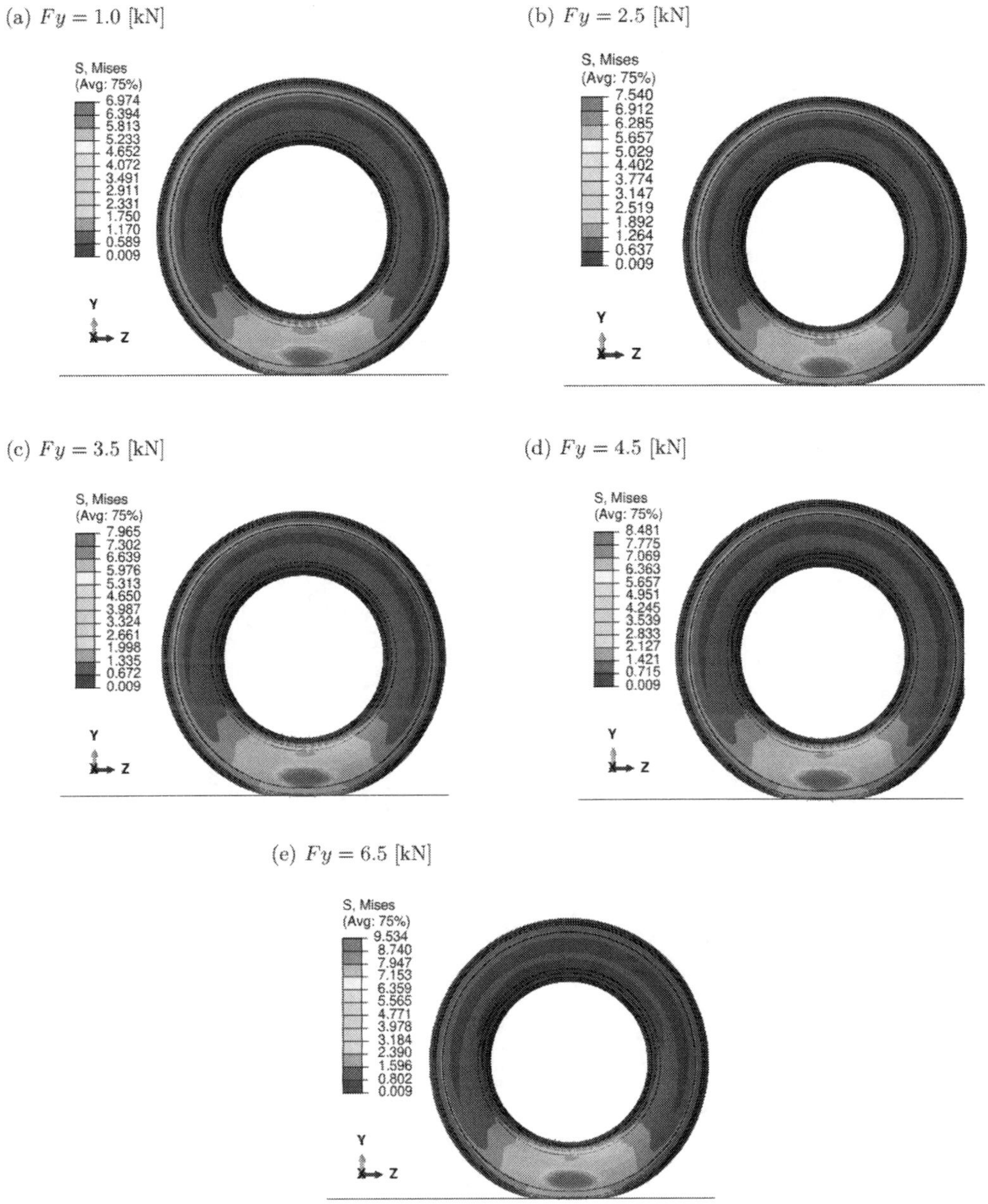

Figure 5: σ_{VM} at the beginning of each creep test.

We see that there is a stress concentration and significant deformation in the red area. So, a sensor was placed at this point to record strain history during the creep test. Thereby, strain history outputs ε_{11} at this point were extracted in the time range $[t_3, t_4]$ and plotted in Fig. 6.

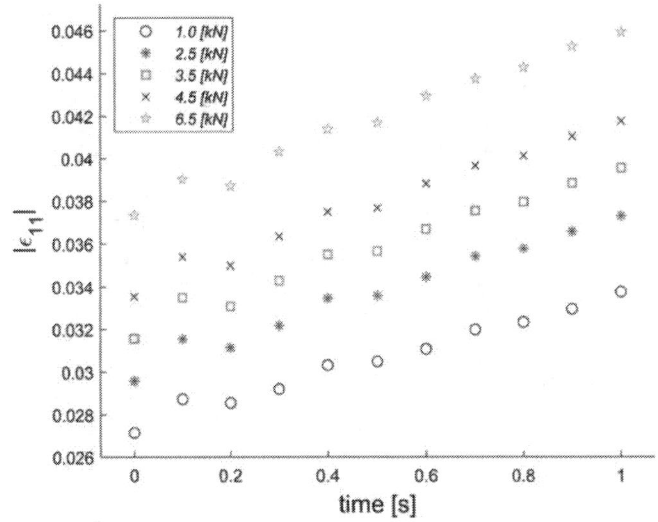

Figure 6: History evolution of ε_{11} with respect to the preconditioning level F_y

It can be seen that there is an increase in strain amplitude throughout the creep test. Also, the material at this point is solicited in tension in the x-direction since ε_{11} is negative. Similarly, we extracted the strain component ε_{22} along the y-axis and plotted it in Fig. 7. The same observation applies, but the deformation type is a compression.

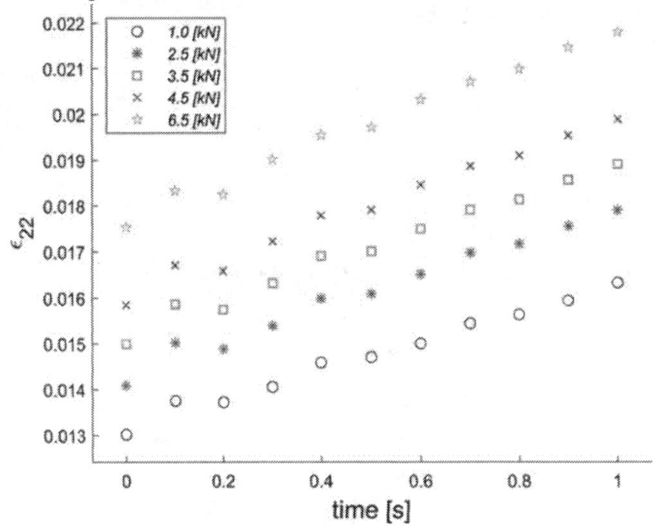

Figure 7: History evolution of ε_{22} with respect to the preconditioning level F_y

3.1 Proposal of a strain history model for tire creep test prediction

In order to define a time history function that can fit the FEM results, we recall that in small strain, the creep response of the differential equation deriving from the Generalized Maxwell model with two dashpots is given as

$$\left(\sum_{i=1}^{2-1}\left(\sum_{j=i+1}^{2}(E_0+E_i+E_j)\,\tau_i\tau_j\right)\right)\frac{\partial^2\varepsilon}{\partial t^2}+\left(\sum_{i=1}^{N}(E_0+E_i)\,\tau_i\right)\frac{\partial\varepsilon}{\partial t}+E_0\varepsilon=\sigma_0 \tag{4}$$

We make the hypothesis that, in hyper-viscoelasticity the same equation holds. Now, let us find a function $\varepsilon(t)$ solution of this differential equation. Basically, $\varepsilon(t)$ can be split into the homogeneous solution $\varepsilon_h(t)$ and particular solution $\varepsilon_p(t)$, such that $\varepsilon_h(t)$ verifies the homogeneous differential equation associated with the left-hand side of Eq.(4). Since the second member is constant, $\varepsilon_p(t)$ reads as

$$\varepsilon_p(t) = a, \quad a \in \mathbb{R} \tag{5}$$

$\varepsilon_h(t)$ reads as

$$\varepsilon_h(t) = be^{r_1 t} + ce^{r_2 t}, \quad b, c \in \mathbb{R} \tag{6}$$

Now, since the characteristic time of the process is driven by that of the rubber compounds, we have $r_1 = -\frac{1}{\tau_1}$ and $r_2 = -\frac{1}{\tau_2}$, such that

$$\varepsilon(t) = a + be^{-\frac{t}{\tau_1}} + ce^{-\frac{t}{\tau_2}}; \quad a, b, c \in \mathbb{R} \tag{7}$$

Using MATLAB[19] built-in nonlinear least square optimization algorithm fit with a tolerance of 10^{-6}, this function was fitted to the FE data. Thus, for the strain component ε_{11}, the obtained values of the constants a, b, c are reported in Table 4

Table 4: Constant parameters for predicted strain ε_{11} for the set of load amplitudes

	F_y [kN]				
	1.0	2.5	3.5	4.5	6.5
a	0.0465	0.0529	0.0557	0.0584	0.0632
b	-5.05×10^{-8}	-2.22×10^{-7}	-4.32×10^{-6}	-1.01×10^{-5}	-4.17×10^{-8}
c	-0.0193	-0.0233	-0.0241	-0.0249	-0.0259
R^2	0.978	0.986	0.986	0.985	0.987

The visualization of the predicted function of ε_{11} is shown in Fig. 8 and 9.

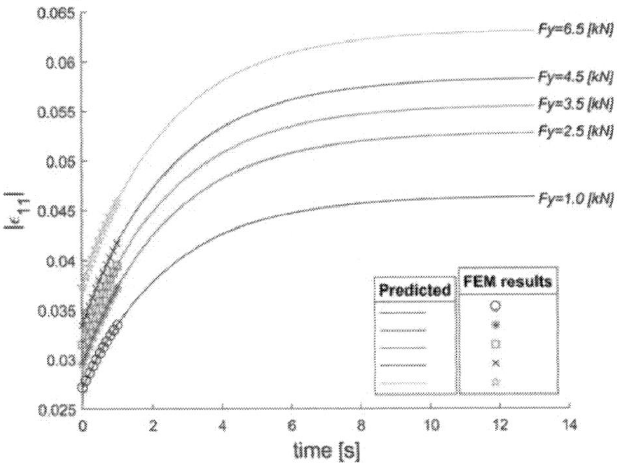

Figure 8: Comparison between the calculated and predicted values of ε_{11}.

Similarly, the values of the constants a, b and c shown in Table 5 were obtained for the predicted function ε_{22} for all the preconditioning levels regarded.

IOP Conf. Series: Materials Science and Engineering **1319** (2024) 012037 doi:10.1088/1757-899X/1319/1/012037

Table 5: Constant parameters for predicted strain ε_{22} for the set of load amplitudes

	F_y [kN]				
	1.0	2.5	3.5	4.5	6.5
a	0.0226	0.0253	0.0268	0.0281	0.0296
b	-2.220×10^{-14}	-7.200×10^{-4}	-1.280×10^{-5}	-1.007×10^{-5}	-1.62×10^{-3}
c	-0.0096	-0.0105	-0.012	-0.012	-0.011
R^2	0.983	0.988	0.987	0.987	0.988

Fig. 9 depicts the visualization of the predicted function ε_{22} for each level of loading.

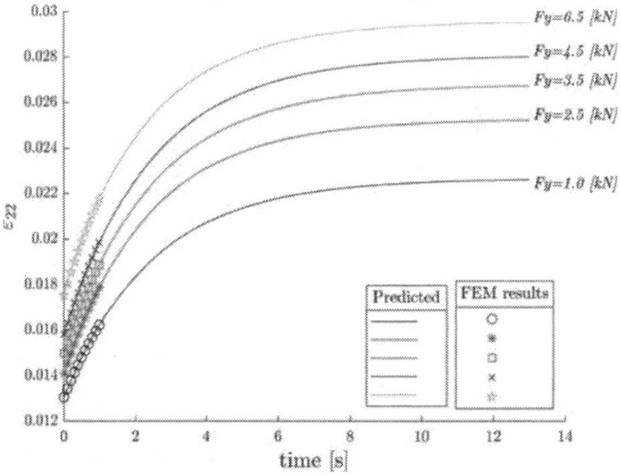

Figure 9: Comparison between the calculated and predicted values of ε_{22}

In view of the prediction, we see that in both cases, the value of the strain at infinity is almost double that at the start of the creep test. Beyond approximately $t > 13$ s, the predicted function ε reaches the ceiling value a.

Conclusion

In this study, we have presented a finite element model to analyze the creep response of a tire, specifically the 175/75R14 model, under various loading conditions using the software ABAQUS. The model incorporates the visco-hyperelastic properties of rubber compounds and linear elastic reinforcements (textile cords, steel cords, and carcass cords), providing a detailed insight into the tire's deformation behavior over time when subjected to a constant vertical force. The results demonstrate the effectiveness of the finite element method to predict tire creep behavior, offering a significant advantage over traditional experimental methods in terms of cost and time efficiency. By analyzing the strain evolution in response to load amplitudes, we were able to identify the critical area of the tire that is most susceptible to creep failure. Then, based on the expression of the function solution of the Maxwell generalized solid with two parameters, we fitted this function to the FE data using nonlinear least square optimization with a very high correlation coefficient (at least 0.98). Afterward, the fitted function was used to predict the maximal strain reachable for a long-term application of the vertical force. This information is invaluable for improving tire design, enhancing durability, and preventing creep failure. Overall, this research contributes to the ongoing efforts to improve tire durability and safety, offering a practical approach to understanding and mitigating creep-related failures. By leveraging advanced finite element analysis, tire manufacturers can better design and test their products, ultimately leading to safer and more reliable tires for consumers. In upcoming work, we will investigate the effect of temperature on tire creep response and validate it with an experimental tire creep test.

Acknowledgement

The authors acknowledge the funding provided by the University of Pardubice through the grant number SGS 2024. This research work had been supported by University of Pardubice grant number SGS 2024 and the Cultural and Educational Grant Agency of the Slovak Republic (KEGA), project No. KEGA 003TnUAD-4/2022 "Simulation of basic and specific experiments of polymers and composites based on experimental data in order to create a virtual computational-experimental laboratory of mechanical tests".

Statements and Declarations
Conflicts of Interest: The authors declare no conflict of interest.

References
[1] B. S. Sadjiep Tchuigwa, J. Krmela, and J. Pokorny, "Literature review on tire component requirements," *Perner's Contacts*, vol. 16, p. 6, Dec. 2021. DOI: 10.46585/pc.2021.2.1740.

[2] A. Gallrein and M. Bäcker, "CDTire: A tire model for comfort and durability applications," *Vehicle System Dynamics*, vol. 45, no. supl, pp. 69–77, 2007. DOI: 10.1080/00423110801931771.

[3] J. Krmela, *Experiments and computational modelling of tires: Textbooks for university students*, Zborov, Czech Republic, 2020.

[4] G. Wang, W. Wang, C. Liang, and L. Cao, "Fatigue life prediction of radial tire bead using a maximum strain energy density range method," *Applied Sciences*, vol. 11, no. 12, p. 5477, 2021, ISSN: 2076-3417. DOI: 10.3390/app11125477.

[5] C. Derham, "Creep and stress relaxation of rubbers—the effects of stress history and temperature changes," *Journal of Materials Science*, vol. 8, pp. 1023–1029, 1973. DOI: 10.1007/BF00756634.

[6] L. Chen, "Research on the relaxation and creep characteristics of tread rubber based on thermal and mechanical coupling," pp. 104–107, 2021. DOI: 10.1109/AIAM54119.2021.00030.

[7] G. Sagar, D. Zheng, A. Suwannachit, M. Brinkmeier, K. Fietz, and C. Hahn, "On the Development of Creep Laws for Rubber in the Parallel Rheological Framework," *Tire Science and Technology*, vol. 47, no. 1, pp. 2–30, Mar. 2019, ISSN: 0090-8657. DOI: 10.2346/tire.18.470104.

[8] A. R. Johnson, J. A. Tanner, and A. J. Mason, "Quasi-static viscoelastic finite element model of an aircraft tire," NASA, Tech. Rep. No. NAS 1.15: 209141, Apr. 1999.

[9] T. Wang, Z. Dong, K. Xu, S. Ullah, D. Wang, and Y. Li, "Numerical simulation of mechanical response analysis of asphalt pavement under dynamic loads with non-uniform tire-pavement contact stresses," *Construction and Building Materials*, vol. 361, p. 129 711, 2022.

[10] H. He, R. Li, Q. Yang, J. Pei, and F. Guo, "Analysis of the tire-pavement contact stress characteristics during vehicle maneuvering," *KSCE Journal of Civil Engineering*, vol. 25, pp. 2451–2463, 2021.

[11] J. Phromjan and C. Suvanjumrat, "A suitable constitutive model for solid tire analysis under quasi-static loads using finite element method," *Engineering Journal*, vol. 22, pp. 141–155, Mar. 2018.

[12] Y. Li, C. Liu, Y. Sun, Y. Li, C. Xu, and M. Xie, "Modeling Methods and Simulation Analysis of Radial Tire with Different Tread Patterns," vol. 9, no. 1, pp. 20–30, 2020.

[13] H. Fathi, Z. El-Sayegh, J. Ren, and M. El-Gindy, "Modeling and validation of a passenger car tire using finite element analysis," *Vehicles*, vol. 6, no. 1, pp. 384–402, 2024, ISSN: 2624-8921. DOI: 10.3390/vehicles6010016.

[14] C. M. Andersen, "Deep anistropic shell program for tire analysis," NASA, Tech. Rep., 1981.

[15] J. Krmela, "The computational modelling of tire," *Recent*, vol. Vol. 10, no. 3(27), Nov. 2009.

[16] Y. Li, X. Sun, J. Song, S. Zhang, and S. Han, "Topological structure and experimental investigation of a novel whole tire bead," *Materials Design*, vol. 203, p. 109 592, 2021, ISSN: 0264-1275. DOI: 10.1016/j.matdes.2021.109592.

[17] Dassault Systemes Simulia Corpia Corp, *ABAQUS/Standard User's Manual, Version 6.9*, 2020.

[18] B. Fazekas and T. J. Goda, "Closed-form and numerical stress solution-based parameter identification for incompressible hyper-viscoelastic solids subjected to various loading modes," *International Journal of Mechanical Sciences*, vol. 151, pp. 650–660, Feb. 2019, ISSN: 00207403. DOI: 10.1016/j.ijmecsci.2018.12.011.

[19] MATLAB, *Release r2023b*, Natick, Massachusetts, 2024.

AUTHOR INDEX